Cisco Multiservice Switching Networks

Carlos Pignataro, CCIE certified No. 4619

Ross Kazemi, CCIE certified No. 2742

Bil Dry, CCIE certified No. 4191

Cisco Press

Cisco Press
201 West 103rd Street
Indianapolis, IN 46290 USA

Cisco Multiservice Switching Networks

Carlos Pignataro, Ross Kazemi, Bil Dry

Copyright© 2003 Cisco Systems, Inc.

Cisco Press logo is a trademark of Cisco Systems, Inc.

Published by:
Cisco Press
201 West 103rd Street
Indianapolis, IN 46290 USA

Printed in the United States of America 1 2 3 4 5 6 7 8 9 0

First Printing September 2002.

Library of Congress Cataloging-in-Publication Number: 2001-092523

ISBN: 1-58705-068-4

Warning and Disclaimer

This book is designed to provide information about multiservice switching networks. Every effort has been made to make this book as complete and as accurate as possible, but no warranty or fitness is implied.

The information is provided on an "as is" basis. The authors, Cisco Press, and Cisco Systems, Inc. shall have neither liability nor responsibility to any person or entity with respect to any loss or damages arising from the information contained in this book or from the use of the discs or programs that may accompany it.

The opinions expressed in this book belong to the authors and are not necessarily those of Cisco Systems, Inc.

Trademark Acknowledgments

All terms mentioned in this book that are known to be trademarks or service marks have been appropriately capitalized. Cisco Press or Cisco Systems, Inc. cannot attest to the accuracy of this information. Use of a term in this book should not be regarded as affecting the validity of any trademark or service mark.

Feedback Information

At Cisco Press, our goal is to create in-depth technical books of the highest quality and value. Each book is crafted with care and precision, undergoing rigorous development that involves the unique expertise of members of the professional technical community.

Reader feedback is a natural continuation of this process. If you have any comments regarding how we could improve the quality of this book, or otherwise alter it to better suit your needs, you can contact us through e-mail at feedback@ciscopress.com. Please be sure to include the book title and ISBN in your message.

We greatly appreciate your assistance.

Publisher	John Wait	
Editor-In-Chief	John Kane	
Cisco Systems Program Management	Michael Hakkert	
	Tom Geitner	
Acquisitions Editor	Amy Lewis	
Production Manager	Patrick Kanouse	
Development Editor	Howard Jones	
Project Editor	Eric T. Schroeder	
Copy Editor	Gayle Johnson	
Technical Editors	Michael Carey	Dave Warren
	Timothy Cricchio	Steve Wisniewski
	Gary Day	
	Mark Newcomb	
	Robin Smith	
Team Coordinator	Tammi Ross	
Book Designer	Gina Rexrode	
Cover Designer	Louisa Adair	
Compositor	Mark Shirar	
	Octal Publishing, Inc.	
Indexer	Heather McNeill	

CISCO SYSTEMS

Corporate Headquarters
Cisco Systems, Inc.
170 West Tasman Drive
San Jose, CA 95134-1706
USA
http://www.cisco.com
Tel: 408 526-4000
      800 553-NETS (6387)
Fax: 408 526-4100

European Headquarters
Cisco Systems Europe
11 Rue Camille Desmoulins
92782 Issy-les-Moulineaux
Cedex 9
France
http://www-europe.cisco.com
Tel: 33 1 58 04 60 00
Fax: 33 1 58 04 61 00

Americas Headquarters
Cisco Systems, Inc.
170 West Tasman Drive
San Jose, CA 95134-1706
USA
http://www.cisco.com
Tel: 408 526-7660
Fax: 408 527-0883

Asia Pacific Headquarters
Cisco Systems Australia,
Pty., Ltd
Level 17, 99 Walker Street
North Sydney
NSW 2059 Australia
http://www.cisco.com
Tel: +61 2 8448 7100
Fax: +61 2 9957 4350

Cisco Systems has more than 200 offices in the following countries. Addresses, phone numbers, and fax numbers are listed on the Cisco Web site at www.cisco.com/go/offices

Argentina • Australia • Austria • Belgium • Brazil • Bulgaria • Canada • Chile • China • Colombia • Costa Rica • Croatia • Czech Republic • Denmark • Dubai, UAE • Finland • France • Germany • Greece • Hong Kong • Hungary • India • Indonesia • Ireland Israel • Italy • Japan • Korea • Luxembourg • Malaysia • Mexico • The Netherlands • New Zealand • Norway • Peru • Philippines Poland • Portugal • Puerto Rico • Romania • Russia • Saudi Arabia • Scotland • Singapore • Slovakia • Slovenia • South Africa • Spain Sweden • Switzerland • Taiwan • Thailand • Turkey • Ukraine • United Kingdom • United States • Venezuela • Vietnam • Zimbabwe

About the Authors

Carlos Pignataro, CCIE certified No. 4619, is a senior engineer on the Escalation Team for Cisco Systems, Inc. In this role he handles difficult and complex escalations, works on critical or stalled defects, and provides input on and tests new products and developments—helping improve their reliability, availability, and serviceability. He has extensive experience in large-scale IP and ATM deployments, and he works with different development teams. He also is an active speaker at networkers' conventions. He has a BS and MS in electronic engineering. Carlos can be reacxhed at cpignata@cisco.com.

Ross Kazemi, CCIE certified No. 2742, is a senior engineer on the System Test Team for Cisco Systems, Inc. He develops and executes test cases that simulate QoS usage in large SP MPLS/VPN networks. He also teaches an internal Cisco class to Cisco support engineers. Previously, he worked as an escalation engineer, dealing with networks that use MPLS over Cisco ATM switches. He also worked as a consultant engineer for Cisco's big enterprise customers. He has a BS in electrical engineering. Ross can be reached at rkazemi@cisco.com.

Bil Dry, CCIE certified No. 4191, manages one of Cisco's Advanced Services teams that offers design consulting, escalation support, and best-practice recommendations to service provider packet telephony customers. His work in wide-area, IP, and ATM networks has helped him achieve double CCIE status in Routing and Switching and WAN Switching. Currently, he is focused on packet voice signaling protocols such as MGCP and SIP. He also speaks on troubleshooting WAN protocols at networkers' conventions. He has a BS in electrical engineering. Bil can be reached at bil@cisco.com.

About the Technical Reviewers

Michael Carey is a technical marketing engineer with the Cisco Multi Service Switching Business Unit (MSSBU), specializing in large scale Frame Relay and ATM network deployments. In several countries, over the past 28 years, Michael has held a variety of product design, sales, and marketing positions in the networking and data communications industry.

Timothy Cricchio is a Cisco field trial engineer in the MSSBU. His responsibilities include demonstrating new products and features on-site to carriers worldwide, providing consultation on how they might integrate and use these new items in their existing network, performing competitive tests between Cisco and other vendors on-site, creating technical documents detailing new feature configurations, and working with Tech Pubs ensuring release documentation is accurate. Timothy has also worked in technical support for over 20 years and for various vendors.

Gary Day, CCIE certified No. 6310, is a project consultant within Cisco Systems, Inc. He is responsible for network designs for EMEA service providers across a wide product range with a core focus on MPLS. He has been with Cisco for five years and has performed a wide range of roles, including technical support of some of the largest MSSBU networks worldwide.

Mark Newcomb, CCNP, CCDP, owns Secure Networks, a consulting firm in Spokane, Wash. Secure Networks' focus is on providing comprehensive networking solutions to clients throughout the Northwest. These services include security audits through network design and implementation. He has 20 years of experience in the networking industry, with the last five years focused on network security issues. He is a frequent author, contributor, and reviewer for books by Cisco Press, McGraw-Hill, Coriolis, New Riders, and Macmillan Technical Publishing. He can be contacted at mnewcomb@wanlansecurity.com.

Robin Smith, CCIE certified No. 4230, is a technical marketing engineer for Cisco's Multi-Service Switching Business Unit (MSSBU). He ensures that the field organization and customer receive technical and commercial marketing information on the MSSBU product line. He came to Cisco via the StrataCom acquisition in 1996. His area of expertise is the integration of ATM and MPLS service provider networks.

Dave Warren, CCNP-WS, CCIP, CCSI, is an independent consultant and trainer. He has worked in the computer and networking industry since the early '80s. For the past five years he has taught and consulted on an array of software and hardware products, both Cisco and beyond. He works with Sequoia Networks in San Jose, Calif., which specializes in training for WANs.

Steve Wisniewski, CCNP, has an MS in telecom engineering from Stevens Institute of Technology. He has more than ten years of networking experience and is employed as a senior engineer for Greenwich Technology Partners. He has authored two books on networking for Prentice Hall and is presently coauthoring a book for Cisco Press. He lives in East Brunswick, N.J. with his wife, Ellen, and their 16 dogs.

Dedications

Carlos Pignataro: I especially dedicate this book to the memory of Vivi. You were a pillar of strength for all around you, and an inspiration and model for me. I would also like to dedicate this book to my wonderful wife, Veronica, with all my love. To my father and mother, Juanca and Chiche, with enormous gratitude and love. Last but not least, to my sister Marie, her husband Gaby, and their two daughters, Maria Eugenia and Maria Victoria.

Ross Kazemi: I dedicate this book to my wife, Nathalie, Alex, and Marie for making my life more interesting and complete.

Bil Dry: I dedicate this book to all the NOC engineers who keep the MPLS and PNNI networks of the world running smoothly.

Acknowledgments

Carlos Pignataro: I would like to thank the MPLS team in Chelmsford, Massachusetts.

Contents at a Glance

Table of Contents

Introduction   xviii

Chapter 1   What Are Multiservice Switching Networks? 3

Ships in the Night 3

Separating the Control and Switching Planes 6

Partitioning, Virtual Switches, and Multiple Control Planes 7

Advantages of the Architecture 10

Completing the Picture 11

Multiservice Switching Forum 13

Summary 13

Chapter 2   SCI: Virtual Switch Interface 15

VSI Functions 15

VSI Master and VSI Slaves 19

VSI Messages 21

End-to-End Connections 24

Controller Location Options 28

Hard, Soft, and Dynamic Resource Partitioning 30
Hard Partitioning 31
Soft Partitioning 31
"Starved" Partitioning 42
Dynamic Partitioning 43

Traffic Service Types, Categories, and Groups 44

Enhanced VSI Protocol Capabilities 45
Slave Discovery 45
Master Keepalive Function 46
Database Synchronization 46
Clock Synchronization 46
Flow and Congestion Control 47
Message Passthrough 47

Summary 47

Icons Used in This Book

Router

Bridge

Hub

DSU/CSU

Catalyst
Switch

Multilayer Switch
without text

ATM Switch

ISDN switch

Communication
Server

Gateway

Access
Server

PC

PC with
Software

Sun
Workstation

Macintosh

Terminal

File
Server

Web
Server

Cisco Works
Workstation

Printer

Laptop

IBM
Mainframe

Front End
Processor

Cluster
Controller

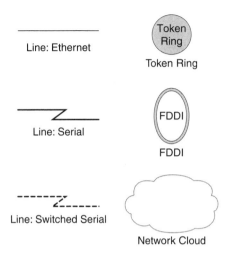

Line: Ethernet

Token Ring

Line: Serial

FDDI

Line: Switched Serial

Network Cloud

Command Syntax Conventions

The conventions used to present command syntax in this book are the same conventions used in the IOS Command Reference. The Command Reference describes these conventions as follows:

- Vertical bars (|) separate alternative, mutually exclusive elements.

- Square brackets ([ ]) indicate optional elements.

- Braces ({ }) indicate a required choice.

- Braces within brackets ([{ }]) indicate a required choice within an optional element.

- **Bold** indicates commands and keywords that are entered literally as shown. In configuration examples and output (not general command syntax), bold indicates commands that are manually input by the user (such as a **show** command).

- *Italic* indicates arguments for which you supply actual values.

Introduction

Service providers (SPs) and large enterprise customers (LECs) must deliver different services over a common infrastructure so that they don't interfere with one another (in other words, so that they pass like ships in the night). Multiservice switching networks achieve this goal. This book's objective is to shed light on the general architecture of multiservice switches and to present case studies on MPLS and PNNI, as well as both protocols running simultaneously. This helps engineers who design, deploy, and troubleshoot networks that use the BPX/IGX/MGX families of multiservice switches.

SPs and LECs offer managed VPN services and circuit emulation to enterprise customers and wholesale DSL services to ISPs and NSPs. In addition, SPs and LECs offer traditional ATM transport services from T1 to OC-48 using the same common infrastructure. They provide not only ATM PVC transport, but also lower-cost ATM SVC services.

Goals and Methods

This book covers the architecture of multiservice switches in which one or more controllers are attached to a controlled switch and the controllers act independently, like ships in the night. In this context, the specific details of MPLS and PNNI implementations are discussed with theory and examples.

Some of the most important details covered include the following:

- How to partition the MSS network between MPLS and PNNI domains and plan for expansion
- How to maintain IP QoS across a cell-based MPLS network
- How to build redundancy into the MSS network and minimize single points of failure
- An explanation of the VSI protocol and how to troubleshoot problems associated with it
- A discussion of PNNI as it relates to the multiservice switching product line, covering limitations on peer group hierarchy and size
- Effective utilization of IISP and AINI routing in a PNNI network
- An explanation of SVC signaling and how it pertains to PNNI
- An explanation of ILMI and how address autoregistration works using ILMI
- A description of cell-based MPLS, including how label distribution occurs and how VC-merge works
- A discussion of POP size and oversubscription
- The theoretical framework and standard bodies that describe the virtual switch and controller, using a switch control protocol for communication

Who Should Read This Book?

This book provides full coverage of MPLS and PNNI on multiservice switching platforms, including the benefits of these technologies and how to build and manage production networks based on them.

In general terms, network engineers, network architects, network design engineers, customer support engineers, and consultants who design, deploy, operate, and troubleshoot multiservice switching networks and who want to provide new-world enabling services on their networks should read this book.

Academics who need to understand this technology model and theoretical framework as part of their research will also find this book useful.

How This Book Is Organized

Although this book could be read cover-to-cover, it is designed to be flexible so that you can easily move between chapters and sections to cover just the material you need more work on.

Consistent with the covered technologies, this book follows a modular design. It is divided into three parts.

The first part includes Chapters 1, 2 and 3. It introduces the architectural framework and delves into virtual switch interface details and realizations. This part describes the controller/controlled switch architecture and the different master and slave models.

The second part, comprised of Chapters 4 through 7, details MPLS architecture, configuration, and design in multiservice switches.

The third part is comprised of Chapters 8 through 11. It discusses PNNI and ATM Forum technologies in multiservice switches, including theory, implementation, configuration, and design.

Chapter 12 presents some general conclusions for the whole book.

The 12 chapters cover the following topics:

- **Chapter 1, "What Are Multiservice Switching Networks?"**—This chapter introduces the multiservice switching architecture, which lets you use a common network infrastructure to support a mixture of services natively and without their interfering with each other. This chapter introduces and defines controllers, partitioning, virtual switches, and the "ships in the night" concept, among other constitutive pieces of the multiservice switching model.

- **Chapter 2, "SCI: Virtual Switch Interface"**—This chapter discusses the virtual switch interface and protocol, which is the switch control interface implemented in Cisco software. The virtual switch interface allows multiple networking protocols to control resources on a virtual switch. This chapter also covers partitioning concepts.

- **Chapter 3, "Implementations and Platforms"**—This chapter applies the ideas from the previous two chapters, which discuss multiservice switching implementations. Different MPLS and PNNI realizations of the multiservice switching architecture are analyzed in great detail.

- **Chapter 4, "Introduction to Multiprotocol Label Switching"**—This chapter introduces the multiprotocol label switching protocol. It covers Label Distribution Protocol (LDP) functionality and procedures including LDP sessions, label distribution methods, and loop detection. It also provides details on quality of service (QoS) using multi-vc, VC-Merge LSR capability, and MPLS Virtual Private Networks (VPNs).

- **Chapter 5, "MPLS Design in MSS"**—This chapter examines design concepts and best practices for MPLS implementations. It also presents scalability concepts including MPLS PoP sizing, how to calculate the required number of label virtual circuits (LVCs), and ATM MPLS convergence. Other sections include IP routing protocols and LSC redundancy.

- **Chapter 6, "MPLS Implementation and Configuration"**—This chapter covers the configuration of MPLS in multiservice switches. A complete MPLS network with three points of presence is implemented from the ground up. This MPLS network is the foundation for the application plane services introduced in Chapter 7. This chapter also presents a generic configuration model that applies to all platforms, as well as some troubleshooting scenarios. A table summarizing all the configuration commands and steps is presented as well.

- **Chapter 7, "Practical Applications of MPLS"**—This chapter covers the configuration of MPLS VPNs and MPLS quality of service.

- **Chapter 8, "PNNI Explained"**—This chapter answers the question "Why PNNI?". Private Network-to-Network Interface is described and analyzed, in both the routing and signaling aspects. This chapter also describes ATM UNI signaling, details ATM addressing concepts, and explains the use of Interim Interswitch Signaling Protocol (IISP) and ATM Inter-Network Interconnect (AINI) to join two PNNI networks.

- **Chapter 9, "PNNI Network Design Goals"**—This chapter examines PNNI network design goals related to providing end-user services, PNNI POP design and scaling, redundancy design, and core bandwidth concerns.

- **Chapter 10, "PNNI Implementation and Provision"**—This chapter discusses the configuration of ATM Forum services in multiservice switching platforms. After discussing a general configuration model, this chapter sets up a PNNI network made up of three Points of Presence (POPs). It also covers the configuration of SVC and SPVC services, NCDP, and filters. A table summarizing all the configuration commands and steps is presented.

- **Chapter 11, "Advanced PNNI Configuration"**—This chapter covers the fundamentals of hierarchical PNNI configuration, achieving the hierarchical abstraction. Other advanced topics and configurations, such as IISP, traffic engineering, and connection admission control, are also covered.

- **Chapter 12, "Virtual Switch Review"**—This chapter provides a closing for the book, reconverging in the multiservice switching architecture and providing general conclusions.

- **Appendix A, "Service Traffic Groups, Types, and Categories"**—This appendix presents the virtual switch interface (VSI) service groups, types and categories used to deliver QoS in Multiservice Switches.

CHAPTER 1

What Are Multiservice Switching Networks?

Suppose you needed one PC for each application you run on your computer: one computer for Web browsing, another for word processing, and so on. Or suppose a city had one set of roads and highways to be used by cars, a different set for trucks, and a third for motorcycles.

From economic and practical points of view, these setups don't seem like the best choices. Service providers and Enterprise customers need to consider the problems they face so that they don't end up with similar scenarios. They should ask why they need a dedicated network to provide voice services, a different infrastructure to carry IP traffic, and a third one to supply ATM virtual circuits. Why should anyone need to design and maintain a network of routers and a parallel ATM switch network? A multiservice switching network uses the same infrastructure to carry any type of traffic, regardless of whether it is encapsulated in IP, ATM, or Frame Relay.

A first approach to a definition of a multiservice switching network is a common transmission and switching infrastructure that can natively provide multiple services so that each service does not interfere with the others. Multiservice switches are therefore network elements that constitute the multiservice switching network and natively provide those different services.

Ships in the Night

One of the requirements for multiservice switching networks is independence between different services. In other words, services do not interfere with each other. An upgrade of the multiprotocol label switching (MPLS) software that controls the switches should not affect the digital subscriber line (DSL) customers who use ATM SVCs from network access or voice customers who are supported with VoATM using AAL2. A first draft of a multiservice switching network might look like Figure 1-1.

Figure 1-1 *Example of a Multiservice Switching Network Providing Three Services*

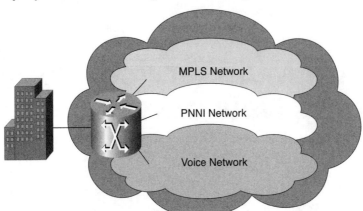

Multiservice Switching Network

A number of things are worth mentioning about this figure. First, note that the only physical network is the big dark cloud. The smaller clouds within the physical network are logical partitions of the physical network by service.

This multiservice switching network provides unified access to multiple services or virtual networks—MPLS, Private Network-to-Network Interface (PNNI), and voice in this example. The links from the multiservice switch into each logical network are then logical links, created from logically partitioning resources in the physical link.

But the most important point to note is that this example has no overlays. All the services are running natively on the multiservice switches. For example, with the MPLS logical network, if you have an underlying ATM network, the multiservice switches run an IGP and LDP for setting up LVCs and not MPLS over PVCs. In other words, every switch in the multiservice network is aware of the services it provides.

NOTE   Resource partitioning is a key element in the architecture. It is revisited several times in this chapter.

A diagram of a network as a cloud is an oversimplification. Let's explore this a bit further. Figure 1-2 represents the physical topology of the ATM network.

Figure 1-2   *Underlying Physical ATM Networks*

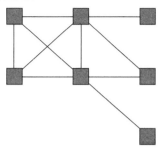

Underlying ATM Network

Resources in the network can be partitioned for the different services to create the logical networks over a common infrastructure. Links and even entire nodes can be made invisible to some control planes in the process of building these virtual networks, as shown in Figure 1-3.

Figure 1-3   *Virtual Networks over the Common Infrastructures*

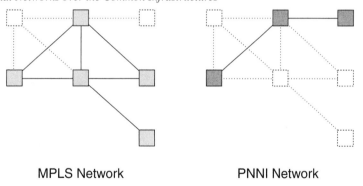

MPLS Network   PNNI Network

From the total of seven nodes and ten links that exist in the physical network, the MPLS network is visible to only five nodes and six links, and the PNNI network "believes" that there are only three nodes and two links. Moreover, from the MPLS network's perspective, the PNNI network does not exist. Two nodes and one link are shared by the two logical networks, but the MPLS control plane in those common nodes or any other node is unaware of the existence of the PNNI network with which it shares resources, and vice versa. This behavior is called *ships in the night*. It means that the different control planes act like the brain of the multiservice switches, operating independently on the same network fabric simultaneously.

Separating the Control and Switching Planes

Let's take a different approach to looking at the characteristics of multiservice switching networks. This section goes one step deeper into explaining these virtual networks by delving into the switch architecture. How does a switch need to be designed in order to build these networks?

We can start by studying a provider that offers different services, such as voice and ATM, over separate networks. Analyzing the telephony switches and ATM switches that make up these infrastructures, we find that typically they are dedicated pieces of equipment in which the control plane in those switches is indivisible from the switching plane. Moreover, in the software architecture of those switches, usually the control logic is aware of the different cards in the switch and the specifics of the interfaces and even drivers. This section investigates the advantages that a modular approach like the one outlined in Figure 1-4 brings.

Figure 1-4   *Modular Architecture of a Multiservice Switch*

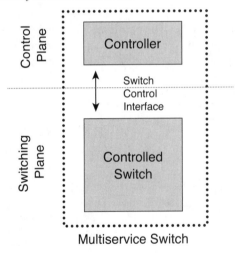

The multiservice switch shown in Figure 1-4 has different modules or components. Their functions are outlined in Table 1-1.

Table 1-1   *Multiservice Switch Modules and Their Functions*

Module	Function
Controlled switch	This is the module responsible for traffic switching.
Controller(s)	This is the module responsible for managing the controlled switch. It acts like the "brain" of the controlled switch.
SCI	Switch Control Interface. The interface that manages the communication between the controller and the controlled switch.

The following parallels can be made:

- The controller is analogous to the system's "brain."
- The controlled switch is comparable to the "muscles" of the system performing the switching.
- The SCI connecting the preceding two items is equivalent to the "nervous system," bringing information from the switch to the controller and commands from the controller to the switch.

This analogy is shown in Figure 1-5.

Figure 1-5 *Functions of Each Constitutive Module in a Multiservice Switch*

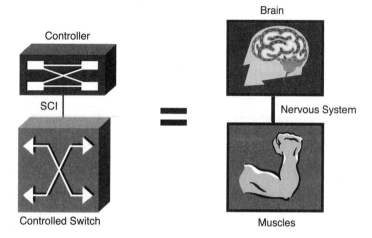

The controller operates independently of the controlled switch. Different controllers operate completely independently of each other. Resource partitioning (discussed later) enables the multiservice functionality of the controlled switch.

This architecture provides *separation of routing and forwarding*, meaning routing running in the controller and forwarding performed by the controlled switch. This gives you the ability to design the best routing software controlling the best switch, which could be practically impossible otherwise. This also presents the framework for implementing technologies that rely on that separation of routing and forwarding, such as MPLS.

Partitioning, Virtual Switches, and Multiple Control Planes

The controlled switch can have different physical cards with a number of interfaces. The network operator can partition the number of connections and the interface resources (by configuration performed in the controlled switch), such as bandwidth, VPI, and VCI ranges (assuming that the switch is an ATM switch, but it does not have to be an ATM switch). These partitioned resources can then be assigned to different controllers.

Examine Figure 1-6, an ATM switch that has five physical interfaces and two different controllers. The first two interfaces can be in a card, and the other interfaces can be in a different card. The controlled ATM switch also has an MPLS controller and a PNNI controller connected to it.

Figure 1-6   *Multiservice Switch with Two Controllers*

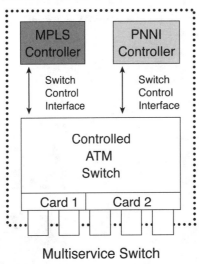

Figure 1-7 shows the resource partitioning and controller assignment in the ATM switch. These partitions of resources from the physical interface are called controlled interfaces from now on.

Figure 1-7   *Resource Partitioning of the Interfaces*

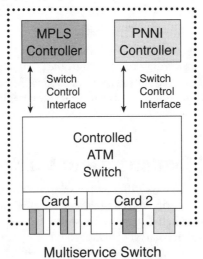

The first interface's resources are partitioned such that one third of the interface resources make an MPLS partition, and another third are assigned to a PNNI partition. The same thing happens with the switch's second interface. From the fourth interface, two-thirds of the resources are set aside for an MPLS partition. The total resources of the fifth interface are given to a PNNI partition.

NOTE   The third interface's resources are not assigned to any controller; therefore, from the controller's perspective, that interface does not exist.

To explicitly demonstrate the "ships in the night" concept, Figure 1-8 shows the visibility of the switch that each controller has. The MPLS controller "perceives" the switch as a box with three interfaces.

Figure 1-8   *Virtual MPLS Switch*

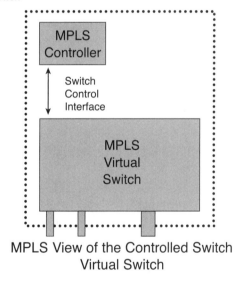

MPLS View of the Controlled Switch
Virtual Switch

The MPLS controller does not make any distinctions as to whether the interfaces are in the same physical card or not. It believes the switch has three interfaces it can control and utilize with the given resources. This architecture relies on the separation between network information and specific hardware information.

This brings us to the next concept. The ATM switch is presented as a *virtual switch* to the controller, with only the interfaces partitioned for it (which are logical interfaces with generic resources).

Likewise, the ATM controlled switch is presented as a virtual switch to the PNNI controller, as shown in Figure 1-9.

Figure 1-9   *Virtual PNNI Switch*

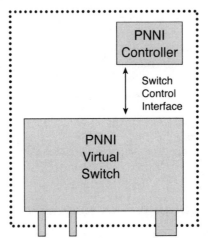

PNNI View of the Controlled Switch
Virtual Switch

In this application we can differentiate three switches:

- The physical switch with resources partitioned in the interfaces to be used by the control logic internal to the ATM switch (the one we mentioned that could not be physically separated). This control plane might or might not exist. If it does not exist, the resources not assigned to any controller are not used.

- A virtual switch presented to the MPLS controller.

- A virtual switch presented to the PNNI controller.

The MPLS virtual switch doesn't know about MPLS. It's completely oblivious. The controller based on the MPLS information controls the operation of the MPLS virtual switch (resources partitioned in the ATM switch), and the pair form a Label Switch Router (LSR) (discussed in more detail in Chapter 3, "Implementations and Platforms").

Advantages of the Architecture

At this point we can step back and take a look at some of the advantages of multiservice switches. From a network operator perspective, several advantages are

- Multiple services can operate on the same platform. This lets providers develop multiple revenue streams from a single investment in packet networking.

- "Ships in the night" functionality. Controllers operate independently from each other; resulting in no intercontroller dependencies. For example, if one controller needs to be upgraded (in software or hardware) or removed, it does not interrupt service for other controllers.

- The capability to choose the control plane that is most suited for the particular application. As a starting point, you can choose PNNI for connection-oriented services and MPLS for IP connectionless services. Also, you can choose the controlled switch based on the switching needs.

The benefits from a networking equipment developer and vendor perspective are

- Platform-independent software. The fact that the controller is unaware of the platform-specific details allows the reuse of controller software. This allows a faster time to market for newer switches.

- The ability to reuse controller software also provides consistency and interoperability/ interworking in operation.

- Simpler controller software because of the elimination of platform-specific code results in higher-quality code and fewer defects.

- Different groups or business units can develop the controlled switch and the different controllers.

As another advantage, the network control module and the switching and forwarding functions can evolve independently from each other.

Completing the Picture

So far we have been under the assumption (sometimes implicit and other times explicit) that the controlled switch is an ATM switch with ATM interfaces. But the architecture of multiservice switches is more general in order to allow multiple realizations of the model. Two more planes need to be added to present the complete framework, as shown in Figures 1-10 and 1-11.

First, an adaptation plane is required to provide the multiple service access and adaptation to the core technology (see Figure 1-10). For example, if the controlled switch is an ATM switch, services such as Frame Relay and voice need adaptation to ATM.

Figure 1-10 *Adaptation Plane*

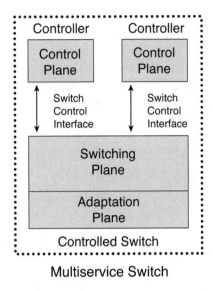

Multiservice Switch

This plane might or might not be present. In some cases, the adaptation is done in the Customer Premises Equipment (CPE).

Finally, a management plane allows the multiservice switches to be configured, controlled, and monitored, as shown in Figure 1-11.

Figure 1-11 *Management Plane*

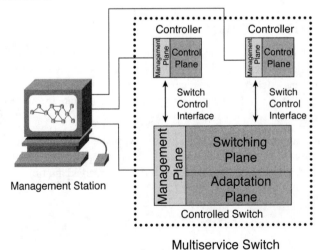

Multiservice Switch

The interface between the control plane and the switching plane is the SCI. The adaptation plane contains the partitioning function that partitions and manages resources in the different interfaces in the controlled switch.

Multiservice Switching Forum

The Multiservice Switching Forum (MSF) develops the implementation agreements for multiservice open architectures. The Multiservice Switching Forum was organized by WorldCom, Cisco Systems, Inc., and Telcordia. It was cofounded in 1998 by Cisco Systems, Inc. The founding members include leading telecom service providers, AT&T, BT, Telia, Telecom Italia, and US West, as well as equipment vendors such as Alcatel, Ascend, Fujitsu, Lucent, Nortel, and Siemens. Membership is open to anyone.

Its mission is to accelerate the deployment of open multiservice switching networks, focusing on the development of architectures and industry agreements.

MSF members are international service providers and equipment suppliers. Both carriers and vendors gain benefits from this global forum.

The MSF also sponsors interoperability events to ensure proof of implementation agreements and true multivendor capability.

The MSF leverages existing agreements and standards from many associations through liaisons with the ATM Forum, IEEE, ETSI, and IETF.

More information about the MSF can be found at http://msforum.org.

Summary

This chapter presented a first exposure to the multiservice switching architectural framework. The modular approach, with separation among the control, switching, adaptation, and management planes, carries a number of associated benefits from both end-user and designer/ vendor viewpoints. This includes economic benefits that stem from the support of a full range of network services and technologies over a single, common infrastructure.

This architecture promotes high-availability networks, with distributed topology and connection information in the controllers.

As a consequence of the multiservice switch decomposition, different control planes acting independently provide "ships in the night" functionality. In this model, multiple protocol controllers utilize a share of a switch's resources, but each controller is oblivious to the fact that others are in operation. Protocol controllers, such as MPLS or PNNI controllers, act as the brains to manage the brawn that is the controlled switch's resources. Resource and partition resource management are the key to true multiservice capabilities.

The next chapter concentrates on analyzing the Virtual Switch Interface and Cisco Systems, Inc.'s implementation of multiservice switches.

CHAPTER 2

SCI: Virtual Switch Interface

Virtual Switch Interface (VSI) is the Switch Control Interface (SCI) that Cisco Systems, Inc. implements in multiservice switches. SCI allows multiple controllers to independently control a single ATM switch, as described in the preceding chapter.

VSI is a mature Switch Control Protocol (SCP) that has proven to be a solid implementation providing true multiservice capability. Other SCIs include IETF's GSPM v.3.

VSI is a message-based interface between the network software and the platform software. It provides a hardware-independent control interface between one or more VSI masters to control a switch (possibly a remote switch). Multiple controller coexistence is possible because of the resource partitioning provided by VSI. This chapter explores the VSI protocol and implementation.

VSI Functions

VSI is a master-slave protocol through which a master can control the operations of one or many slaves. VSI was invented by Cisco Systems. The current implementation is VSI version 2.4.

The main functions of VSI are as follows:

- **Discovery of the switch configuration**—VSI learns about the controlled switch configuration. It can discover the available logical controlled interfaces, the resources allocated in those interfaces, and their status. The VSI slave informs the master of a set of resources associated with a logical interface.

- **Control the switch cross-connects**—The setup and teardown of cross-connects in the controlled switch allow data forwarding, according to what the network protocol mandates. This includes QoS, different service types, and oversubscription.

- **Monitoring of port and channel statistics**—VSI allows the VSI master to gather statistics or counters about ports and connections from the VSI slaves.

- **Resynchronization**—If the VSI master and the slaves get out of sync as far as the cross-connect databases, the resynchronization capabilities allow them to resynchronize without having to tear down and rebuild every connection from scratch. This is done with checksums.

A controller-controlled switch pair (that comprises the multiservice switch) is shown in Figure 2-1.

Figure 2-1   *Multiservice Switch Modules*

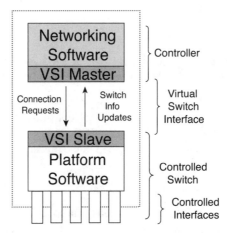

The different modules and components shown in Figure 2-1 can be classified as follows:

- **Controller**—Controls the operation of the controlled switch.

  - **Networking software**—At this layer are the software modules and APIs that control the cross-connect decisions. In an MPLS application, these are the IP routing protocols, such as Interior Gateway Protocol (IGP) (OSPF, IS-IS, and so on) and Label Distribution Protocol (LDP). In a PNNI application, these are UNI signaling and PNNI signaling and routing.

  - **VSI master**—The master portion of VSI is implemented in the controller. Using VSI commands, the controller sets up the cross-connects in the switch. One VSI master is responsible for the entire switch. It interfaces to the networking software.

- **Controlled switch**—The ATM switch under the direction of the controller.

  - **VSI slave**—The implementation of VSI in the switch executes the commands the master sends and updates the master with the switch information. To enable the controllers to act independently, the VSI slave process(es) in the switch must allocate resources to the different control planes.

  - **Platform software**—The platform software handles general switch management and switch-specific functions.

  - **Controlled interfaces**—As soon as resources are allocated in the interface for a specific control plane, that interface is made visible from the controller and thus becomes a controlled interface. This is a purely logical concept, as is the virtual switch to which the interface belongs.

From these architectural modules, an important point discussed in the preceding chapter is made explicit: The VSI slaves present a view of a virtual switch to the VSI master and therefore to the networking software module. A physical interface and a logical interface (for example, a virtual interface that uses only one VPI in the interface, or an Inverse Multiplexing ATM [IMA] interface with multiple T1s) are the same from the VSI master's perspective, and thus from the networking software's perspective. The controller view of controlled interfaces is shown in Figure 2-2.

Figure 2-2   *Controller View of Controlled Interfaces*

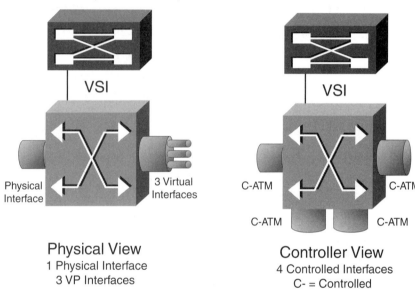

Physical Interface

3 Virtual Interfaces

C-ATM   C-ATM

C-ATM   C-ATM

Physical View
1 Physical Interface
3 VP Interfaces

Controller View
4 Controlled Interfaces
C- = Controlled

As a consequence, a physical interface with multiple virtual path logical interfaces is seen as multiple virtual interfaces from the controller. On the other end of the spectrum, multiple physical interfaces in an IMA group in the controlled switch are presented to the controller as a single interface. This is summarized in Table 2-1.

Table 2-1   *Physical-to-Logical Ratio of Controlled Interfaces*

Logical Interface	Physical-to-Logical Ratio
Physical interface	1:1
Virtual interface	1:N
IMA interface	N:1

Another direct consequence of this is the fact that VSI's Application Programming Interface (API) hides the platform hardware-specific parameters, software, and firmware from the networking software, cleanly separating them. VSI provides primitives for generic ATM cross-connect creation, deletion, and modification. Typically, platforms have more parameters (usually hardware-specific) to configure for a connection than the networking protocol (such as PNNI or MPLS). Those parameters are called *extended parameters*. Figure 2-3 shows this separation concept.

Figure 2-3   *VSI Hides Platform Specifics from the Networking Software*

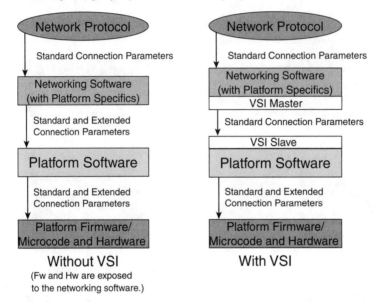

In switches that do not implement this modular approach, the mapping from standard protocol parameters to switch-specific extended parameters is typically done at the networking software, making the networking modules difficult to port to another platform with a different set of extended parameters. These parameters might be as important as different QoS queues.

In multiservice switches, the mapping is done at the platform software (eliminating platform-specific code from the networking software). This allows platform-independent networking software and results in each platform's optimally mapping standard parameters to extended platform-specific parameters. VSI achieves this by defining *service types,* letting the controller specify a category of traffic management when setting up a connection.

NOTE   The switch selects *logical interface numbers* (LINs) to be used by VSI while updating the controller. Those numbers need to be unique and persistent, but this is done switch-implementation-specific. (LINs are assigned by the slave in a platform-specific way. They need to be unique even if the controlled switch is a multishelf node.) The mapping between physical interface name and LIN is maintained by the VSI slave(s) and is hidden from the VSI master. Moreover, the VSI master does not assume any structure in the number and does not presume the range to be compact (even if sometimes a structured approach is desirable).

The LIN is a 32-bit value that the switch needs to advertise before the controller can use it. Valid LINs range from 0 to 0xFFFFFFFE. The VSI master uses the 0xFFFFFFFF value to indicate to the VSI slave that it wants to start walking the slave's LIN values.

VSI Master and VSI Slaves

The VSI separates the networking application from the switch-specific software. The networking application includes a VSI master module, and the switch includes one or more VSI slave modules.

Although there's only one VSI master per controller, there can be one or more VSI slaves in the controlled switch. The VSI slaves can be classified as centralized, distributed, and hybrid.

In the centralized model, shown in Figure 2-4, a single VSI slave process is running on the switch control card. It controls all the interfaces in the switch. The VSI master interfaces with only one VSI slave process. The switch's resource management is centralized. This model is also called the proxy VSI slave because it handles the slave functionality for the interface cards.

Figure 2-4   *VSI Slaves: Centralized Model*

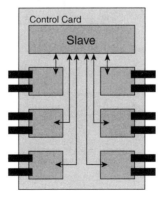

In the distributed model, shown in Figure 2-5, each card has its own VSI slave process, so several VSI slave processes are running in the switch. These cards have more intelligence than the cards in the centralized model. Resource management and connection setup and teardown run on the processors on the service cards.

Figure 2-5 *VSI Slaves: Distributed Model*

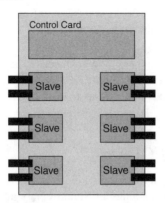

The controller card does not take part in connection control; it performs configuration management operation.

Connection requests from the VSI master might need to be performed with two slaves to set up the two *legs* of the *cross-connects*.

The last model, shown in Figure 2-6, is a combination of the previous two models. It can be seen as a special kind of distributed model. The control card controls interfaces on some cards, and other cards run their own VSI slave process.

Figure 2-6 *VSI Slaves: Hybrid Model*

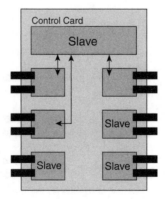

From a different angle, a switch controlled by multiple controllers using VSI looks like Figure 2-7.

Figure 2-7   *Multiple Controllers*

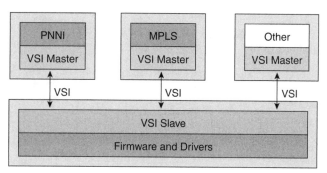

Each controller uses a different controller-id when communicating with the slave(s).

There is no requirement for master-to-master communication. Resources in the switch are managed by the platform software, reinforcing the "ships in the night" concept.

VSI Messages

The information flow between VSI master and VSI slave has two directions. The VSI master sends connection requests to the VSI slave(s), and the VSI slave(s) update the VSI master with switch information.

Figure 2-8 shows the format of the VSI message. It allows the future definition of new parameter group formats without having to redefine the message format.

Every VSI message has a version-id field in the header. The VSI master and VSI slave are expected to support multiple old versions to allow upgrades and backward compatibility. The VSI master is required to select the VSI protocol version used by the slaves. The master chooses the highest version that is common to both the VSI slave and VSI master, performing *VSI version negotiation*. Also, in multiservice switching networks, a slave is expected to be able to handle different versions, because different VSI masters can select different versions for the same VSI slave.

Figure 2-8   *General VSI Message Format*

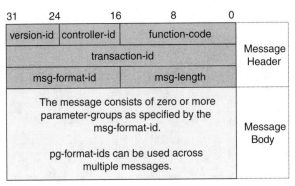

VSI version negotiation happens at the VSI session level. In other words, a VSI master can use VSI version X with some slaves and VSI version Y with other slaves. Moreover, a VSI slave can use VSI version A with one master and VSI version B with a second master.

The parameter-group format is shown in Figure 2-9.

Figure 2-9   *Parameter-Group Format*

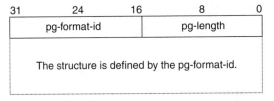

In a VSI message, a given function code can have more than one valid message-format-id. A parameter-group structure can be used in different message formats (and function codes). A given parameter group type can have more than one valid parameter-group format.

VSI messages can be divided into the following main groups:

- **Connection messages**—There are two subtypes of connection messages:
    - **Connection requests**—The controller requests connection-related commands (setup reserve, setup commit, teardown, modification, or get information).

— **Connection request responses**—The VSI slave acknowledges the connection requests. The slave is required to reply with either ACK or NACK, and the master is required to get an explicit reply for every connection request.

- **Interface information**—The controller gets interface information from the switch using these messages. The VSI slave(s) can also send notification of changes asynchronously with VSI traps. VSI is designed such that it eliminates the need for the controller to know the details of the interface.

- **Switch information**—The controller gets switch-related information with these messages. The switch can also send updates asynchronously (such as changes in synchronization state [session-id] and changes to logical interfaces). These messages are also often used as keepalive messages between the master and all the slaves.

- **Miscellaneous information**—These messages include the generic error response message, the generic debug command and response, and passthrough (up and down, command and response).

- **Bulk statistics**—Messages (commands, responses, and traps) related to bulk statistics.

It's worth mentioning that in a distributed model, the slaves have specific functions apart from the VSI slave functions, such as performing resource management operations and call acceptance, responding to connection requests, notifying the controller of changes in the interfaces, and interfacing with the hardware drivers. These functions are to set up VSI communication channels with other VSI slaves in the controlled switch, communicate with other slaves for connection setup, maintain individual session-ids per slave, and maintain an interfaces-to-slave mapping. The switch is also responsible for setting up a full mesh of VSI interslave communication channels, transparently to the VSI master.

NOTE   Each VSI master maintains individual and independent sessions with each slave (each session has its own flow-control configuration). The VSI slaves, therefore, are also called *sessions*.

The VSI protocol specifies that the slave modules communicate with each other in setting up connections that span multiple slaves, as shown in Figure 2-10.

Figure 2-10 *Cross-Connects: VSI Slave Distributed Model*

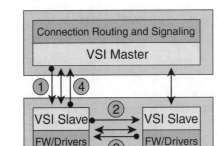

In Figure 2-10, the networking module (connection routing and signaling) decides to set up (reserve) a cross-connect between interfaces A and B in the controlled switch. Two different VSI slaves manage the two interfaces' resources:

Step 1 The VSI master sends a **connection request** command to the VSI slave that manages interface A's resources.

Step 2 After passing the CAC (Connection Admission Control), VSI slave A allocates resources locally and sends the **connection request** to VSI slave B for setting up the second leg of the cross-connect.

Step 3 VSI slave B checks the CAC, allocates resources, and acknowledges the request with a **connection request response** to VSI slave A.

Step 4 When the two legs or endpoints of the cross-connect are set up (reserved), VSI slave A sends the **connection request response** to the VSI master. The response also includes updated loading information for the interfaces. This approach also eliminates any central bottleneck and allows performance scalability using distributed processing.

Note that this example only requests that a connection segment be set up—that is, reserved. The activation of the segment is performed with a **connection commit**.

End-to-End Connections

This section analyzes the steps involved in the creation of an end-to-end connection. As you know, the multiservice switching architecture is based on the separation of the switching plane and the control plane. A general three-node network might look like Figure 2-11.

Figure 2-11 *Multiservice Switching Network: End-to-End Connections*

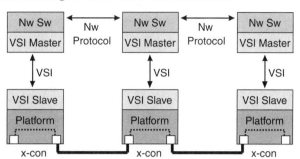

Based on the networking protocol running in the controllers (at the networking protocol layer), the VSI master instructs the controlled switch to set up the cross-connects that are the segments of the end-to-end connections. The controlled switch does not know the end-to-end connection information or topology, only the local cross-connect.

If the controllers are MPLS controllers, the networking protocol is LDP and an IGP, such as OSPF or IS-IS. The label distribution mechanism is Downstream on Demand, and will be discussed in detail in chapters 4 "Introduction to Multiprotocol Label Switching" and 6 "MPLS Implementation and Configuration." Figure 2-12 shows a multiservice switching MPLS network.

Figure 2-12 *Multiservice Switching Network: MPLS Example*

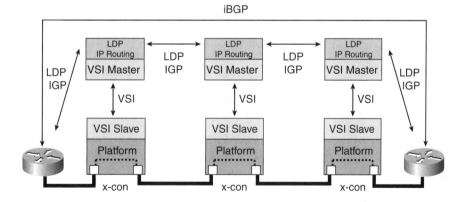

NOTE	The label distribution technique used in ATM LSR has some variations that are detailed in Chapters 4 and 6. For ATM-LSRs that are not VC-merge-capable, the distribution mechanism must be Downstream on Demand, and optionally is the same for ATM LSRs that support VC-merge. Downstream on Demand has two control modes: independent (optimistic label allocation) and ordered (conservative label allocation). These are defined in RFC 3031 ("Multiprotocol Label Switching Architecture") and RFC 3036 ("LDP Specification"). These variations determine how end-to-end connections are signaled and set up, and thus at which point the cross-connects are installed in the controlled switches.

NOTE	Here is the definition of VC-merge in RFC 3035 ("MPLS Using LDP and ATM VC Switching"), Section 3: "VC-merge is the process by which a switch receives cells on several incoming VCIs and transmits them on a single outgoing VCI without causing the cells of different AAL5 PDUs to become interleaved."

NOTE	A VPN MPLS network also has edge-to-edge communication. The PE (Provider Edge) devices have an iBGP neighboring relationship.

Let's explore a complete example of how a controller can set up and tear down a connection segment of an end-to-end connection in a multiservice switching PNNI network. See Figure 2-13.

Figure 2-13 *Multiservice Switching Network: PNNI Example*

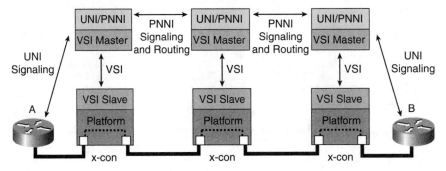

The first part of the example details the steps that occur in a Connection SETUP:

Step 1 Router A sends a **SETUP** message to router B to establish an SVC.

Step 2 The controller in the first node receives the **SETUP** message and sends a **CALL PROCEEDING** back to the router.

Step 3 The controller requests a connection segment using VSI to be set up (reserved) at the local controlled switch.

Step 4 The controller sends the **SETUP** signaling message to the next node.

Step 5 The VSI slave in the local controlled switch parses the VSI message from Step 3, checks the CAC, allocates resources (message execution), and sends the response back to the controller (message responding). The response includes updated loading information for the interface.

Step 6 Router B receives the **SETUP** message and replies with a **CALL PROCEEDING**.

Step 7 Router B sends a **CONNECT**, and the **CONNECT** message reaches the controller in the first node.

Step 8 The controller sends a **CONNECT** message to router A.

Step 9 The controller requests the connection segment to be activated (committed) using VSI at the local switch.

Step 10 The local switch configures the hardware to enable data flow and sends an acknowledgment back to the controller using VSI.

The preceding is a simplified example from a Q.2931 point of view, trying to emphasize VSI messaging in every step. The details of PNNI signaling will be covered in Chapters 8 "PNNI Explained" and 10 "PNNI Implementation and Provision."

The second part of the example describes the stages in a Connection RELEASE process:

Step 1 The controlled switch detects a failure from the hardware on a specific interface, such as a loss of signal (LOS).

Step 2 The controlled switch notifies the controller of the logical interface failure using VSI.

Step 3 The controller sends **CALL RELEASE** messages for every call on that interface to other controllers and routers.

Step 4 The controller requests that the cross-connects be removed from the local switch.

Step 5   The controlled switch removes connection legs and releases resources. The VSI slave sends a **connection request response** VSI message back to the controller with updated loading information for the corresponding interfaces.

Controller Location Options

There are multiple realizations or implementations of the controller/controlled switch architecture. Figure 2-14 summarizes the controller location options.

Figure 2-14 *Controller Location Options*

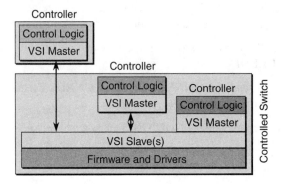

You can see that three different locations exist for the controller with respect to the controlled switch:

- On the external platform—a separate "controller shelf"
- On the switch on a separate card, cocontroller card, or coprocessor
- On the switch controller card

Because the controller can reside on either the switch or off-switch, the choice for lower layers is left for individual implementations. In the first case, where you have a separate controller, there's also a physical interface between the controller and the control switch. That physical interface is not part of the VSI specification, even though there is an appendix in the VSI specification that states how LLC/SNAP/AAL5 encoding over ATM can be used. VSI across Frame Relay, LAN, or a reliable message transport such as TCP or SSCOP is not currently defined, but it might be defined in future versions of the VSI specification.

When using ATM, the controller and switch use ATM VCs to communicate with each other. As shown in Equation 2-1, there is one ATM PVC per slave using these virtual path identifiers (VPIs) and virtual circuit identifiers (VCIs) by default:

Equation 2-1 *Default VPIs/VCIs Used by VSI*

$$
\begin{cases}
VPI = 0 \\
VCI = 40 + SlaveID
\end{cases}
$$

These are the default VPI and VCI values used. They can be changed in some implementations by mutual agreement between the master and the slaves (in other implementations, they are fixed). In the case of a centralized VSI slave, the **slave-id** is 0, so the VCI is 40.

The encapsulation used is AAL5 LLC/SNAP, as shown in Figure 2-15.

Figure 2-15 *ATM Encapsulation of VSI Messages*

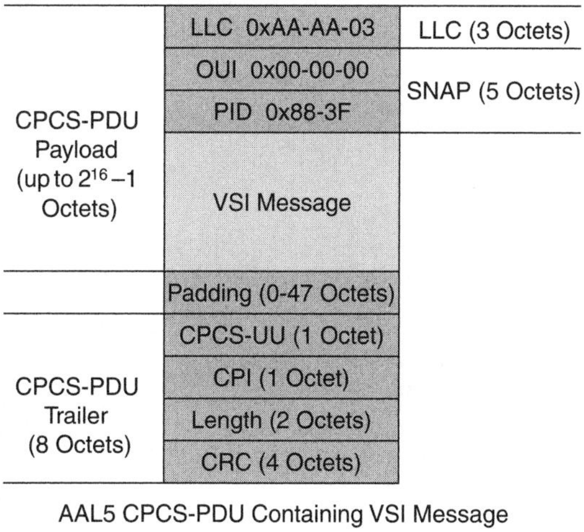

	LLC 0xAA-AA-03	LLC (3 Octets)
	OUI 0x00-00-00	SNAP (5 Octets)
CPCS-PDU Payload (up to $2^{16}-1$ Octets)	PID 0x88-3F	
	VSI Message	
	Padding (0-47 Octets)	
	CPCS-UU (1 Octet)	
CPCS-PDU Trailer (8 Octets)	CPI (1 Octet)	
	Length (2 Octets)	
	CRC (4 Octets)	

AAL5 CPCS-PDU Containing VSI Message

The VSI message is directly encapsulated in AAL5 using Common Part Convergence Sublayer Protocol Data Unit (CPCS-PDU), LLC/SNAP header. The presence of a SNAP header is indicated by the LLC header value of 0xAA-AA-03. A SNAP header has two fields: OUI (Organizationally Unique Identifier) and PID (Protocol Identifier). An OUI value of 0x00-00-00 states that the following PID is an Ethertype. A PID of 0x883F indicates that the PDU is a VSI message.

NOTE   The LLC/SNAP header described in Figure 2-15 corresponds to the VSI specification that was submitted to the Multiservice Switching Forum. As a Cisco Systems extension to VSI, the implementation of VSI over AAL5 LLC/SNAP uses a proprietary header value: a Cisco OUI of 0x00000C with a PID of 0x0103, indicating VSI.

Hard, Soft, and Dynamic Resource Partitioning

Even though resource partitioning is not required for VSI, it's of paramount importance to support multiservice switching networks. The Resource Management Module (RMM) in the VSI slaves is the enabler of the multiservice functionality.

As you know, interfaces have resources associated with them (such as VPI/VCI ranges, bandwidth, and number of channels). A multiservice switch supporting multiple controllers must assign some resources to each controller (at least, it must assign interfaces to different controllers). Some of those resources must be used exclusively by one controller, such as VPI/VCI ranges, and others can be exclusive or shared with other controllers, such as bandwidth. A partition is an allocation of resources for use by a specific controller, such as MPLS or PNNI.

After assigning resources to different partitions on the same interface, the switch must then advertise only the resources allotted to a given partition to the controller assigned to the partition. This defines the creation of the virtual switch, and the controller does not need to know that the interface has resources partitioned. The resources might also include pooled resources.

In Figure 2-16, resources are partitioned in the interface and are assigned to two controllers—PNNI and MPLS. The resources are channel or logical connection number (LCN) space, bandwidth, and VPI/VCI space.

Figure 2-16 *Resource Partitioning*

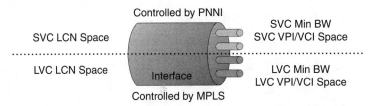

A mechanism must exist to share those resources among various partitions assigned to different controllers. There are two ways to do that: hard partitioning and soft partitioning.

A few key points to remember when it comes to assigning resources to a partition during configuration are

- No partition can utilize the minimum bandwidth or LCN space allotted to another partition for static resource allocation, even if the other partition is not currently using the resource.

- All partitions configured within a physical interface share that interface's bandwidth.

- All partitions in a port group share the LCNs allotted to that port group. A port group is a set of interfaces that share LCN resources. Port groups can contain multiple interfaces.

Hard Partitioning

VPI/VCI space partitioning is strict, meaning that one VPI/VCI range is assigned to one and only one owner controller, and it cannot be shared with other controllers. VPI/VCI ranges do not overlap among partitions. The term to describe these kinds of partitions is *hard partitioning*.

In a hard partitioning method, each controller gets a fixed amount of resources (bandwidth or channels). Although this is a straightforward scheme, the drawback is that the resources can be used inefficiently.

NOTE   In the case of AutoRoute-to-PNNI migration, soft partitioning of VPI/VCI resources is implemented with some restrictions. In other words, different control planes can have an overlapping VPI/VCI space.

Soft Partitioning

In contrast to hard partitioning, *soft partitioning* of resources has two implications:

- It provides resource guarantees for some types of resources (for example, LCN and bandwidth).

- It provides a pool of shared resources that are available to all partitions in addition to the guaranteed resources.

One way of complying with both requirements of soft partitioning is to specify a minimum and maximum amount for each resource when soft partitioning those resources. The minimum is the guaranteed value, and the maximum is the upper boundary provided that the extra resources are available and are not used by other partitions. In other words, this allows for a common pool of shared resources.

Bandwidth Soft Partitioning

This list defines bandwidth soft-partitioning concepts:

- **Partition_Minimum_BW**—The amount of bandwidth a partition is guaranteed.

- **Partition_Maximum_BW**—The upper limit on the amount of bandwidth a partition can use.

- **Interface_VSI_reserved_BW**—The total amount of bandwidth reserved for VSI in an interface.

- **Interface_VSI_PoolSize_BW**—The total amount of bandwidth that constitutes a *pool* that is shared among VSI partitions in the same interface.

- **Partition_VSI_PoolSize_BW**—The amount of bandwidth that a given partition on an interface can use out of the Interface_VSI_PoolSize_BW.

Users only directly configure the minimum and maximum partition bandwidth. The rest of the values are calculated from these two.

To make this point completely clear, let's analyze two examples of bandwidth partitioning between three partitions in the same interface. See Figures 2-17 and 2-18.

Figure 2-17 *Soft Partitioning of Bandwidth Resources*

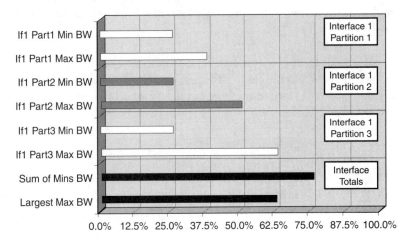

Figure 2-18 *Soft Partitioning, Part 2*

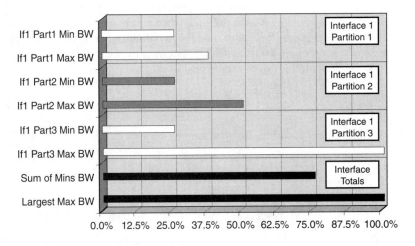

For each partition in each interface, you define the Partition_Minimum_BW and Partition_Maximum_BW. The first requirement is to guarantee the minimum bandwidth for all the partitions in the interface. The bottom sections in Figures 2-17 and 2-18 specify the sum of the minimum bandwidth for all bandwidth partitions on the interface. So the bandwidth reservation for the entire interface must be at least equal to that sum, which is 75 percent of the bandwidth in the interface.

As shown in Figure 2-17, reserving Interface_VSI_reserved_BW equal to the sum of the minimums (75 percent) is enough to guarantee the minimum bandwidth for each resource partition in the interface. In this case, when the bandwidth reserved for VSI for the interface equals the sum of the minimums, there is no pool of shared bandwidth (Interface_VSI_PoolSize_BW = 0), so no partition can reach its maximum bandwidth. The important concept is that only the minimum must be guaranteed.

As shown in Figure 2-18, if partitions 1 and 2 are transmitting at their minimum bandwidths, but the reservation of bandwidth for the interface is 100 percent, partition 3 can go beyond its minimum. In this case, Interface_VSI_reserved_BW needs to be set to the largest maximum bandwidth value. Note that partition 3 can never achieve its maximum bandwidth of 100 percent, because the minimum bandwidth allotted to partitions 1 and 2 must be maintained.

The cases shown in Figure 2-18 can be summarized using the formula shown in Equation 2-2:

Equation 2-2 *Per Interface VSI Reserved Bandwidth*

$$Interface\_VSI\_reserve\_BW = \max[\sum(MinBW), \max(MaxBW)]$$

The bandwidth reserved for VSI in the interface (Interface_VSI_reserved_BW) is equal to the sum of all the minimum bandwidths, or the largest of all the maximum bandwidths of all the partitions in the interface, whichever is greater.

If the bandwidth is soft-partitioned, the difference between the largest maximum bandwidth value and the sum of all the minimum bandwidths of all the partitions is a common resource to be used by any partition that needs it, provided that the partition does not exceed its own maximum bandwidth. This is what we defined earlier as Interface_VSI_PoolSize. In the case shown on the right side of Figure 2-17, 25 percent of the interface bandwidth is shared between the three partitions. If partitions 1 and 3 are transmitting at their minimum bandwidths, partition 2 can reach its maximum bandwidth of 50 percent. On the other hand, if partitions 2 and 3 are transmitting at their minimum bandwidths, partition 1 can transmit at its maximum of 37.5 percent, and 12.5 percent of the interface bandwidth is left for partitions 2 and 3 to share.

Consider the formula in Equation 2-3:

Equation 2-3 *Per Interface VSI Bandwidth Poolsize*

$$Interface\_VSI\_PoolSize\_BW = \max[0, \max(MaxBW) - \sum(MinBW)]$$

You subtract the sum of the minimum bandwidths from the largest maximum bandwidth value. If the difference is greater than 0, an interface bandwidth pool (Interface_VSI_PoolSize_BW) exists. Otherwise, there is no interface bandwidth pool.

To determine a particular partition's pool size (Partition_VSI_PoolSize_BW), you compare the interface pool size to the difference between the maximum and minimum bandwidths for that partition, and then you assign it the smaller value. For the *i*th partition, you have Equation 2-4:

Equation 2-4 *Per Partition VSI Bandwidth Poolsize*

$$Partition\_VSI\_PoolSize\_BW_i = min(Interface\_VSI\_PoolSize\_BW, (MaxBW_i - MinBW_i))$$

This means that the partition does not use more than its configured maximum value if there is shared bandwidth.

LCN or Channel Soft Partitioning

From a different viewpoint, resources can belong to the interface, to a set of interfaces, or to the entire card. VPI/VCI space and bandwidth clearly belong to the interface. In the case of bandwidth, we just pointed out how the resource pool applies at the interface level.

But in some cases, the LCN (the number of connections) might be a resource belonging to a set of interfaces called a *port group*. To refresh the definition, a port group is a set of interfaces that share LCN resources and that might contain multiple interfaces. The concept of soft partitioning has significance only in the interfaces that can actually share resources among them. When bandwidth is soft-partitioned, it's meaningful only within the interface. Likewise, when LCN is soft-partitioned, it is significant in the port group.

The corresponding LCN or channel soft-partitioning concepts are defined in the following list:

- **Port group**—A set of interfaces that share LCN resources.
- **Partition_Minimum_LCN**—The number of channels a partition is guaranteed.
- **Partition_Maximum_LCN**—The upper limit on the number of channels a partition can use.
- **Port_Group_VSI_reserved_LCN**—The total number of channels reserved for VSI in a port group.
- **Port_Group_VSI_PoolSize_LCN**—The total number of channels that constitute a *pool* that is shared among VSI partitions in the same port group.
- **Partition_VSI_PoolSize_LCN**—The number of channels that a given partition on a port group can use out of the Port_Group_VSI_PoolSize_LCN.

The user configures the Partition_Minimum_LCN and Partition_Maximum_LCN values. The rest of the values are derived from these two.

The same rules as with bandwidth resources apply to LCN resources, taking into account that LCN resources generally span multiple interfaces because of chip implementations. For LCN resources, the reservation is done at the port group level, including different interfaces.

Given the following partition information, let's see how to calculate the partition and port group LCN pool sizes, the port group reserved LCNs, and the partition-available LCNs.

Table 2-2 shows the configured minimum LCN and maximum LCN values for each partition in every interface and in each port group.

Table 2-2   *Minimum and Maximum LCN Partition Configuration*

Port Group	Interface	Partition	Minimum LCNs	Maximum LCNs
1	1	1	500	1000
1	1	2	500	2000
1	2	1	1000	3000
2	1	1	300	1000
2	2	1	300	1000
3	1	1	800	1500
3	1	2	800	1500

Using the formulas for port group, you first calculate the port group VSI reserved LCN and port group VSI pool size LCN.

For LCN resources as well, the number of LCNs reserved for use by VSI in the port group is the greater value of the sum of the minimum LCN values and the largest maximum LCN value for every resource partition in the port group. See Equation 2-5:

Equation 2-5 *Per Portgroup VSI Reserved LCNs*

$$PortGroup\_VSI\_reserved\_LCN = \max[\sum(MinLCN), \max(MaxLCN)]$$

The LCN pool size is the difference between the largest maximum LCN value and the sum of the minimum LCN values if it is greater than or equal to 0. It is 0 otherwise. See Equation 2-6:

Equation 2-6 *Per Portgroup VSI LCN Poolsize*

$$PortGroup\_VSI\_PoolSize\_LCN = \max[0, \max(MaxLCN) - \sum(MinLCN)]$$

Table 2-3 shows the reserved LCN and LCN pool size per port group from comparing the sum of the minimum LCN to the largest maximum LCN.

Table 2-3   *Sum of Minimum LCN Versus Largest Maximum LCN*

Port Group	sum(Mins)	max(Maxs)	Port Group Reserved LCNs	Port Group Pool Size
1	2000	3000	3000	1000
3	600	1000	1000	400
3	1600	1500	1600	0

For the first and second port groups, the largest maximum LCN is greater than the sum of the minimum LCNs, meaning that there will be a pool size. The third port group has a null shared pool of LCN resources, because sum(Mins) is greater than max(Maxs).

Finally, you can calculate the partition LCN pool size and available channels for each partition using the following two formulas.

This first formula shows that you determine the LCN pool size for a given partition by comparing the port group pool size to the difference between the maximum and minimum LCN values for a given partition and choosing the smaller. This formula ensures that you do not use more than the maximum LCN allotment. See Equation 2-7:

Equation 2-7 *Per Partition VSI LCN Poolsize*

$$Partition\_VSI\_PoolSize\_LCN_i = min(PortGroup\_VSI\_PoolSize\_LCN, (MaxLCN_i - MinLCN_i))$$

Now that you know the partition pool size, you can easily determine the partition's available LCNs using the formula in Equation 2-8:

Equation 2-8 *Available LCNs*

$$Available\_LCN_i = MinLCN_i + Partition\_VSI\_PoolSize\_LCN_i$$

Table 2-4 provides the final objective of this exercise. It shows the pool size and available LCNs on a per-partition basis.

Table 2-4   *Per-Partition Pool Size and Available LCNs*

Port Group	Interface	Partition	Partition Pool Size	Available LCNs
1	1	1	500	1000
1	1	2	1000	1500
1	2	1	1000	2000

Table 2-4   *Per-Partition Pool Size and Available LCNs (Continued)*

Port Group	Interface	Partition	Partition Pool Size	Available LCNs
2	1	1	400	700
2	2	1	400	700
3	1	1	0	800
3	1	2	0	800

For partition 1 in interface 1 and port group 1, the difference between the maximum LCN and minimum LCN (from Table 2-3) is less than the port group pool size (from Table 2-4). This means that the partition can reach its maximum, provided that LCN resources are available.

For the rest of the partitions in the same port group, the port group LCN pool size (from Table 2-4) is the lower value, so those partitions can't reach their configured maximum LCN value. If LCN resources are available (channels not used), the partitions can use only 1500 LCNs and 2000 LCNs, respectively. This ensures that the minimum LCN is guaranteed for the other partition in the port group.

Finally, partitions in port group 3 have a null partition pool size, meaning that at most they can use their minimum configured LCN.

More Soft-Partitioning Examples

This section discusses two examples of VSI resource partitioning. It introduces a graphical approach to resource partitioning. The first example describes a scenario with equal maximum partition sizes, and the second example presents a case with different maximum partition sizes. The two examples strengthen the concepts of resource partitioning with figures that demonstrate how the partitioning interrelationships work. The figures also help you understand the virtual switch concept covered later in this chapter.

Example 1: Equal Maximum Partition Sizes

Figure 2-19 represents the soft partitioning of resources for use by two VSI controllers. These resources can be bandwidth in an interface or LCNs in a port group. Remember that, in soft partitioning, you specify a minimum and a maximum for each partition. In this particular case, the maximum for both partitions is the same.

Figure 2-19 *VSI Resource Partitioning Where Max_1 = Max_2*

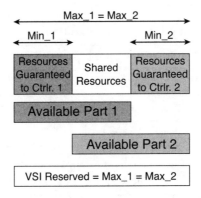

The first concept to mention is that the guaranteed resources in each partition are equal to the minimum configured for the partition. Because both Max_1 and Max_2 are larger than the sum of the minimums (Min_1 + Min_2), you set VSI Reserved equal to the larger of Max_1 and Max_2. See Equation 2-9:

Equation 2-9 *Soft Partition Example 1*

$$Max\_1 = Max\_2 > Min\_1 = Min\_2$$

The difference between VSI Reserved (in this example, equal to Max_1 and Max_2) and the sum of the minimums (Min_1 + Min_2) represents a pool of shared resources. These two definitions (VSI reserved and shared resources) are meaningful within the scope of the specific type of resources. If the resources being partitioned are bandwidth resources, the scope is the physical interface. If the resources being partitioned are LCN resources, the VSI reserved and shared resources apply at the port group level.

In both partitions, the difference between the maximum and the minimum is larger than the shared resources, so either partition can fully utilize the shared resource pool. Consequently, the corresponding available resources for each partition are the sum of the guaranteed resources (the minimum resources) plus the shared resources. The partition pool size for both partitions equals the VSI pool size of shared resources.

Example 2: Different Maximum Partition Sizes

On the contrary, if the *maximums* of both partitions are not the same, you can use the case shown in Figure 2-20.

Figure 2-20 *VSI Resource Partitioning Where Max_1 <> Max_2*

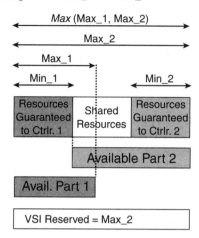

In this case, the VSI reserved resources are still based on the maximum of the maximums. However, Max_1 is smaller than Max_2. Although the shared resources are the same as before, some differences in the partition behavior exist.

The available resources for partition 2 are the same as before and are equal to the minimum resources plus the shared resources. On the other hand, for partition 1, the difference between Max_1 and Min_1 is smaller than the shared resources, so the available resources for partition 1 are equal to the maximum for the partition, because it cannot use beyond its configured maximum. The partition pool size for partition 1 is Max_1 minus Min_1.

Soft Partitioning and Virtual Switches

It is interesting to see the virtual switches that are "created" by partitioning the controlled switch. For the special case in which the maximums are equal to or larger than the sum of the minimums, Figure 2-21 shows different guaranteed and shared resources.

Figure 2-21   *Virtual Switch Representations for the Case Where Max_1 = Max_2*

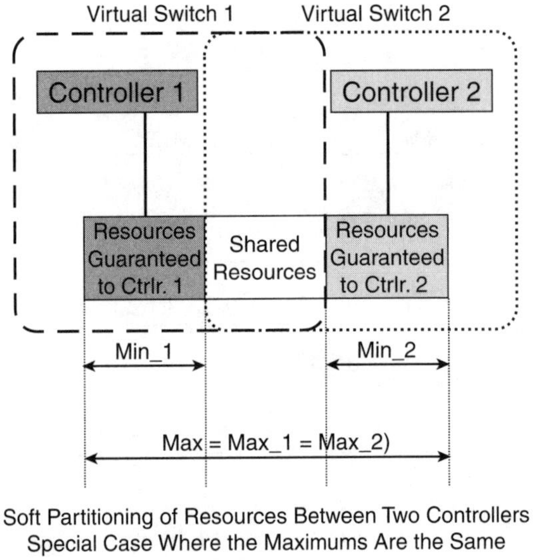

Soft Partitioning of Resources Between Two Controllers
Special Case Where the Maximums Are the Same

Conversely, if the partition pool size for one partition is smaller than the shared resources, we have the scenario shown in Figure 2-22.

Note that virtual switch 1 is now "smaller," meaning that it has fewer resources available. As a consequence, virtual switch 2 has more resources of its own that it does not share with virtual switch 1.

Finally, if the VSI reserved comes from the sum of the minimums (when the sum of the minimums is greater than or equal to the largest of the maximums), there are no shared resources, as shown in Figure 2-23.

Figure 2-22 *Virtual Switch Representations for Restricted Partition Pool Size*

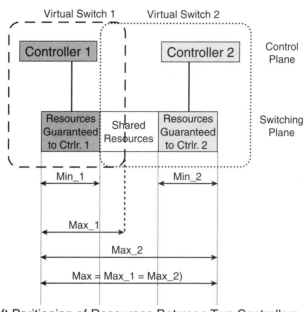

Soft Paritioning of Resources Between Two Controllers

Figure 2-23 *Virtual Switch Representations Where sum(Mins) ≥ max(Maxs)*

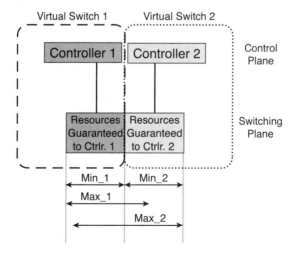

"Starved" Partitioning

There's one case in which a deficit condition for minimum resources can occur. Suppose you have a port group with four interfaces in which the first three interfaces have one VSI partition each with the channel resources shown in Table 2-5.

Table 2-5   *Port Group Before a Starved Condition*

	Min LCN	**Max LCN**
Interface 1	1000	4000
Interface 2	1000	4000
Interface 3	1000	4000

The port group to which all four interfaces belong has 4000 LCNs reserved for VSI.

The port group pool has 1000 LCNs, as well as the partition pool for each partition; therefore, each partition has 2000 LCNs available.

Suppose that, currently, interfaces 1 and 2 are using 1000 LCNs and interface 3 is using 2000 LCNs, using up all the resources available. At this point, you add a new resource partition in interface 4 with 1000 guaranteed LCNs. This new partition is shown in Table 2-6.

Table 2-6   *Port Group at a Starved Condition*

	Minimum LCN	**Maximum LCN**
Interface 1	1000	4000
Interface 2	1000	4000
Interface 3	1000	4000
Interface 4	1000	4000

Figure 2-24 shows the use of LCN resources in the existing partitions as well as in the new partition.

Figure 2-24 *Starved Partition*

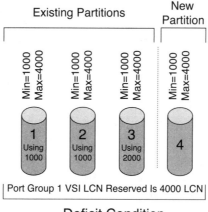

Figure 2-24 shows that this new partition does not have enough LCNs to satisfy the minimum, because the third interface's partition is using all the LCNs from the previous pool of 1000 LCNs. This newly added partition is marked as starved, meaning that not enough free LCN resources are present to satisfy this partition's minimum requirement. Those LCN resources are used by the existing partitions because they belong to a common pool. For the purposes of not disrupting service, it's desirable that those LCNs are not taken away from interface 3, but when interface 3 frees one LCN, that LCN is automatically assigned to interface 4. When the deficit state is over, there's no way of reaching the same condition again, because now the port group pool size is 0 LCNs, so each partition's pool size is 0 LCNs as well.

Dynamic Partitioning

Although the term *dynamic partitioning* is used in industry literature, it is actually a misnomer; it should be called *dynamic repartitioning*. It applies to hard or soft partitioning and allows the operator to increase or decrease a partition's resource allocation with minimal impact to service. Without dynamic partitioning, a user would have to delete the resource partition and re-add it with the new values.

The partitioning function resides in the VSI slave modules. Therefore, the implementation of the dynamic partitioning capability is both platform-dependent and release-dependent.

Traffic Service Types, Categories, and Groups

VSI provides support for many different traffic *service types*. The switch is expected to support as many traffic types as possible (including a signaling traffic type), but it is also expected to provide a default traffic type.

Service categories combine related service types into a single category to ease load advertisement.

Service categories are used in interface policy configuration. They are intended to accommodate ATM Forum service types. (For example, service types CBR.1, CBR.2, and CBR.3 are grouped into the CBR service category.) For other *service groups,* the service categories and service types are the same.

The controller first selects a service group in the Switch Configurable Info parameter group in a VSI **switch set configuration** command.

Valid service groups are shown in Table 2-7.

Table 2-7   *VSI Service Groups*

Service Group	Description
0	MPLS service categories
1	ATM Forum service categories
2	IP service categories (obsolete)
3 through 10	Vendor-specific service categories (used by local agreement between switch and controller)

NOTE   The IP service group (2) and its corresponding service types and service categories are obsolete, because DiffServ has redefined the meaning of the Type of Service (ToS) bits of the IP packet header.

In the Interface Configuration Trap or Get/Get-more Response messages, after specifying the LIN in the Interface Identifier parameter group, the switch advertises the number of configurable service categories in the Interface General Info parameter group.

Another parameter group that must be included in the interface configuration is Interface Load Info. It can also be included in the response to a connection request specifying updated load information for the partition. The interface load info is composed by a number of load parameter structures, which advertise load info to the controller. It advertises the values that apply to all service categories (common values) and exceptions for service categories that differ from the common values.

NOTE   Exception values are reported whenever the value reported for the service category is less than the common value.

Service types are used in the Connection Bandwidth Info parameter group and the Service Type Identifier parameter group. The first parameter group is included in connection-related messages (Reserve and Commit commands, Get/Get-more Configuration responses, and Configuration traps), and the second is used in messages dealing with getting interface statistics per service type (Interface Get Service Type Statistics command and response).

NOTE   VSI allows the controller to provide per-service-type policy configuration. In that case, you can further subdivide a partition's bandwidth among different service types. Policy parameters are specified in percentages to eliminate reconfiguration if the partition bandwidth changes.

Appendix A, "Service Traffic Groups, Types, and Categories," lists all the service groups, service types, and service categories defined in VSI.

Enhanced VSI Protocol Capabilities

This section covers some further details and enhanced capabilities of the VSI protocol.

Slave Discovery

The VSI master continuously sends poll messages to all possible VSI slaves. The VSI message used in slave discovery polls is the **switch get configuration** command. When the VSI master receives a response to the discovery poll message or gets a Switch Get Configuration Trap from a VSI slave, it stops the discovery polling and starts the keepalive polling. It also performs a resynchronization. The keepalive and resynchronization functions are described in the following sections.

Master Keepalive Function

The keepalive is a simple request message that the master sends. It expects a response from the slave to detect when the slave cannot respond. The preferred VSI message for this function is the **switch get configuration** command. When the keepalive timeout is exceeded, the slave is declared unavailable, and subsequent connection messages for that slave are NACKed in the master.

To recover from a slave-unavailable condition, the master keeps sending keepalive messages, and the slave can also send traps to the master.

Database Synchronization

The master is responsible for keeping or reestablishing connection segment synchronization, in which database information is exchanged in order to enter a known state. In the event of a loss of communication between master and slave, synchronization is lost, but traffic is unaffected. After a loss of synchronization (for example, a LOS between controller and switch), there is an optional state of resynchronization isolation after which the resynchronization process starts.

Resynchronization isolation is an optimized method of identifying connections that need to be resynchronized. The master groups connections and associates a checksum with each group. When synchronization is lost, the slave is queried for the checksum associated with each group. Only the ones that do not match the calculated checksum are passed to the resynchronization process. The groups of connections are called checksum blocks.

Resynchronization can be triggered by a slave or controller rebuild or switchover, response timeouts, or new slave attachment.

Clock Synchronization

VSI provides messages so that a network clock application can select the clock source that the switch synchronizes itself and its interfaces to in order to synchronize the network to a single clock source. Such a network clock application is Network Clock Distribution Protocol (NCDP).

The switch advertises to the controller the clock quality, the interfaces that can be clock sources, and which interfaces can pass clocking information. External clock interfaces are also advertised as logical interfaces supporting no connections.

In a distributed slave model, the controller can send clock-related commands to any slave.

Flow and Congestion Control

VSI messages involve a transaction (and consequent transaction-id) for each message sent. As mentioned earlier, a response is expected for every command sent. As another flow-control mechanism, a windowing function prevents sending more messages than the number of outstanding (unacknowledged) messages subtracted from the transmit window. Flow-control windows in both the controller and the switch might be configurable.

In a distributed environment, when a VSI command is sent to a VSI slave, the VSI response is expected to be returned from the same slave to maintain the flow-control window properly.

An example of congestion control within VSI is the response code REJ (reject). It indicates that the VSI request was denied because of congestion control on the slave. The REJ response code has three possible reasons: master-detected VSI congestion, slave-detected VSI congestion, and slave-detected VSI congestion at another slave.

Message Passthrough

VSI provides a mechanism to pass messages transparently from the switch to the controller and vice versa. To that extent, four VSI messages are provided: the passthrough pass up command, the passthrough pass up response, the passthrough pass down command, and the passthrough pass down response (in which the direction *up* refers to the direction from switch to controller). The Passthrough Information Type parameter group contains the pg-format-id, the pg-length, a protocol-id, the info-length, and the info-data padded to be an integer number of bytes.

For example, in a PNNI application, the VSI slave can pass through ILMI messages for address registration to be processed by the controller. The slave can process ILMI keepalive messages in a distributed ILMI implementation, where the ILMI Managed Entity (IME) is in the controlled switch.

Summary

VSI is an open interface between a controller and a controlled switch. It's a robust SCI implementation that has proven to be solid and versatile in deployments worldwide.

This chapter has delved into the Cisco Systems SCI implementation called the Virtual Switch Interface (VSI) and analyzed VSI master and slave functions and their relationship.

The VSI master module residing in the protocol controller discovers, controls, monitors, and resynchronizes the controlled switch. Taking instructions from the VSI master, the VSI slaves in the controlled switch translate those instructions into platform-specific directives that tell the switch to set up, manage, and teardown connections and manage queues. Because multiple VSI masters can control a single switch, we also discussed resource allocation. A switch's resources are partitioned using hard or soft allocation. Hard allocation forces each partition's resource allotment to a fixed range or value and does not allow sharing. Conversely, soft allocation provides a shared pool of resources for use by all partitions, but it still guarantees resources for every partition. We also examined dynamic partitioning, which is a platform-dependent way to allow service modules within the switch to expand and contract a partition's resource allocation on-the-fly and with minimum service impact.

The next chapter deals with the different implementations of multiservice switches and VSI in the Cisco Systems equipment.

CHAPTER 3

Implementations and Platforms

This chapter applies all the concepts from the previous two to actual implementations. We will analyze the details of the MPLS and PNNI implementations, as well as draft other Cisco Systems, Inc. proprietary implementations, such as AR and PAR. Also, we will explore Virtual Switch Interface (VSI) slave-specific implementations, such as logical interface number (LIN) encoding and resource management.

This is a product-oriented chapter, and as such, it is subject to modifications and aging. The implementations are studied from an architectural standpoint, so the following pages probably won't be outdated in the near future.

This chapter assumes that you have some knowledge of the switch platforms. Up-to-the-minute switching product information can be found at Cisco Connection Online (CCO) UniverCD at http://www.cisco.com/univercd/home/home.htm. Select Cisco WAN Switching Solutions under Cisco Product Documentation.

Many of the features and characteristics covered in this chapter are software release-specific. This chapter does not contain release or baseline information.

Before diving into the details of the specific MPLS and PNNI implementations, we'd like to repeat one of the most important concepts about multiservice switches: Networking software is decoupled from platform specifics by the VSI. Both MPLS and PNNI software is common in all these implementations. PNNI and MPLS software manage logical interfaces, whereas each platform manages the physical interface specifics.

MPLS Implementations

This section describes in detail all the different MPLS implementations in multiservice switching devices. It starts by defining the label switch controller (LSC) and the general MPLS switch architecture and concepts, as well as summarizing the MPLS switch realizations. The rest of the section explores each implementation.

Label Switch Controller

The controller in an MPLS application is called a *label switch controller*. It contains both the networking modules (routing protocols, LDP, and so on) and the VSI master modules. The LSC and the ATM-controlled switch together constitute a single *Label Switch Router* (LSR), as shown in Figure 3-1.

Figure 3-1 *Multiservice Switching LSR*

ATM Switch   Label Switch   ATM Label
                    Controller   Switch Router

ATM Switch Plus LSC Constitute an MPLS ATM LSR

In essence, the LSC is a router. Moreover, it is an MPLS router, with all the functions of an LSR plus extra functionality to control the switch using VSI. The LSC's primary function is to set up the ATM switch so that the LSC is not in the data path.

Internally, each interface is represented with a Cisco IOS interface descriptor block or IDB (a data structure in Cisco IOS software that represents the hardware interface). Within the LSC, each interface in the controlled switch is represented as a new type of virtual interface called the *extended MPLS ATM* (XmplsATM IDB) interface. This concept allows the existing MPLS applications to control and interact with an ordinary ATM interface.

Two different implementations were considered for the representation of the ports in the slave switch: implementations in which the XmplsATM IDBs were all subinterfaces of the control interface, and implementations in which the XmplsATM IDBs were an entirely new type of IDB. The former was discarded mainly because interfaces on the switch have one address space per interface and can have overlapping VPIs/VCIs. Each LC-ATM interface has its own label space. As such, XmplsATM interfaces were implemented as a new type of full (hardware and software) IDBs. XmplsATM interfaces expose the LC-ATM interface in the controlled switch.

The XmplsATM interfaces are similar to ATM router interfaces in most respects. However, XmplsATM interfaces support only MPLS, not Q.2931 signaling, LANE, or Classical IP (see RFC 1577). In fact, most ATM functions and commands are disabled in an XmplsATM interface. Internally, it is not seen as an ATM interface.

The following functions are present in an ATM interface but not in an XmplsATM interface:

- Modify ATM parameters
- Modify LANE parameters
- Configure internal loopback on an interface

- Manually set the interface MAC address
- Configure a static map group
- SSCOP interface subcommands
- MPLS Switch Control Protocol (SCP) commands

Functionally, when a packet is given to the XmplsATM device for transmission, it is redirected to the control interface. The XmplsATM interface has no hold queue; the packet might be put in the control interface's hold queue if that interface is congested.

Introducing the extended ATM interface is not the only different concept we need to review concerning MPLS implementation in multiservice switches. The next idea we need to explore is the categorization of the virtual circuits (VCs) in the LSC/switch picture.

From the LSC perspective, we can divide the VCs into three categories:

- **VSI VCs**—The LSC has one VC from the VSI master to each VSI slave. These VCs do not leave the LSR. They are visible only to the VSI master modules. They are invisible to the MPLS application. By default, VSI VCs use a base-vc of VPI = 0 and VCI = 40.

- **Control-vcs**—VCs for untagged packets. By default, these use VPI = 0 and VCI = 32. They are used by LDP and IP routing. These VCs always terminate on the LSC.

- **Label virtual circuits (LVCs)**—LVCs carry labeled traffic implicitly in the VPI/VCI fields and are set up with LDP. They can be subdivided into terminating LVCs and transit LVCs:

  - **Terminating LVCs**—Data flows between a port in the switch and the LSC.

  - **Transit LVCs**—These are implemented as a single cross-connect. Data flows between two ports in the switch so that the LSC does not touch the data. (It is responsible only for setting up the cross-connect.) From the LSC's perspective, traffic flows between an XmplsATM interface to another XmplsATM interface.

NOTE   Other VSI cross-connects in the controlled switch are invisible to the LSC—the VSI slave-slave cross-connects in a VSI distributed model.

Given the similarities between the control-vcs and terminating LVCs, they are often called *terminating VCs*. There are differences, though. For example, control-vcs are bidirectional, and LVCs are unidirectional.

All data between the LSC and the controlled switch (only for terminating VCs) traverses the control interface. An additional port on the router and a corresponding port on the switch serve this purpose.

One important thing to mention is that all terminating VCs have one leg of the cross-connect in the controlled switch. These cross-connects can all terminate in different switch interfaces with the same VPI/VCI pair. (The control-vc can have the default VPI/VCI values of 0/32, and for terminating LVCs, each LC-ATM interface has its own *label space*.) The other leg of the cross-connects terminates on the LSC passing through only one control interface (between the LSC and the switch) so that its cells are segmented and reassembled (SARed) and processed by the LSC. A problem arises with the need to avoid collisions of VPIs/VCIs on the control interface. To solve that problem, terminating VCs are remapped inside the switch cross-connect on the control interface. These remapped VCs are called *private VCs*. Inside the LSC software, private VCs are mapped back to their original values so that the MPLS control plane observes the correct values (values used in the actual interface in the switch, usually called *external VPI/VCI*).

To support the notion of private VCs, a new encapsulation type is used in XmplsATM interfaces, indicating that a packet received in this VC is actually destined for an XmplsATM VC. See Figure 3-2.

Figure 3-2   *Mapping of Terminating VCs to Avoid VPI/VCI Collisions in the Control Interface*

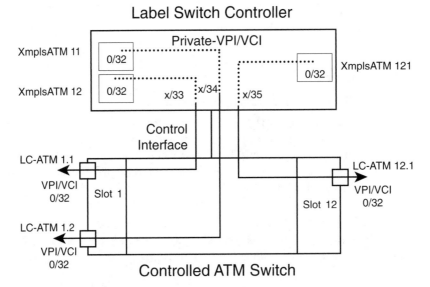

The cross-connects shown in Figure 3-2 are set up using VSI. Therefore, the VPI used in the *control interface* (x in the figure) is within the VPI range specified in the corresponding *partition* in that interface.

NOTE	VSI runs between the LSC and the controlled switch through the control interface. The LSC-switch pair forms the MPLS LSR. The control interface is invisible to neighboring LSRs. From their perspective, the control interface can be thought of as a bus that is internal to the LSR. LDP and IP routing see a single LSR, not the LSC-switch pair.

The LSC is the controller used on the implementations of MPLS multiservice switches. Before we get into the details, look at Table 3-1, which summarizes the different MPLS LSRs, specifying the controlled switches and the corresponding LSC implementation.

Table 3-1 *MPLS LSR Implementation Summary*

Controlled Switch	VSI Slave Model	VSI Slave Location	VSI Master Platform	Controller Location
BPX-8600	Distributed slaves	BXM cards	Cisco 72xx	External platform
IGX-8400	Distributed slaves	UXM and URM cards	Cisco 72xx	External platform
IGX-8400	Distributed slaves	UXM and URM cards	URM	Co-controller card on switch
MGX-8850	Hybrid slaves	AXSMs, RPM-XF, and PXM-45s	RPM-PR or RPM-XF	Co-controller card on switch
MGX-8950	Hybrid slaves	AXSM/B, RPM-XF, and PXM-45/B	RPM-PR or RPM-XF	Co-controller card on switch

Normally, the edge Label Switch Router (eLSR, also called a Label Edge Router [LER]) is an external router connected to an LC-ATM interface in the LSR. An eLSR also has two special implementations, in which the edge functionality resides in a router blade inside the controlled switch. In these cases, the router card can act as either an LSC or an eLSR. One last implementation exists in which the eLSR resides in a router card inside an ATM switch without LSR functionality. These three cases are summarized in Table 3-2.

Table 3-2 *MPLS eLSR Implementation Summary*

Platform	eLSR Card	LSR Functionality?
IGX-8400	URM card	Yes
MGX-8850 (PXM-45-based)	RPM-PR or RPM-XF	Yes
MGX-8250 (PXM-1-based)	RPM/B and RPM-PR	No

BPX-8600 MPLS Implementations

The BPX-8650 IP + ATM switch is an MPLS-enabled switch. It consists of a BPX-8620 ATM switch with distributed VSI slaves in BXM or BXM/E cards (processor-based service cards) and a Cisco 72xx-based LSC including the VSI master modules. In the MPLS architecture, the BPX-8620 switch together with the LSC acts as an ATM LSR.

NOTE   The 7200 router-based LSC also can act as an eLSR concurrently with the LSC functionality. However, it is strongly recommended that the eLSR functionality on a 7200 router not be used when it serves as an LSC. The eLSR functionality is enabled by default. You will see how to disable it in the sections named "Disabling Headend LVCs" and "Disabling Tailend LVCs" in Chapter 6, "MPLS Implementation and Configuration."

The control link uses an ATM port adapter in the LSC and an ATM port in the BPX switch. As you know, all communication between the switch and the LSC is performed over ATM VCs. The following figures help you better understand the ATM channels involved. Figures 3-3 through 3-7 show the control and data paths in an LSC-controlled BPX-8600 MPLS LSR.

Figure 3-3 shows the controlled switch configuration over comm-bus. In the BPX platform, the BCC card runs the switch software. It communicates with the different cards that have comm-bus messaging.

Figure 3-3   *BPX-8600 LSR—Controlled Switch Configuration Path*

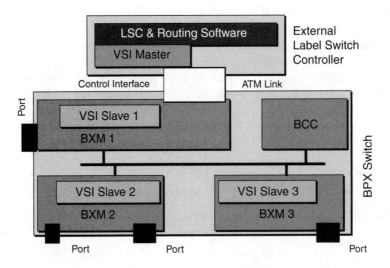

Figure 3-4 shows the VSI master-slave VCs, part of the VSI control path. These channels are used for master-slave VSI communication. There is one master-slave VC from the VSI master in the LSC to each slave in active BXM cards. In the control interface, the default VPI/VCI is 0/40.

Figure 3-4   *BPX-8600 LSR—VSI Master-Slave Channels*

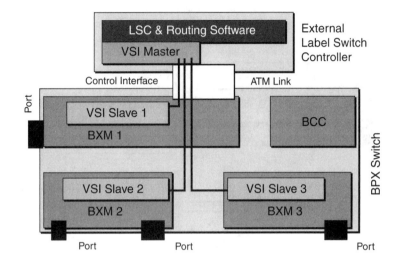

Figure 3-5 shows the VSI interslave cross-connect, completing the VSI control path. A distributed slave model has a full mesh of cross-connects among the slaves, going from each slave in an active BXM card to every other slave in all other active BXM cards.

Figure 3-5   *BPX-8600 LSR—VSI Slave-Slave Cross-Connects*

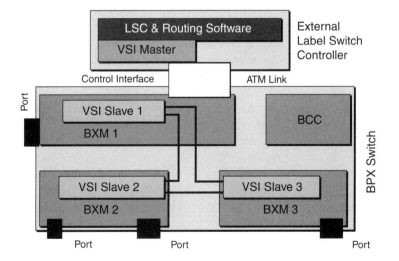

Figure 3-6 shows LDP routing signaling VCs, which are the MPLS control path. Using VSI, the LSC creates the control-vcs. They are used for unlabeled IP traffic, such as IGP updates and TDP/LDP. They are bidirectional, and by default, the VPI/VCI used is 0/32 in the LC-ATM interfaces. In the control interface, private VCs are used.

Figure 3-6 *BPX-8600 LSR—Unlabeled (non-MPLS) Traffic Connections*

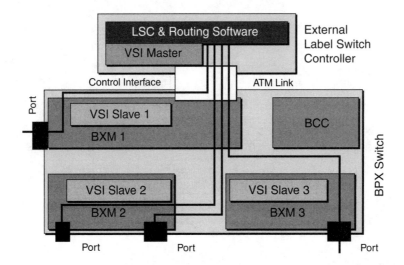

The networking software uses the information gathered with these VCs to create LVCs.

Figure 3-7 shows the label-controlled cross-connects, which are the data path. Creating LVCs is the objective. Using VSI, the LSC creates the unidirectional cross-connects in the controlled switch. Figure 3-7 clearly shows that as soon as the LVCs are set up, the LSC does not touch the data.

NOTE   Figures 3-3 through 3-7 help you understand the VCs and cross-connects involved in the LSR. They are not for studying a BPX-8600 from a switching perspective. The BCC card in the BPX-8600 not only runs the switch software but also houses the switch fabric (19.2 Gbps peak cross-point matrix). Every cross-connect inside the BPX-8600 is switched in the active BCC.

Figure 3-7 *BPX-8600 LSR—LVCs*

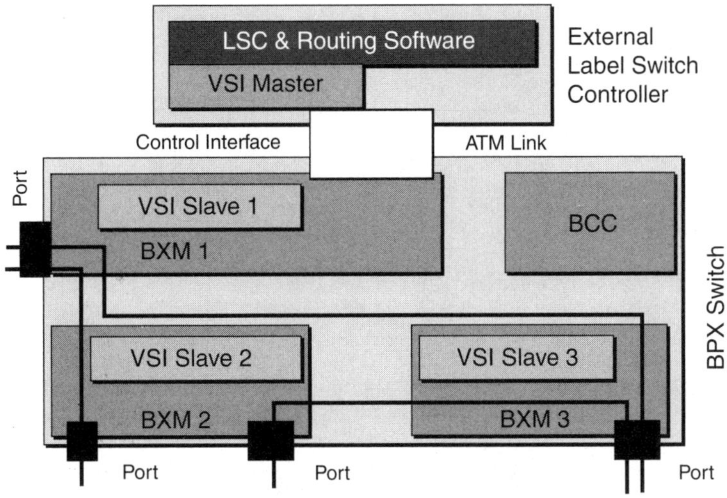

IGX-8400 MPLS Implementations

The LSC is the controller used in MPLS applications. The IGX family of switches has two different LSC implementations: an external LSC and a co-controller card LSC. In both cases, the slave model is distributed. Every UXM, UXM-E, and URM card has one VSI slave implementation.

IGX-8400 with an External LSC

The IGX 8400 family of switches (IGX-8410, IGX-8420, and IGX-8430, with 8, 16, and 32 slots, respectively) can be MPLS-enabled with an external Cisco 72xx-based LSC in the same way a BPX-8600 is. From a multiservice switch perspective, the LSC-controlled IGX architecture is the same as the LSC-controlled BPX discussed earlier. However, the IGX and BPX platforms have different switching technologies and capacities, slots, port densities, and so on.

IGX-8400 with a Co-Controller Card LSC

The LSC controlling an IGX switch has another implementation. For this other implementation, the LSC is located on the switch on a separate card, with the URM co-controller card.

NOTE   The URM card can be configured to act as an LSC controlling the IGX or as an eLSR. This feature enables the deployment of MPLS networks without deploying separate MPLS-capable routers. LSC and eLSR functionality in the same URM card is strongly discouraged.

From a very high-level perspective, the Universal Router Module (URM) card looks like Figure 3-8.

Figure 3-8   *Universal Router Module*

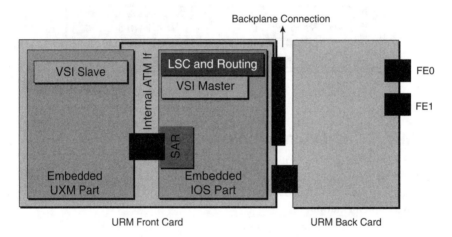

Seen as a whole, the URM hosts the networking protocols (IP routing, dynamic MPLS, and traffic engineering), a VSI master module, and a VSI slave module.

The following figures illustrate the channels used in a URM-LSC-controlled IGX. For completeness, two URM cards are present in the figures:

- URM 1 acts as an LSC. In the figures, it is decomposed into the two main constitutive modules (UXM and router modules).

- URM 2 acts as an eLSR. In that case, the VSI master module in the URM-acting-as-LSC controls the VSI slave in the URM-acting-as-eLSR. The router portion of the URM acts as an ordinary eLSR.

Figure 3-9 shows the configuration path over the C-bus, which is the platform control path. In the IGX platform, the NPM card runs the switch software. It communicates with the different cards with C-bus messaging.

Figure 3-9 *IGX-URM LSR—Controlled Switch Configuration Path*

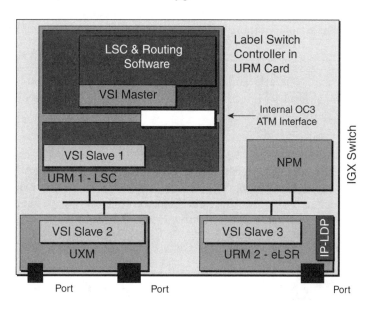

Figure 3-10 shows the VSI master-slave VCs. These channels are used for master-slave VSI communication. There is one master-slave VC from the VSI master in the URM's LSC to each slave in active UXM or eLSR-URM cards.

Figure 3-10 *IGX-URM LSR—VSI Master-Slave Channels*

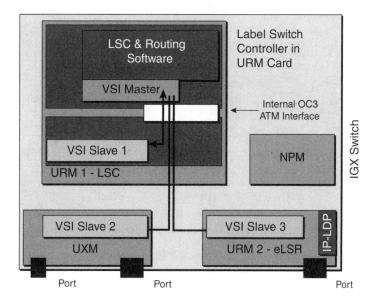

There is also master-slave VSI communication inside the URM that acts as the LSC. The communication between the UXM portion and IOS portion inside a URM card uses the IPC (interprocess communication) VC (VPI/VCI = 0/1023), not a specific VSI VC.

NOTE   In a URM card, VPI/VCI = 0/1023 cannot be used for controller addition in an MPLS LSC application or in a URM port because it's used for IPC. It can be used in a UXM with an external controller.

Figure 3-11 shows VSI interslave cross-connects. Together with the VSI master-slave VCs, they form the VSI control path. These VSI slaves also follow a distributed model. Therefore, there's a slave-to-slave cross-connect from every slave in an active UXM or URM card to every other slave in active UXM or URM cards.

Figure 3-11 *IGX-URM LSR—VSI Slave-Slave Cross-Connects*

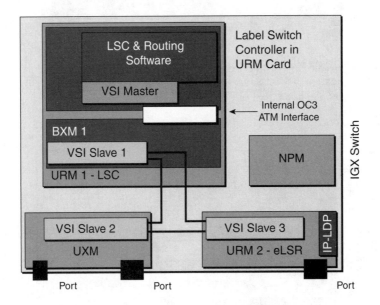

Figure 3-12 shows the LDP routing signaling VCs, which are the MPLS control path. Using VSI messages, the LSC creates the control-vcs that are used for untagged IP traffic (IGP updates and TDP/LDP). The VPI/VCI used by default is 0/32 in the LC-ATM interfaces.

In UXM LC-ATM interfaces, those control-vcs terminate in the next LSR or eLSR. In a URM-based eLSR, the control-vc terminates in the embedded router.

Figure 3-12 *IGX-URM LSR—Unlabeled (non-MPLS) Traffic Connections*

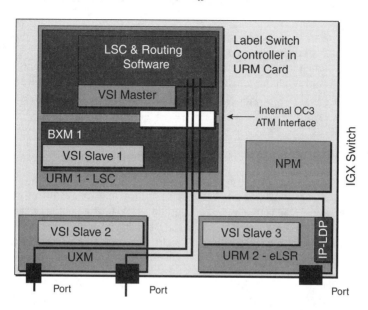

Figure 3-13 shows the label-controlled cross-connects that make up the data path. Using VSI, the URM-LSC creates LVC cross-connects. No transit LVCs traverse the URM-eLSR, only headend or tailend VCs.

Figure 3-13 *IGX-URM LSR—LVCs*

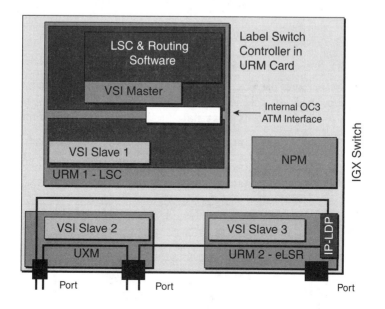

After the LVCs are set up, the URM-LSC does not touch the data.

<table>
<tr>
<td>**NOTE**</td>
<td>An IGX-8400 like the one shown in Figures 3-9 through 3-13 with one URM card with LSC functions and another URM card with eLSR functions is seen from other LSRs as two different pieces of equipment: one LSR (which is made up of the URM-LSC and UXM cards with LC-ATM interfaces and VSI slaves) and an eLSR (which is the URM-eLSR).</td>
</tr>
</table>

MGX-8850 and MGX-8950 MPLS Implementations

PXM-45-based MGX multiservice switches natively support MPLS LSR functionality in two different ways that provide a one-box solution in a very efficient way: by placing the LSC functionality in route processor module (RPM) cards, with the same networking modules and VSI master modules as in the BPX and IGX implementations. Both RPM-PR and RPM-XF cards can perform LSC or eLSR functions in an MGX-8850 and LSC functionality in an MGX-8950.

To fully understand the MPLS LSR and eLSR implementations in MGX-8850 and MGX-8950 IP + ATM switches, you need to understand not only the LSC implementation but also the VSI slave architecture. From a switch perspective, PXM-45-based MGX switches present a VSI hybrid model. Each BASM (Broadband ATM Service Module, such as AXSM cards and RPM-XF) presents a VSI slave inside the card. Alternatively, the PXM-45 and PXM-45/B cards implement a VSI slave acting as a proxy VSI slave for CBSMs (Cell Bus Service Modules, such as RPM-PR). Details on the VSI slave architecture are given in the later section "MGX-8850 and MGX-8950 Controlled Switches and VSI Slaves."

MGX-8850 and MGX-8950 with an RPM-PR-Based LSC

This section explores the VCs, cross-connects, and relationships among the different functional modules. This section's figures are based on an RPM-PR-based LSC and eLSR, so the PXM-45 card acts as a proxy slave for the LSC as well as for the eLSR.

Figure 3-14 depicts the VSI master-slave VCs. These channels are used for master-slave VSI communication. There is one master-slave VC from the VSI master in the LSC to each slave in active AXSM cards, as well as to the active PXM-45 card.

Figure 3-14 *MGX-8850 and MGX-8950 LSR—VSI Master-Slave Channels*

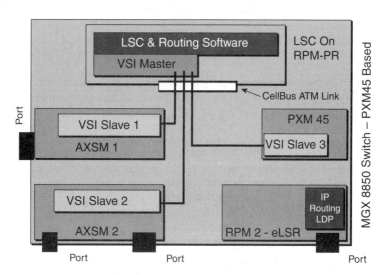

LSC controller-to-slave communication is performed using virtual circuit connections (VCCs) for VSI sessions, as shown in Figure 3-14. These cross-connects have well-known VPI/VCI pairs, on both the RPM-PR side and the BASM side. (PXM-45 acts as a proxy VSI slave for Narrow Band Service Modules [NBSMs].)

- On the RPM-PR, VCCs with VPI/VCI = 0/65507 through 0/65519 are added for communication with slots 1 through 6, active PXM-45, and slots 9 though 14, for a total of 13 VCs. This range is fixed.

- On the BASM, a VCC with VPI = 0 and VCI = controller-id is established for master-slave communication.

- The PXM-45 sets up the cross-connects for each VCC through the serial bus, assuming that the LSC is on a CBSM. These cross-connects are created when the controller is added to the switch. The VCC to the VSI slave residing on the PXM-45 terminates at the PXM-45.

Table 3-3 summarizes the well-known VPI/VCI values for RPM-PR and RPM-XF.

Table 3-3 *Well-Known VPI/VCI Values for RPM Cards*

Connection Type	VPI	VCI
Management VCs	0	65520 to 65535
VSI master-slave VCs	0	65507 to 65519

Management VCs are used for IPC. The IPC protocol introduced the concept of IPC zones to accommodate the ability to house multiple active RPM cards.

Figure 3-15 shows VSI interslave communication. This implementation uses a hybrid slave model, with distributed VSI slaves in all BASMs as well as a centralized slave in the PXM-45 for all NBSMs (such as the RPM-PR acting as an eLSR).

Figure 3-15 *MGX-8850 and MGX-8950 LSR—VSI Slave-Slave Path*

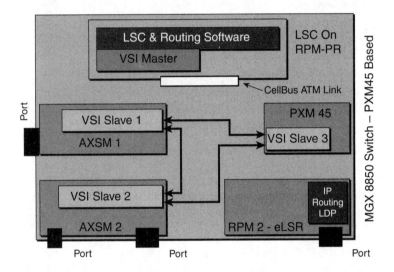

As such, communication among all slaves is needed. But this case is different from the previous two because there are no slave-to-slave VCs or cross-connects. Slave-to-slave communication is accomplished using IPC (which is why Figure 3-15 has an arrow).

For slave-to-slave VSI communication, each VSI slave (including the proxy VSI slave) creates an IPC endpoint. This is completely transparent to the VSI master located in the RPM-PR. The name bound to this endpoint is formed from the string **vsis** concatenated with the slave ID (in hexadecimal format). For example, an AXSM card in slot 3 binds the IPC endpoint to the name **vsis3**, and an AXSM-E in slot 12 uses the name **vsisC**. So the inter-slave communication can be seen as VSI over IPC.

Figure 3-16 shows the LDP routing signaling VCs. As with the previous cases, the first VCs that are created using VSI are the control-vcs.

Figure 3-16 *MGX-8850 and MGX-8950 LSR—Unlabeled (non-MPLS) Traffic Connections*

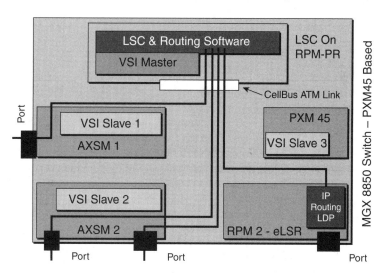

In AXSM LC-ATM interfaces, those control-vcs terminate in the next LSR or eLSR. In an RPM-based eLSR, the control-vc terminates in the RPM.

Figure 3-17 shows the label-controlled cross-connects, which as data paths are the ultimate goal. Using VSI, the RPM-LSC creates LVCs.

Figure 3-17 *MGX-8850 and MGX-8950 LSR—LVCs*

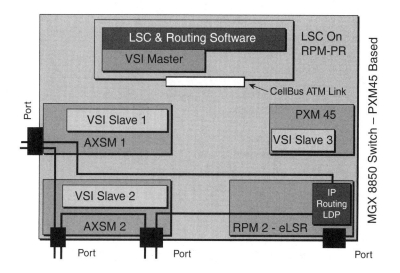

For the eLSR endpoints, the VSI master sends the setup messages to the PXM-45 slave.

After the LVCs are set up, the RPM-LSC does not touch the data.

MGX-8850 and MGX-8950 with an RPM-XF-Based LSC

RPM-XF uses a different architecture and technology than RPM-PR, providing much higher performance. RPM-XF is based on Parallel Express Forwarding (PXF), a new technology, and a Toaster Micro-Controller (TMC) processor. RPM-XF is a BASM (using the serial bus for data), but also with access to the cell bus for control traffic. So RPM-XF manages two interfaces to the PXM-45. RPM-XF separates VSI traffic (in the cell bus) from TDP signaling and LVCs (in the serial bus), providing complete separation of control and user traffic. RPM-XF supports the OC-12 Packet Over Sonet (POS) back card and the Gigabit Ethernet (GigE) back card (as well as a console back card with two Fast Ethernet ports for management).

As mentioned earlier, RPM-XF hosts both the LSC and VSI slave. The in-card VSI slave manages the partition in the serial bus interface. It is responsible for all setup/teardown operations in the serial bus interface. The RPM-XF architecture is shown in Figure 3-18.

Figure 3-18 *RPM-XF*

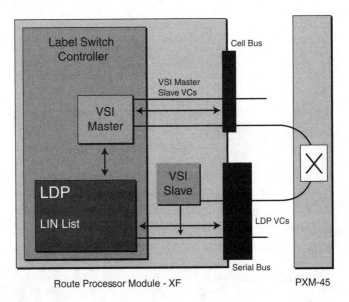

In Figure 3-18, the VSI master-slave VCs use the cell bus. LDP VCs as well as LVCs use the serial bus. The communication between the VSI master in the RPM-XF and the embedded

VSI slave controlling the interface to the serial bus is cross-connected in the PXM-45. The LDP VC shown in the figure was created by the VSI slave in the RPM-XF, as indicated by the arrow.

From an operation/configuration point of view, the differences between RPM-PR and RPM-XF are minimal. RPM-XF has a superset of commands.

The RPM-XF is the first RPM module that supports LSC and eLSR functionality coexistence on the same RPM. This is achieved by separating control and user data in different hardware paths, protecting control data during user congestion in the switching paths, creating the control-vc with the highest-priority service type, and other mechanisms.

Figures 3-19 and 3-20 show the channels used in an RPM-XF-based LSC.

The complete VSI control plane comprised of VSI master-slave and VSI slave-slave communication is shown in Figure 3-19.

Figure 3-19 *RPM-XF—VSI Communication*

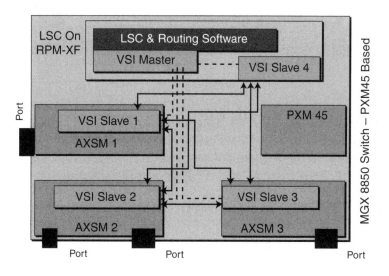

It is important to note in Figure 3-19 that there is VSI communication within the RPM-XF card.

Figure 3-20 shows the MPLS control plane VCs used by unlabeled traffic such as LDP signaling and IP routing, as well as label VCs that create the data path.

Figure 3-20 *RPM-XF—LDP and Routing VCs and LVCs*

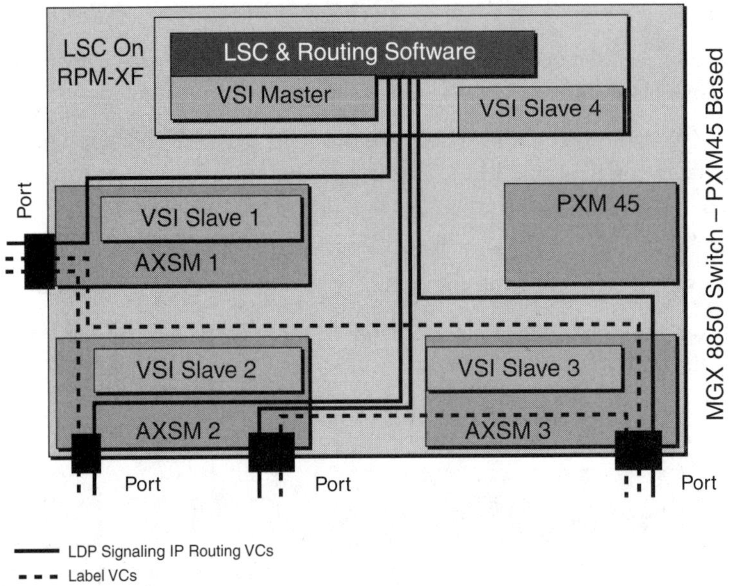

LSC On RPM-XF

LSC & Routing Software

VSI Master

VSI Slave 4

Port

VSI Slave 1

AXSM 1

PXM 45

VSI Slave 2

AXSM 2

VSI Slave 3

AXSM 3

MGX 8850 Switch – PXM45 Based

Port   Port   Port

———— LDP Signaling IP Routing VCs
- - - Label VCs

eLSR Functionality in an MGX-8230 and MGX-8250

The MPLS functionality in PXM-1-based MGX-8230 and MGX-8250 is different from all the previous platforms. These two switches can act only as eLSRs; they cannot provide LSR functionality. The eLSR capabilities are accomplished with the RPM/B or RPM-PR cards. RPM-XF is not supported on MGX-8250 because the PXM-1 card can switch cells only from CBSMs. Edge MPLS capability in MGX-8230 and MGX-8250 is shown in Figure 3-21.

Figure 3-21 *Multiservice Switching eLSR*

Edge Switch   +   Edge LSRs   =   Multiservice Edge Label Switch Routers

ATM Edge Switch Plus eLSRs Constitute a Multiservice MPLS eLSR

NOTE   RPM-PR (RPM-premium) offers significant performance improvements over RPM/B, including an OC-6 rate to the cell bus, more than double the forwarding rate and almost three times the number of virtual interfaces or IDBs (2000 IDBs in RPM-PR).

The main reason for the MGX-8230 and PXM-1-based MGX-8250 not having LSR functionality is that you cannot partition resources in their interfaces, so an LSC cannot label-control them. The target for MGX 8230 and MGX-8250 is an edge concentrator, and the eLSR functionality is consistent with that.

An example of the MGX-8250 eLSR functionality is shown in Figure 3-22.

Figure 3-22 *eLSR Functionality in an MGX-8230 and MGX-8250*

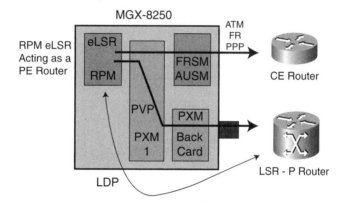

In a typical application, you can tunnel an LDP session as well as the created LVCs in a Permanent Virtual Path (PVP). To re-emphasize the concept, in the LSR, you partition resources (for example, VPI range) in the controlled switch, and the LSC learns about those resources (by means of VSI) and uses them. In contrast, in the eLSR, the switch has no re-sources to partition, and the MPLS application has no VSI messaging; consequently, you need to specify in the RPM acting as the eLSR the VPI to use as a tunnel.

MPLS Implementation Summary

As a summary, an MGX-8850 PXM-45-based LSR connected to a BPX-8600 LSR running MPLS looks like Figure 3-23.

In the MGX-8850, the PXM-45 acts as a slave for connection setups/commits for the RPM-PR acting as the eLSR. From the LSC perspective, those resource partitions assigned to it are seen as XmplsATM interfaces.

Figure 3-23 *BPX-8600 and MGX-8850 LSRs*

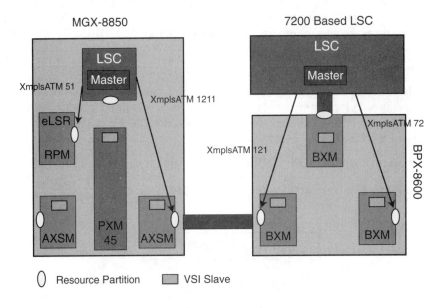

Finally, you can add eLSRs at both ends to complete the picture, as shown in Figure 3-24.

Figure 3-24 *BPX-8600 and MGX-8850 LSRs and MGX-8250 and Cisco 7200 eLSRs*

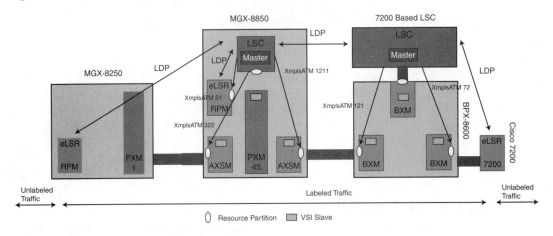

The PXM-1-based MGX-8250 can act only as an eLSR. The labeled traffic and unlabeled traffic references do not consider the eLSR in the RPM-PR card in the MGX 8850.

PNNI Implementations

This section explores multiservice switching PNNI implementations, focusing on different multiservice switches that natively run PNNI. First we will go through the service expansion shelf (SES) controlling a BPX-8600 switch. Then we will analyze the PXM-45-based MGX-8850 and MGX-8950. Finally, we will review PXM-1E-based PNNI nodes.

The three implementations share the same PNNI software, ensuring picture-perfect interoperability. The PNNI controller interfaces with the controlled switch using VSI. VSI slaves advertise interfaces—which are partitions of resources of physical or virtual interfaces managed by those VSI slaves—to the VSI master. The PNNI software calls these logical interfaces PnPorts. This concept (equivalent to XmplsATM interfaces in an MPLS implementation) allows PNNI software to control those logical interfaces.

Table 3-4 summarizes the multiservice switching PNNI implementations.

Table 3-4   *Multiservice Switching PNNI Implementations*

Controlled Switch	VSI Slave Model	VSI Slave Location	VSI Master Platform	Controller Location
BPX-8600	Distributed slaves	BXM cards	SES controller	External platform
PXM-1E-based MGX-8850	Centralized slave	PXM-1E	PXM-1E	Controller card
PXM-45-based MGX-8850	Hybrid slaves	AXSMs and PXM-45s	PXM-45	Controller card
MGX-8950	Hybrid slaves	AXSM/B and PXM-45/B	PXM-45/B	Controller card

Currently, no PNNI support is planned for the IGX platforms. PNNI support for UXM cards using a SES controller is a feasible development, but support for all other IGX cards (such as Universal Frame Relay Module [UFM]) would entail developing a more powerful NPM controller card.

SES Controller: BPX-8600 PNNI Implementation

The SES enables PNNI networking on a BPX-8600. The SES uses PXM-1 redundant cards with PNNI and VSI software, and it has the same form factor as an MGX-8230.

The architecture of a BPX-8600/SES PNNI node is equivalent to the LSC controlling the same BPX-8600 in an MPLS implementation: The slaves reside in BXM and BXM-E cards, in a distributed model, and the same master-slave and slave-to-slave channels are used. Signaling channels terminate in the controller (UNI and PNNI in this case; LDP in MPLS), and that allows the creation of end-to-end connections (SVCs and SPVCs in this case; LVCs in MPLS).

The SES controller offers fewer degrees of freedom than an LSC. The base-vc for master-slave communication is fixed and is equal to VPI/VCI = 0/40. In addition, the controller-id has a fixed value of 2. Both parameters can be configured in the BPX-8600, and they need to be set to the SES fixed values. On the LSC, both parameters can be configured.

Other differences exist as well. One of the differences is that apart from VSI, Annex.G runs between the BPX-8600 and the SES controller using VPI = 3 VCI = 31. By means of Annex.G (the polling mechanism between the BPX and SES), the BPX-8600 and the SES discover each other's names and IP addresses. Also, IP Relay communication exists between the controller and the controlled switch. IP Relay uses VPI = 3 and VCI = 8. The current network time is also propagated from the BPX node to the SES shelf using Cisco extensions to baseline Annex.G functionality.

Figure 3-25 shows the VSI channels and the feeder VCs in a SES-controlled BPX PNNI node. The VSI channels allow VSI master-slave and interslave communication. The feeder VCs consist of Annex.G and IP Relay VCs.

Figure 3-25 *VSI and Feeder Channels in a SES-Controlled BPX-8600*

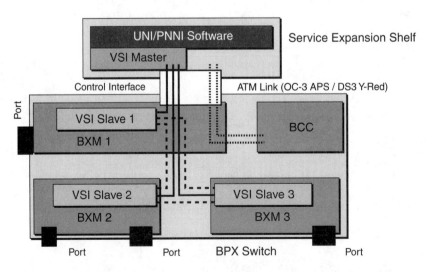

Using VSI, UNI signaling and PNNI signaling and routing connections are created, terminating in the SES PNNI software, as shown in Figure 3-26. That PNNI control plane enables the creation of SVCs and SPVCs forming the data plane.

Figure 3-26 *Signaling, Routing, and Data Channels in a SES-Controlled BPX-8600*

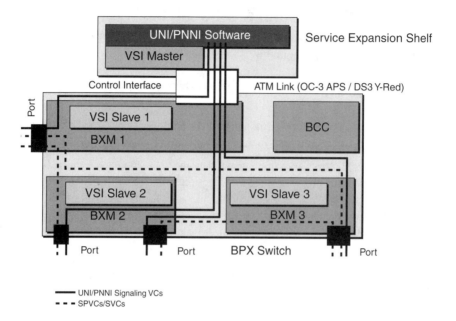

Figure 3-26 summarizes all the VCs and cross-connects present in a PNNI node using a BPX-8600 controlled by a SES. UNI and PNNI signaling VCs use VCI = 5 for SSCOP signaling and VCI = 18 for PNNI routing control channels or RCC, where the VPI can be 0 or some other value for virtual interfaces. Remember that a virtual interface is used to build a trunk across a purchased virtual path connection. The new additions are the feeder VCs, used for Annex.G and IP Relay.

Also, because the PNNI software runs on a system with PXM-1 redundancy, the link between the controller and the switch has APS 1+1 or Y-red redundancy features.

Compared to a PXM-45 MGX-8850 supporting PNNI, one difference is that the SES/BPX pair cannot be a Network Clock Distribution Protocol (NCDP) client. This is because there is no VSI slave in the BCC card where the external clock inputs reside. This fact does not impose a limitation because clock distribution and selection are provided by AutoRoute in the BPX (as discussed in the section "AutoRoute").

For time-of-day synchronization, BPX nodes distribute the network time among themselves, and each BPX node propagates that time to its feeder shelves, including the SES if present. Where MGX-8850/PXM-45 nodes also exist in the PNNI network space, the BPX/ SES can be enabled as a Simple Network Time Protocol (SNTP) server, allowing it to propagate the network time to the MGX-8850/PXM-45 nodes (SNTP clients).

PXM-45-Based MGX-8850 and MGX-8950 PNNI Implementation

From a controller perspective, the PXM-45-based MGX-8850 and MGX-8950 PNNI implementation is the first multiservice switch implementation where the controller is located on the switch controller card. Therefore, some of the communication paths are internal software interfaces. Yet the communication between the controller and the controlled switch is VSI.

Both the networking software (including ATM signaling, PNNI, SSCOP, and so on) and the VSI master reside in software in the PXM-45 card.

On the other side, from a switch perspective, the slave model is hybrid. The details of the slave model are covered in the later section "MGX-8850 and MGX-8950 Controlled Switches and VSI Slaves." They apply to all applications, including PNNI. To introduce some of these concepts, each BASM card (such as AXSM) has its own VSI slave implementation, and the PXM-45 card houses a VSI slave that manages interfaces in CBSM cards as well as the internal virtual port and clocking interfaces. In particular, the first CBSM supported in a PXM-45-based MGX-8850 is RPM-PR. Future support for NBSMs might include VISM, FRSM, and AUSM. MGX-8950 does not support single-height NBSMs.

NOTE	The terms CBSM and NBSM are used interchangeably.

Figure 3-27 summarizes some of these ideas.

As opposed to the previous PNNI implementation, where VCs are set up for master-slave communication and cross-connects for slave-slave communication, the PXM-45-based MGX-8850 uses IPC to achieve this goal. This is true of both master-slave and slave-slave communication.

The VSI master creates an IPC endpoint to receive messages from the VSI slaves. The predefined global name to bind to this endpoint is created by concatenating the string **VSIM** with the controlled ID of the VSI master in hexadecimal notation.

The VSI slaves (including the PXM-45 proxy slave) create two IPC endpoints each. The first endpoint is to receive messages from the VSI master. Each slave binds the string **vsim** concatenated with the slave ID (which is the slot number) in hex to this endpoint.

Figure 3-27 *VSI Communication Paths in a PXM-45-Based MGX-8850 PNNI Implementation*

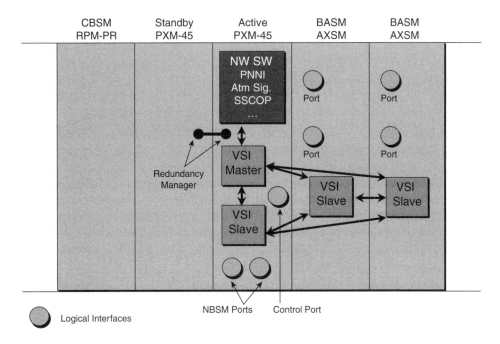

The second endpoint is used for slave-to-slave messaging and is transparent to the VSI master and the networking software. The name bound to this endpoint is created by concatenating the string **vsis** with the slave ID in hexadecimal.

NOTE   IPC provides the name registration and resolution facilities. IPC also provides infrastructure for messaging between tasks in the same card or tasks in different cards. There are pre-established card-to-card ATM VCs for IPC communication. IPC Global Name Registry and Name Resolution services are present in the PXM-45 card, and every card has an IPC Name Resolution Client.

The VSI slave module in each BASM is responsible for the logical interfaces in that card. The VSI slave in the PXM-45 is responsible for managing NBSM logical interfaces as well as PXM-45 internal logical interfaces. In Figure 3-27, the RPM-PR has two logical interfaces in the PXM-45: one for VCCs and the other for Virtual Path Connections (VPCs). The logical interface for the control port is also shown in Figure 3-27. The control port is where control-vcs (connections that need to be SARed and analyzed by the networking software) terminate. Those connections are signaling VCs for UNI and NNI interfaces, routing VCs for NNI interfaces, and IP connectivity SVCs. The control port is reported to the networking software (with a VSI interface trap from the PXM-45 VSI slave to the VSI master) and is treated as any other logical interface. The control port and control-vc details are shown in Figure 3-28.

Figure 3-28 *Control Port and Control-vcs*

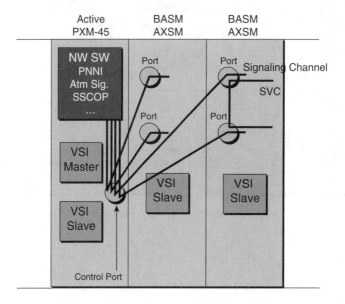

The control port is the primary interface for control-vcs. This means that the VSI master sends the connection setup message to the PXM-45 VSI slave (managing the control port), and that slave communicates with the AXSM VSI slave.

NOTE   The clock input interfaces are also represented as logical interfaces in the PXM-45 (not shown in the figure). More details about PXM-45's logical interfaces can be found in the later section "VSI Slaves: Logical Interface Numbers."

PXM-1-E-Based MGX-8830 and MGX-8850 PNNI Implementation

The last implementation of a multiservice switching PNNI node for routing and signaling is a PXM-1E-based system. The PXM-1E processor card is an enhanced processor card that uses PXM-45 and AXSM technology, providing higher port density and better performance than a PXM-1 processor card. It has the same switching capabilities as a PXM-1 by means of the shared memory (PXM-1E does not have a crossbar switching fabric), targeting the PNNI node for edge PNNI functionality. A PXM-1E-based MGX supports only CBSMs (FRSM, AUSM, VISM, CESM, and RPM-PR), not BASMs (AXSM cards).

An MGX system cannot share different types of processor cards in the same chassis, even though the MGX chassis is the same for PXM-1 (it is not PNNI-capable because of processor power), PXM-1E, and PXM-45 processor cards. That is, a mix of PXM-1, PXM-1E, and PXM-45 boards in the same node is not supported. PXM-1E is also supported in the small-factor MGX-8230 chassis with horizontal slots. PXM-45 is not currently planned to be supported.

In this PNNI implementation, as in a PXM-45 implementation, the PNNI software modules as well as the VSI master reside in the PXM-1E card. A centralized VSI slave also lives in the PXM-1E processor card.

The uplink is in the back card of the processor card. One of the features that makes a PXM-1E-based MGX node a versatile PNNI edge node is its support for multiport combinations with a daughter card/back card combo in the PXM-1E flexible uplink (a combo daughter card supports 8xDS3/E3 + 4xOC3), as well as a 16xT1/E1 daughter card and back card for ATM links and IMA links and a 2xOC12 daughter card and back card.

There is no graceful upgrade path from a PXM-1-based system in a feeder environment to a PXM-1E-based MGX acting as a PNNI routing node.

PNNI Implementation Summary

As a summary, Figure 3-29 details a PXM-45-based MGX-8850 and the BPX-8600 PNNI node part of a PNNI network.

The resources partitioned for PNNI are advertised to the PNNI controller. In the controller, they are represented as PnPorts.

If the CPEs are SVC-capable, they can place SVC calls using UNI signaling. The three PNNI nodes run PNNI signaling and routing, providing the ability to set up SPVCs. This is useful when the CPE is not SVC-capable.

Figure 3-29 *SES-Controlled BPX-8600 and MGX-8850 Running PNNI*

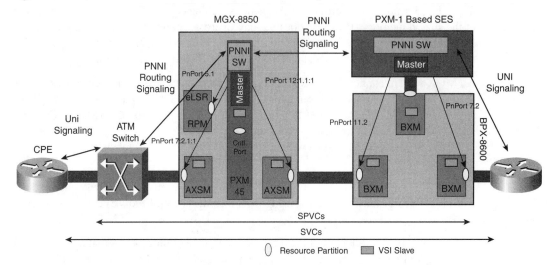

Other Implementations

This section discusses two Cisco Systems, Inc. proprietary network protocols and control-plane implementations. The following sections cover a monolithic version and a portable version of AutoRoute.

AutoRoute

AutoRoute (or Automatic Routing) is a very optimized and robust Cisco Systems, Inc. proprietary networking protocol that runs on BPX-8600 and IGX-8400 switches. AutoRoute (AR) makes use of a distributed topology database called the Network Information Base (NIB) that is used by the distributed Connection Management (CM) protocol. Processor cards in the nodes (BCC [Broadband Controller Card] in BPX-8600 and NPM [Network Processor Module] in IGX-8400) communicate with one another using network messages that need to be acknowledged by the receiving node. Each node maintains its own representation of the network topology and is aware of the current network state (including alarm conditions and the amount of traffic routed over a trunk, also called trunk-based loading). AutoRoute allocates bandwidth and routes. It reroutes connections over optimal paths through the network following either a hop-based route or a cost-based route. Route selection is source-based and is based on Dijkstra's Shortest Path Algorithm and the distributed database.

AutoRoute offers a wide range of added features, such as automatic connection management, automatic alternate routing, optional route cache, the ability to specify preferred paths or to avoid certain trunks for some connections, priority bumping, network clock distribution, time of day and date distribution, and in-band management. With its distributed intelligence, AR eliminates central points of failure. A connection's master endpoint is responsible for rerouting it, not a centralized route processor.

As far as scalability, the AR limit is 223 routing nodes. AR has proven itself in large networks. Each routing node can support up to 16 feeder nodes (which do not participate in the distributed topology), expanding the network node count to more than 1000. However, the long-term solution to "infinite scalability" in routing nodes is to migrate to PNNI.

NOTE   PNNI nodes also support feeder nodes that are not topology-aware.

AutoRoute is included in the software image that BCC and NPM processor cards run, called Switch Software (SwSW). The first implication is that ATM cards (BXM and UXM) always have an AutoRoute partition, controlled by SwSW and not VSI. Apart from the AR partition, using SwSW you can configure partitions to be controlled by the VSI slave in those cards.

NOTE   One of the differences between the AR partition and VSI partition management is that VSI connection setups are always acknowledged by the VSI slave, whereas normally AR connections are not. Switch Software sends the connection setup command to the line or trunk card and assumes it is successful. Switch Software checks Connection Admission Control (CAC). (In AutoRoute, Switch Software can set **overlap checking flag** and **ACK required flag** in a connection setup, specifically for VPI = 0.)

Portable AutoRoute

Portable AutoRoute (PAR), as the acronym clearly indicates, separates AutoRoute from the platform software. It is implemented in PXM-1-based MGX-8230 and MGX-8250. It uses VSI as the interface between the networking software (PAR) and the platform software.

The PXM-1 card houses the centralized VSI slave for all service modules in the MGX-8230 or MGX-8250, as well as the PAR application and VSI master. The VSI slave also manages logical ports in the PXM-1 for external clock interfaces and IP connectivity. All connections and resource management go through the centralized slave.

In a PXM-1-based MGX-8250 implementation, PAR has four functions:

- **Connection provisioning**—Performs feeder connections and local connections (DAX connections). The current implementation does not support AutoRoute routing functionality. It also maintains database synchronization.

- **Connection alarm management**—PAR sends messages to condition connections when it receives a failure event.

- **Annex.G**—In a feeder configuration, PAR manages Annex.G LMI communication between the MGX-8230 or MGX-8250 and the BPX-8600.

- **Clocking**—PAR handles the clocking selection in the edge switch.

An MGX-8230 or MGX-8250 node can be connected as a feeder node to a BPX-8600 or a PXM-45-based MGX-8850 or can act as a standalone node. A feeder node does not share topology information. In both cases, it supports single-height service modules such as FRSM, AUSM, CESM, and VISM and also double-height modules such as RPM/B and RPM-PR.

PXM-45-based MGX-8850 also implements PAR as an internal controller in the PXM-45 card.

Multiple Controllers

At this point, we have explored MPLS and PNNI implementations as well as AR and PAR implementations in Cisco Systems, Inc. multiservice switches. This section is intended to glue some of those concepts together by analyzing a complete BPX-8600 and MGX-8850 implementation.

Starting with a BPX-8600, you know that it supports AR (PVCs), MPLS (LVCs), and PNNI (SVCs and SPVCs) coexistence. The "ships in the night" concept (SIN) can be seen in the protocol stack shown in Figure 3-30. The networking software is platform-independent, ensuring consistent features and operation. Also, the controllers act independently. So the SES PNNI controller can be upgraded or removed without affecting the service for the MPLS controllers.

Figure 3-30 *Ships in the Night: Different Control Planes in a BPX-8600*

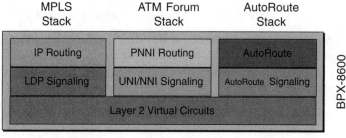

A closer look at the BXM card (see Figure 3-31) shows the interfaces between the networking software and the platform software.

Figure 3-31 *Resource Partitioning for Multiple Controllers in a BPX-8600*

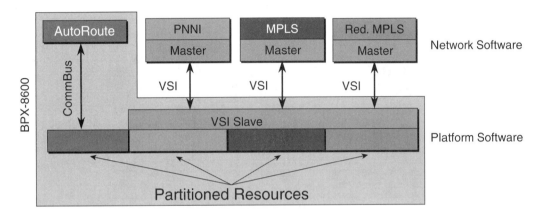

AutoRoute uses commbus messaging as an interface, and all the other controllers (PNNI and redundant MPLS controllers) use VSI.

The resources in a BXM port or trunk are partitioned for AR and VSI. VSI resources are partitioned for different controllers. The end result looks like Figure 3-32.

Figure 3-32 *Partitioning of Resources in a BXM/BXM-E*

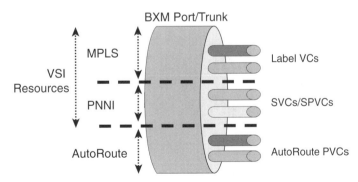

Resource Partitioning in a BXM/BXM-E

On the PXM-45-based MGX-8850, AutoRoute is not present, and all resources are under the control of the VSI slaves. Some controllers have fixed controller ids: 1 for PAR, 2 for PNNI, and 3 for LSC. Controller-ids 4 to 20 can be used by other instances of LSCs. Except for controller-ids 1, 2, and 3, VSI slaves do not make assumptions about the controller type based on the controller-id. Figure 3-33 shows a multiple-controller implementation in an MGX-8850.

Figure 3-33 *Multiple Controllers in a PXM-45-Based MGX-8850*

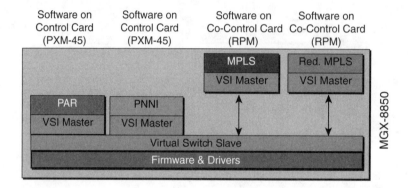

The PNNI controller and MPLS controller use the default control channels shown in Table 3-5.

Table 3-5 *Default Control Channels*

VPI	VCI	Control Channel	Controller
0	5	SSCOP	PNNI
0	16	ILMI	PNNI
0	18	PNNI	PNNI
0	32	Unlabeled traffic (LDP and IP)	MPLS
0	34	NCDP	PNNI

For MPLS applications, the minVCI in the resource partition defaults to 32. For PNNI, it defaults to 35.

VSI Slave Implementations

The VSI implementation in BPX-8600 and IGX-8400 switches follows a distributed model. Each BXM and BXM-E card in a BPX-8600 and each UXM, UXM-E, and URM card in an IGX-8400 houses a VSI slave managing the interfaces in that card.

MGX-8850 and MGX-8950 Controlled Switches and VSI Slaves

The multiservice switching implementations in PXM-45-based MGX-8850 and MGX-8950 have some variations. They are a bit more complex than the controlled switch architecture, presenting a hybrid VSI slave model. To completely understand the MPLS and PNNI implementations on these switches, you need to understand the switch platform.

In a PXM-45-based MGX-8850, the switching is performed in the PXM-45 card. Figure 3-34 shows a simplified hardware overview.

Figure 3-34 *PXM-45-Based MGX-8850—Hardware Overview*

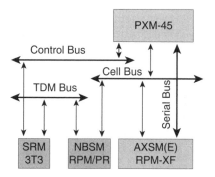

The TDM bus is used for bulk distribution. The redundancy bus is not shown in Figure 3-34.

For the purposes of switching, NBSMs use the cell buses. (The backplane contains eight cell buses.) The PXM-45 card uses shared memory to switch cells from the cell buses, such as a PXM-1. Examples of NBSMs include all single-high service modules (such as VISM, FRSM, and AUSM) and RPM-PR.

BASMs use sets of serial lines called serial buses running to every double-high slot. The PXM-45 card uses cross-point switches to switch cells from the serial buses. Examples of BASMs include all the AXSM and AXSM-E cards and the RPM-XF.

The following list shows the switching path in the PXM-45 card for intranode connections:

- NBSM-to-NBSM connections use the cell bus switching fabric.
- BASM-to-BASM connections use the crossbar switching fabric.
- NBSM-to-BASM connections use both switching fabrics.

The following list shows the switching path in the PXM-45 card for internode connections:

- NBSM-terminated connections use the cell bus switching fabric.
- BASM-terminated connections use the crossbar switching fabric.

On an MGX-8950 platform, the switching of cells from the cell buses is also performed in the PXM-45/B cards, but the switching of cells from the serial buses is performed in the XM-60 cards. This adds another degree of separation between data traffic and control traffic. The control bus also connects the XM-60 and the PXM-45/B cards.

NOTE	The MGX-8950 does not support SRM cards or single-high service modules. The only NBSM it supports is RPM-PR. The MGX-8950 also supports only AXSM/B modules (as opposed to AXSM modules, which fail in an MGX-8950).

Why is this quick overview important for an understanding of the multiservice switching implementation on these two platforms? The answer is very straightforward: BASMs (AXSM, AXSM-E, and RPM-XF) implement the VSI slave in the card, whereas NBSMs (RPM-PR, VISM, AUSM, FRSM, and CESM) use a centralized slave in the PXM-45 card as a proxy VSI slave. The end result is that the slave model is hybrid; BASMs (also known as Serial Bus Service Modules [SBSMs]) provide the distributed part of the model, and NBSMs (also known as CBSMs) provide the centralized portion of the model.

NOTE	The preceding paragraphs do not imply that all NBSMs operate in an MPLS or PNNI application. For example, the Circuit Emulation Service Module (CESM) does not fit in any MPLS application. Nevertheless, when a controller such as the LSC, PNNI controller, PAR controller, or any future controller uses an NBSM, it communicates with the VSI slave located in the PXM-45. In an MPLS application, RPM-PR is the NBSM that is used. The PNNI controller can theoretically use FRSM, AUSM, VISM, RPM-PR, and CESM. RPM-PR is the first CBSM with support for SPVCs in PNNI applications. PNNI controller support for other CBSMs on PXM-45 is under investigation.

VSI Slaves: Logical Interface Numbers

As you know, the VSI slave is responsible for assigning unique and persistent 32-bit-long LINs that identify each logical port.

Even though the VSI master should not make assumptions about the encoding of the LINs, or assume any logic behind it, it is useful for you to understand LIN encoding for low-level troubleshooting. The fact that the VSI master does not associate the LIN with a physical port keeps the networking code platform-independent. A physical descriptor is advertised as an ASCII value from the slave to ease user configuration.

Different VSI slaves encode the LIN differently based on the specifics of each platform.

In BPX and IGX switches, VSI slaves in BXM and UXM cards, respectively, use a straight-forward coding scheme: $LIN_{UXM-BXM}$ = **0x00**$AABBCC$, where AA is the slot number, BB is the physical port number, and CC is the virtual trunk number or 0 if it is not a virtual trunk. The VSI slaves in AXSM and PXM-45 cards in MGX platforms use a different encoding. Figures 3-35 through 3-37 summarize the LIN assignment in all VSI slaves.

The BXM and UXM VSI slave LIN encoding is shown in Figure 3-35.

Figure 3-35 *BXM and UXM LIN Encoding*

Figure 3-35 contains the following field:

- If the partition resides in a physical interface, the virtual trunk number is 0x00.

The AXSM cards and RPM-XF VSI slave LIN encoding are shown in Figure 3-36. RPM-XF is a serial line-based (as opposed to cell bus-based) RPM for MGX-8850 and MGX-8950. RPM-PR is cell bus-based and does not include a VSI slave module.

Figure 3-36 *AXSM and RPM-XF LIN Encoding*

```
       3                 2                 1
1 0 9 8 7 6 5 4 3 2 1 0 9 8 7 6 5 4 3 2 1 0 9 8 7 6 5 4 3 2 1 0
┌───────────────┬─────────────┬────────┬─────────────────────┐
│Shelf # = 0x01 │   Slot #    │ ifType │    Port Number      │
└───────────────┴─────────────┴────────┴─────────────────────┘
```

Figure 3-36 contains the following fields:

- The shelf number always equals 0x01.
- The slot number is the AXSM/RPM-XF slot.
- ifType equals 3 (00011b), meaning atmVirtual.
- The port number is the logical port number.

Figure 3-37 shows the PXM-45 and PXM-1E proxy VSI slave LIN encoding. This encoding is used for service modules for which PXM-45 or PXM-1E is a centralized slave. Those include RPM/B, RPM-PR, and NBSM (that is, all cell bus-based service modules [SMs]), as well as internal logical ports.

Figure 3-37 *PXM-45 and PXM-1E LIN Encoding*

```
        3                   2                   1
1 0 9 8 7 6 5 4 3 2 1 0 9 8 7 6 5 4 3 2 1 0 9 8 7 6 5 4 3 2 1 0
┌──────────────┬──────────────┬──────────┬──────┬──────────────┐
│ Shelf # = 0x01│ Pri PXM Slot#│ SM Slot# │ifTyp │ Interface #  │
└──────────────┴──────────────┴──────────┴──────┴──────────────┘
```

Figure 3-37 contains the following fields:

- Primary PXM slot number is the PXM logical slot number. It is 7 for PXM-45 and PXM-1E-based MGX-8850 systems and 1 for MGX-8830 switches.
- Service module slot number is the slot in which the SM resides.
- ifType equals 3 (00011b), meaning atmVirtual.
- For RPM-B and RPM-PR proxy partitions, the interface number is 1 for VCC partitions and 2 for VPC partitions.
- For PXM-45 and PXM-1E internal ports, interface number 34 is used as the control port and IP connectivity and numbers 35 through 38 are used for building integrated timing supply (BITS) clocking.

The high-speed Frame Relay Service Module (FRSM-12-T3E3) uses the same LIN encoding as AXSM cards and RPM-XF.

As mentioned earlier, the PXM-45 or PXM-45B card in an MGX-8850 platform acts as a centralized slave for cell bus-based service modules. Those service modules include mid-high SMs (such as VISM, FRSM, and AUSM) and RPM-PR cards. The only CBSM supported in the MGX-8950 platform is RPM-PR.

The PXM-45 or PXM-45/B has other virtual ports. (In the LIN, the SM slot number is 7.) They are summarized in Tables 3-6 and 3-7.

Table 3-6 *PNNI Logical Control Port*

Port ID	Name	Description
7.34	Control port	This logical interface is the endpoint for connections that pass through the SAR process and terminate on the PXM-45.

NOTE   The logical interface 7.34 is a logical or virtual port. It does not exist physically in the PXM-45 but is presented as another port to the VSI master application. The need for such a port is as follows: Signaling connections such as UNI and PNNI need to terminate in the PXM-45 for SAR and processing. Those connections are created using VSI messages. Therefore, VSI needs a LIN to reserve and commit those endpoints. Another example is the IPCONN application. This logical interface is the endpoint for SVCs for IP management connectivity to the MGX switch, terminating in the IP stack.

Table 3-7 *BITS Clocking Logical Ports*

Port ID	BITS Input	Description
7.35	Upper slot E1	Represents the upper physical clocking input port where an E1 BITS clock is connected.
7.36	Upper slot T1	Represents the upper physical clocking input port where a T1 BITS clock is connected.
7.37	Lower slot E1	Represents the lower physical clocking input port where an E1 BITS clock is connected.
7.38	Lower slot T1	Represents the lower physical clocking input port where a T1 BITS clock is connected.

Some clarifications are necessary regarding logical ports 7.35 through 7.38. The PXM-45-UI-S3 back card has two physical clock ports, but four LINs are necessary to uniquely identify the BITS mode of both clock ports. In Chapter 2, "SCI: Virtual Switch Interface," the "Clock Synchronization" section described the mechanism used between the slave (advertising the clock input as a LIN with no connection capabilities) and the NCDP application. The PXM-45 slave understands only LINs, and the VSI **set clock** commands do not have a field to indicate clock type (BITS T1 versus BITS E1). Four LINs are needed to overcome this limitation where the user can configure four combinations (CLK1 E1, CLK1 T1, CLK2 E1, and CLK2 T1).

Emphasizing the ideas presented before, the VSI master does not need to know the logical interface encoding. It treats the LIN as a 32-bit number, not assuming a coding scheme. But the user (who sees the whole picture) needs to know it. LINs are used in different contexts. The most visible example is the construction of the SPVC endpoint AESA, where the switch prefix is prefixed to the LIN to create a unique AESA for SPVCs. See Figure 3-38.

Figure 3-38 *SPVC Endpoint AESA Construction Using LIN*

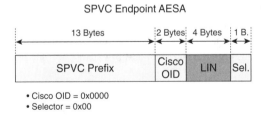

• Cisco OID = 0x0000
• Selector = 0x00

VSI Slaves: Resource Management

One of the most important modules in the VSI slave to ensure true multiservice functionality is the Resource Management Module (RMM). The Resource Manager implements the VSI slave's platform-specific or dependent functionality. It is logically divided into different submodules, as shown in Figure 3-39.

Figure 3-39 *VSI Slave Resource Manager Components*

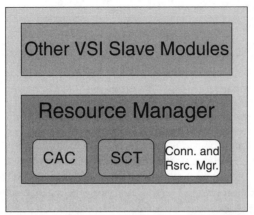

VSI Slave Module

The three constitutive modules of the resource management module are explained in the following list:

- **CAC**—CAC is responsible for accepting or rejecting a connection configuration. This ensures that the partition's requirements can be fulfilled.

- **Service Class Templates (SCTs)**—These provide default values for connection-configurable parameters. The SCT is applied on an interface basis. Its defaults are used if the parameters are not provided, determining a connection's traffic characteristics. SCT details are covered in the later section "Service Class Templates and QoS."

- **Connection and Resource Manager**—This programs the connections on hardware (by means of an API) and manages LCN and bandwidth resources. LCN resources are managed at the partition and port group levels. Bandwidth resources are managed at the partition, logical interface, and physical interface levels.

NOTE   CAC is performed by the VSI slave. In a PNNI application, the PNNI node originating the source route performs a General Connection Admission Control (GCAC) when creating the Designated Transit List (DTL). GCAC is performed by the PNNI controller. Neither call-acceptance mechanism should differ. If the GCAC algorithm is too permissive compared to the CAC algorithm, unnecessary crank-backs result. If GCAC is too restrictive, it prevents the use of routes with enough resources.

NOTE   A physical interface consists of one or more logical interfaces. Each logical interface has one or more resource partitions.

As an example, Figure 3-40 shows some of the practical differences between a centralized slave and a distributed slave in a PXM-45-based MGX-8850 that uses a hybrid slave model.

Figure 3-40   *VSI Hybrid Model in a PXM-45-Based MGX-8850*

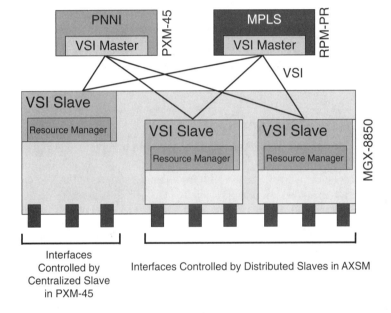

Resources in BASMs are managed by the VSI slave in the card, and resources in NBSMs are managed by the centralized VSI slave in the PXM-45 card, resulting in a VSI hybrid model.

VSI Slaves: Connection and Resource Manager

The Connection and Resource Manager is the third module on the Resource Manager implemented as part of the VSI slave. Different VSI slaves manage and partition different resources—in particular, BXM and BXM-E cards in a BPX-8600, UXM and UXM-E cards in an IGX-8400 partition VPI space, and bandwidth and LCN resources per VSI resource partition. Indirectly, the QBins (or Class of Service [CoS] buffers) are also partitioned.

The AXSM card implementation, however, has finer granularity. The resources that are partitioned per VSI partition are VPI space, VCI space, egress bandwidth, ingress bandwidth, and LCN resources (as well as indirectly partitioning CoS buffers). Compared to BXM and UXM implementations, bandwidth partitioning has one more degree of freedom (direction; the ingress and egress can be different), and it also can partition the VCI space.

At first glance, VCI space partitioning might seem unimportant or trivial, but it has significant consequences. Allowing partitioning of resources in a virtual trunk (VP-Tunnel) and assigning those partitions to two different controllers is one of the outcomes. A customer can buy a VPC from a service provider and run MPLS and PNNI over it—for example, MPLS using VCIs between 33 and 32767 and PNNI using VCIs between 32768 and 65535.

NOTE   If the objective is to limit the VCI space for only one partition, that can be accomplished from the controller, in an MPLS controller or a PNNI controller.

Service Class Templates and QoS

SCTs are extremely important in multiservice switch implementations because they are the means to provide QoS support. SCTs provide a way of mapping protocol connection parameters (such as MPLS or PNNI traffic parameters) passed in VSI connection setup primitives to extended parameters, usually platform-specific. SCTs provide support for different CoSs. Refer to Figure 2-3 and the paragraph following it in Chapter 2.

SCTs are structured in two different templates: VC descriptor templates, with the necessary parameters to establish connections, and Class of Service Buffer (CoSB) descriptor templates, employed to configure the queues that service connections with similar QoS requirements. See Figure 3-41.

Figure 3-41 *SCT Structure*

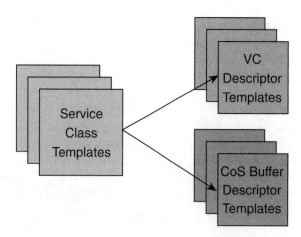

NOTE	CoSB is a general designation that sometimes goes by a different name in a different implementation. In BXM cards, CoSBs are also called QBins. The CoSB descriptor template has 16 entries because each virtual interface has 16 CoSBs or QBins.

When a controller wants to set up a cross-connect in the controlled switch, the VSI message contains optional traffic parameters, along with a service type specified for the connection, as discussed in Appendix A, "Service Traffic Groups, Types, and Categories." That service type is used as the index into the VC descriptor template to find the extended parameters and program the connection. This process is illustrated in Figure 3-42.

Figure 3-42 *SCT—Simplified Connection Setup Process*

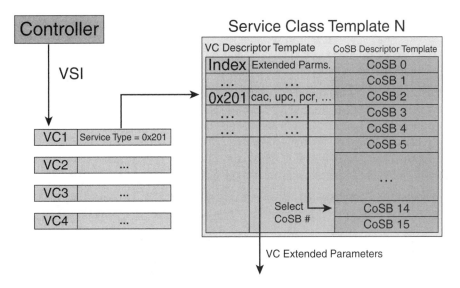

The SCTs also include a default service type entry in the VC descriptor template in case an unsupported service type is requested. The VC descriptor templates can have parameters or "columns" that are scaling factors or multipliers of VSI-provided parameters. Other parameters are independent of VSI-provided parameters. The CoSB templates configure the CoSBs. Some of the parameters included in the CoSB template are CLP, EPD, and EFCI thresholds, and EPD and WFQ enable/disable status. In general, SCTs are valid only for VSI-enabled partitions.

One of the extended parameters in the VC descriptor template is the CoSB that the traffic type uses. This fact is of great importance because it allows the multiservice switch to use different queues for different service types. In an MPLS multi-vc application, for example (as discussed in Chapters 4 and 7), this allows an LSR to use one queue for "standard" traffic and a different one for "premium" traffic. In the multiservice switching LSR, the forwarding is performed by the controlled switch based on the switching label in part of the VPI/VCI in the cell's header. Yet the queues in the controlled switch are IP-aware. In a PNNI scenario, different queues are used for CBR traffic and UBR traffic, allowing true constraint-based routing.

Generalizing this concept, SCTs give the multiservice switch the tools to implement the Core Differentiated Services (DiffServ) architecture (see RFC 2475). The DiffServ architecture defines edge behaviors (classification, marking, policing, and so on) and core behaviors (Per-Hop Behaviors [PHBs], in which the same treatment is applied to all packets with the same DiffServ service class). The building blocks of PHBs are queuing and dropping. With the SCTs, a multiservice switch can implement per-class queuing and dropping policies, as well as assign a subpartitioned bandwidth allocation to each class.

SCTs also allow class type-aware admission control, a building block of Guaranteed Bandwidth Traffic Engineering (GB-TE) or DiffServ-Aware Traffic Engineering (DS-TE), where a connection's class type is signaled. In essence, the CoSB templates in the SCT indirectly partition queue resources.

Different SCTs are suited to different applications (typical interface uses). There are nine predefined SCTs, covering MPLS-only interfaces, PNNI-only interfaces with and without policing, and different combinations of simultaneous MPLS and PNNI interfaces. SCT characteristics are summarized in Table 3-8.

Table 3-8 *SCT Characteristics*

SCT Number	SCT Name	Controller Support	Description	ATMF UPC
1	MPLS1	MPLS	MPLS DiffServ-only template with MPLS ABR support	N/A
2	ATMF1	PNNI	PNNI-only template	On
3	ATMF2	PNNI	PNNI-only template	Off
4, 6, and 8	ATMF_MPLS_MPLSABR_1	MPLS and PNNI	Support for ATMF, DiffServ MPLS, and MPLS ABR	On
5, 7, and 9	ATMF_MPLS_MPLSABR_2	MPLS and PNNI	Support for ATMF, DiffServ MPLS, and MPLS ABR	Off

NOTE   Not all SCTs are implemented in all platforms.

SCT 1 supports only MPLS service. The rest of the SCTs, from 2 to 9, are identical pairs, except that the even SCTs (2, 4, 6, and 8) have policing on, and the odd SCTs (3, 5, 7, and 9) do not have policing for ATMF service types. SCTs 2 and 3 support only ATM Forum types (such as PNNI), and the rest support some combination of MPLS and PNNI service types.

On BPX and IGX slaves, AutoRoute service types use some QBins. Therefore, some of the SCTs implement reduced support for a certain service category, such as MPLS or ATMF, meaning that all service types use only two CoSBs or QBins. Conversely, on AXSM slaves, the SCTs that simultaneously support MPLS and PNNI map the service categories to different CoSBs, providing full support for MPLS and PNNI simultaneously. In MPLS-only SCT (SCT 1) or PNNI-only SCTs (SCTs 2 and 3), no support is provided for the other service categories. This means the service types in those service categories use the same CoSB or QBin as the default service type, all mapped to one single CoSB. The VSI specification requires that unknown service types be mapped to the default service type.

Different platforms implement SCTs differently:

- In BPX-8600 and IGX-8400 switches, SCTs are bundled into a Switch Software image at compilation time. Therefore, SCTs reside in the controller card (BCC and NPM) and are downloaded to the BXM and UXM cards using comm-bus and C-bus messages.

- In PXM-45-based MGX-8850, the SCTs reside in separate files on the PXM-45 hard drive in the directory C:/SCT/. They are downloaded to and processed by AXSM cards upon configuration. In the event that the SCT files are not present, a default SCT numbered 0 is used. SCT 0 is bundled into a PXM-45 image and cannot be modified.

Differences also exist from a configuration perspective. BPX-8600 and IGX-8400 have very few configuration parameters using SwSW commands. The VC descriptor template cannot be configured. MGX-8850 SCTs, on the other hand, can be modified offline using Cisco WAN Manager (CWM) and uploaded to the PXM-45 hard drive.

Summary

This chapter provided architecture descriptions of the various VSI master and slave implementations running in Cisco Systems, Inc.'s BPX, IGX, and MGX product lines. The MPLS LSC can exist in a standalone platform that is physically linked to a BPX or IGX. Otherwise, an RPM or URM service module resides in an MGX or IGX, respectively, and contains the LSC implementation. Similar to MPLS, PNNI controllers either are implemented in external platforms such as the SES, which directs PNNI operation in the BPX, or reside in an MGX switch's PXM control card.

After discussing protocol controller and VSI master implementations, we focused on VSI slaves. VSI slaves are categorized into hybrid models, such as the MGX, or distributed models, found in the BPX and IGX. Remember that hybrid or distributed classification depends on the location of the VSI slaves.

Additional topics in this chapter included a description of the legacy AutoRoute and Portable AutoRoute connection routing algorithms, as well as a detailed explanation of and reason for LINs.

The next chapters delve more deeply into MPLS and PNNI multiservice switching networks.

CHAPTER 4

Introduction to Multiprotocol Label Switching

Multiprotocol label switching (MPLS) represents the next level of standards-based evolution in combining Layer 2 (data link layer) switching technologies with Layer 3 (network layer) routing technologies. The primary objective of this standardization process is to create a flexible networking fabric that provides increased performance and scalability. MPLS also enables advanced features such as traffic engineering capabilities, Quality of Service (QoS) and Class of Service (CoS) awareness, and facilitates the deployment of Virtual Private Networks (VPNs).

MPLS is designed to work with a wide variety of transport mechanisms; however, the initial implementations focus on leveraging ATM, IP, and Frame Relay, which are already deployed in large-scale service provider networks. This chapter highlights the details of IP + ATM-based networks and their capabilities as they apply to the deployment of MPLS and the requirements on a multiservice switch.

MPLS Background

MPLS is defined by the Internet Engineering Task Force (IETF) as a standards-based approach to applying label-swapping technology to large-scale networks. The MPLS Working Group was established in early 1997 and has since defined a large set of working documents that are currently being massaged into standards. Prior to the formation of the MPLS Working Group, a number of vendors had announced and/or built a proprietary version of a label-switching implementation. This widespread interest in label-switching technology initiated the formation of the MPLS Working Group.

The IETF is defining MPLS in response to numerous interrelated problems that need immediate attention. These problems include the following:

- Scaling IP networks to meet the growing demands of Internet traffic
- Allowing differentiated levels of IP-based services to be provisioned
- Merging voice, video, and different data applications onto a single IP network
- Scaling VPNs
- Improving operational efficiency to cut costs

It is important to note that many service providers are active in the MPLS Working Group. This ensures that MPLS's capabilities will have a direct correlation to customer problems. Many of the issues MPLS seeks to address have already been recognized by a number of equipment manufacturers. In fact, many of these vendors have already developed standards-based solutions that address these problems. You can find more information about this in the IETF MPLS charter at www.ietf.org/html.charters/mpls-charter.html.

It is interesting to analyze the etymology of the MPLS acronym. This chapter covers the label-switching paradigm in great depth. However, what does multiprotocol refer to? This question might seem basic, but the answer goes to the heart of the MPLS concept. Multiprotocol means multiple network layer (Layer 3) protocols such as IPv4 and IPv6 and multiple data link layer (Layer 2) protocols such as Asynchronous Transfer Mode (ATM), Frame Relay, Point-to-Point Protocol (PPP), Ethernet, and Fast Ethernet. For this reason, MPLS does not fit into the OSI model.

Although MPLS, as we described, can conceptually support multiple Layer 2 and Layer 3 protocols, the initial work is focused on the integration of IPv4 with ATM and Frame Relay. Current feature developments include IPv6 and AToM (Any Transport over MPLS), just to name a couple.

MPLS Overview

ATM was once envisioned as the best technology for multiservice-based core and edge networks. Many large enterprise customers (LECs) thought ATM would extend from the desktop through the core of the network and terminate at another desktop. This speculation was based on the fact that ATM already had the capability for different classes of service, such as constant bit rate (CBR), variable bit rate (VBR), available bit rate (ABR), and unspecified bit rate (UBR).

Today, these features have been built into an IP-based network and IETF standards included as well. However, economics will play an important role in determining the adoption rate of these next-generation IP networks. Upgrading entire networks by swapping out hundreds or even thousands of pieces of equipment is not cost-effective. Therefore, many service providers will continue to maintain ATM in their networks for the foreseeable future. Less risk is also involved in incrementally designing an IP + ATM multiservice switching network because ATM QoS is proven on a large scale in live deployments.

Consequently, next-generation networks will be built using new technologies that leverage existing technologies such as IP + ATM that are already deployed in existing networks.

A new paradigm is also presented in this book for service providers and large enterprise networks intending to provide existing services such as point-to-point ATM and Frame Relay virtual circuits (VCs), ATM switched virtual circuits (SVCs), and circuit emulation services (CES) over cell-based networks while adding IP services to their portfolio so that those services don't interfere with each other and maintain inherent ATM QoS. For incremental service management and network investment protection, multiservice switching networks will play a very important role in driving next-generation services. Many of these access technologies can be IP-enabled so that customers can choose between a pure Layer 2 or an enhanced Layer 3 service with Layer 2 guarantees.

Finally, the proven reliability, availability, and serviceability (RAS) characteristics inherent in ATM networks can be leveraged to provide similar availability in IP networks.

Label-switching technology is the result of the desire to combine the benefits of switching technologies that live in the core of the network with the benefits of IP routing technologies that live at the edge of the network, getting the best of both worlds. A blended network using both of these technologies creates a problem best described as "how to make IP and ATM interoperate." The IETF and the ATM Forum initially took on this challenge and defined standards such as Classical IP over ATM (RFC 1577 and RFC 2225) and multiprotocol over ATM (MPOA), which allow IP to work over an ATM network. However, MPLS is tackling a different problem best described as "how to integrate the best components of traditional Layer 2 and Layer 3 technologies." As previously mentioned, multiprotocol label switching seeks to combine the best features of Layer 2 switching technologies as they exist in the ATM and Frame Relay world with the best features of Layer 3 routing technologies as they exist in the IP world. This is Layer 3 plus Layer 2 as opposed to Layer 3 over Layer 2.

MPLS, as the standards-based approach to label switching, specifically defines a set of protocols and procedures that allow the fast-switching capabilities of ATM and Frame Relay to be used by IP networks. The key concept in MPLS is identifying and marking IP packets with labels at the edge and forwarding them to a modified switch or router, which then uses the labels to switch the packets through the network. This new equipment with router and switch characteristics is a label switch router (LSR). The labels are created and assigned to IP packets based on the information gathered from existing IP routing protocols. This classification happens in the edge in edge Label Switch Routers (eLSRs), also called Label Edge Routers (LERs).

The upcoming sections cover the following topics:

- **The role of routing**—Gathering information necessary to get to an IP prefix destination
- **The role of switching**—Forwarding packets in a simple and efficient way
- **The role of MPLS**—Separating routing and forwarding, gathering the best of Layer 2 and Layer 3 to then bring them together

The Role of Routing

Internet Protocol (IP) packets contain a header with sufficient information that lets them be forwarded through a network. Packet forwarding has traditionally been based on packet routing. The packet-routing technique used in IP networks is a destination-based routing paradigm. This means that an IP packet is routed through the network based on the destination address contained in the packet. An IP packet is shown in Figure 4-1.

Figure 4-1 *IP Packet*

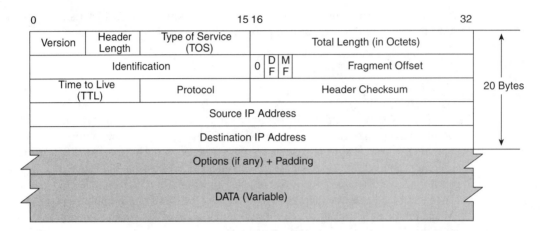

A note on the Type-of-Service (ToS) field shown in Figure 4-1 is included in the upcoming section "Quality of Service and Multi-vc" as well as Figure 4-17.

The forwarding mechanism used by IP networks is hop-by-hop routing, which means that every packet entering a router is examined, and a decision is made as to where to send the packet (in other words, what the packet's next hop is). In this manner, a packet is routed through a network from its source to its destination. Because the packets are individually routed through a network and don't follow a predetermined path, the network is said to be connectionless. Routers exchange information by establishing an adjacency (in other words, a conversation) with every directly connected peer.

To properly route a packet, a router must be able to determine the packet's next hop. Routing protocols, such as Open Shortest Path First (OSPF), allow each router to learn the network's topology. Using the information provided by the routing protocols, the routers build forwarding tables that identify the appropriate next hop for all known IP destination addresses.

NOTE   Routers typically store IP prefixes rather than complete IP addresses in their forwarding tables; however, the details of IP routing are outside the scope of this book.

The Role of Switching

Switching technologies based on ATM and Frame Relay use a much different forwarding algorithm than monolithic routers, which essentially use a label-swapping algorithm. Because this forwarding algorithm is so simplistic, it is typically done in hardware, yielding a better price/performance advantage than traditional IP routing that takes place in a software path.

ATM and Frame Relay are connection-oriented technologies, meaning that traffic is sent between two endpoints only after a connection (in other words, a predetermined path) has been established. Because traffic between any two points in the network flows along a predetermined path, connection-oriented technologies make the network more predictable and manageable. The combination of these attributes helps explain why large networks have been built with a switching fabric in the network's core.

The Role of MPLS: Bringing Routing and Switching Together

MPLS is designed to meet all the mandatory characteristics and requirements of large-scale carrier-class networks. It is evolutionary in the sense that it uses existing Layer 3 routing protocols as well as all the widely available Layer 2 transport mechanisms and protocols, such as ATM, Frame Relay, PPP over leased lines, and Ethernet. For large-scale public networks, Frame Relay and particularly ATM are of great interest, primarily because they support the underlying concepts of QoS and CoS.

MPLS solves the problem of how to integrate the best features of traditional Layer 2 and Layer 3 technologies. It does this by defining a new operating methodology for the network. The key component within an MPLS network, the LSR, can understand and participate in both IP routing and Layer 2 switching. By combining these technologies into a single operating methodology, MPLS avoids the problems associated with methods that define a way for Layer 2 and Layer 3 to interoperate while maintaining two distinct operating paradigms.

Even though MPLS requires LSRs to participate in IP routing, MPLS's forwarding aspect differs significantly from hop-by-hop routing. The LSRs participate in IP routing to understand the IP reachability and network topology from a Layer 3 perspective. This routing knowledge is then used to assign labels to packets. Labels are analogous to the VPI/VCI used in ATM and the DLCI used in Frame Relay. When viewed on an end-to-end basis, the labels combine to define the path between endpoints. These paths, called label-switched paths (LSPs), are intentionally very similar to the VCs used by switching technologies because they provide benefits such as predictability and manageability. In addition, the connection or LSP lets a Layer 2 forwarding mechanism be used. As described earlier, a label-swapping mechanism is typically very fast and can be implemented very easily in hardware.

How Label Switching Works

The multiprotocol label-switching architecture is defined in RFC 3031.

In general terms, label switching in the data path is similar in concept to VC switching in traditional ATM switches. At each switching juncture, the label value is rewritten with an outgoing label, similar to ATM cell switching. The main difference is in the control plane, because LDP is used for label setup as opposed to ATM forum signaling.

The general label that is used for label swapping, also called a shim header, resides in the label stack, as defined in RFC 3032, "MPLS Label Stack Encoding." The shim header is a 32-bit header added in front of the IP header, as shown in Figure 4-2. All link types, including ATM and Frame Relay, insert this special shim header.

Figure 4-2 *Label Stack Entry*

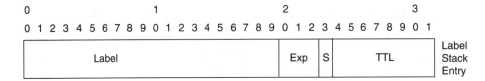

Label: Label Value, 20 Bits
Exp: Experimental Use, 3 Bits
S: Bottom of Stack, 1 Bit
TTL: Time to Live, 8 Bits

The fields in the shim header shown in Figure 4-2 are as follows:

- **Label**—The actual 20-bit label value.

- **Exp**—Experimental use. Although this 3-bit field is defined as reserved for experimental use in RFC 3035, it is used in MPLS QoS support in the DiffServ model.

- **S**—Bottom of stack. This bit is set to indicate the last entry label stack. In a multiple label stack application, such as MPLS VPN, only the innermost label stack entry has this bit set.

- **TTL**—Time to Live. This field encodes the time to live.

The 20-bit label is used for label swapping. This MPLS header or stack of entries appears after the data link layer headers and before the network layer headers. For this reason, the MPLS header is normally called Layer 2-and-a-half (Layer 2 1/2). It is used in frame-based data link layers such as PPP, Ethernet, and Fast Ethernet.

However, ATM and Frame Relay already implement a label-switching forwarding mechanism, swapping identifiers (VPI/VCI and DLCI) on a link-by-link basis. Therefore, it makes sense to use these identifiers as the labels swapped in MPLS. On ATM MPLS, the VPI and VCI fields or the VCI field alone are used as the label. This is defined in RFC 3035, "MPLS

Using LDP and ATM VC Switching." In Frame Relay-based MPLS, as defined in RFC 3034, "Use of Label Switching on Frame Relay Networks Specification," the DLCI is used as the label. Figure 4-3 shows the different label stack entries using different Layer 2 protocols.

Figure 4-3   *Encapsulation in Different Layer 2 Protocols*

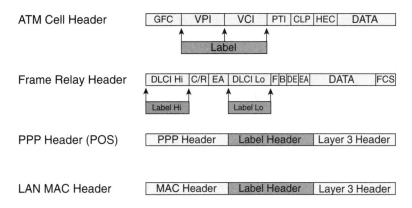

VPI/VCI fields or VCI fields alone are used as the label. Labels are applied to each cell, which makes label swapping look like ATM switching from a forwarding perspective. Both Label Distribution Protocol (LDP, as defined in RFC 3036) and Tag Distribution Protocol (TDP, the Cisco proprietary prestandard distribution protocol) are used for label distribution in ATM on the control plane.

On label switching controlled ATM (LC-ATM) interfaces, label packets are sent using null encapsulation, as defined in Section 6.1 of RFC 2684, in which there is no need for protocol identification. The label stack header is always attached preceding the network layer header before segmenting the packet into cells. In this case, the label stack header carries a null label because, as you know, the top label is carried encoded in the VPI/VCI or VCI only fields. Why carry this shim header if the label is encoded in cell headers? The main reason is to allow a label stack of arbitrary depth, just as on non-ATM links. Another function of the label stack is to carry the TTL and Exp bits. You need both of these fields if a packet is to be label-switched further by a non–ATM-LSR. This function also provides transparency for the Exp bits. In summary, labeled packets must always have a shim. The complete encapsulation of ATM MPLS labeled packets is shown in Figure 4-4.

The ATM MPLS packet shown in Figure 4-4 contains a stack of one label entry, the MPLS label. This would be a packet terminating in an LSR or eLSR. In an MPLS VPN application, the MPLS packet inside the AAL5 PDU has at least two entries in the label stack even when it is carried on a label virtual circuit (LVC): an implicit-null label (since the MPLS label is carried in the VPI/VCI fields), and the BGP VPN label. This will be discussed in detail in the upcoming section "IP Virtual Private Network Services."

Figure 4-4 *MPLS Encapsulation on LC-ATM*

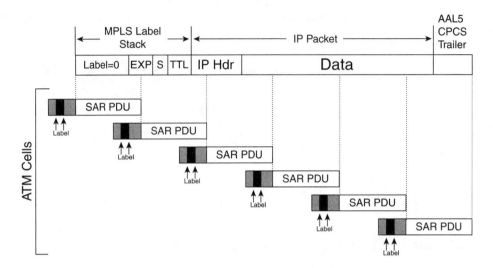

MPLS Operation Example

Let's look briefly at a very basic and general MPLS operation example, as shown in Figure 4-5. At this stage, this example applies to both frame-based and cell-based MPLS.

Figure 4-5 *How MPLS Works*

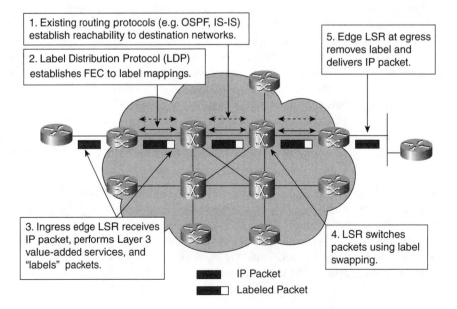

We can divide the process in two planes well-separated by MPLS:

- **Control plane**—It starts with a routing protocol such as OSPF, EIGRP, or IS-IS, establishing reachability across the network and thus determining the Layer 3 topology. As soon as reachability is established when the routing tables are built, the LDP assigns or binds labels to each route and propagates this information to its neighbors.

- **Data plane**—A packet comes into the network. The packet is a normal IP packet. It hits the edge of the MPLS network (an eLSR). A router does a lookup based on the IP address. As soon as a match is found, a label is identified. That label is affixed to the packet, and the packet is forwarded to the appropriate interface. The packet then travels to the next-hop LSR. That LSR does a lookup based on the label only. It finds a match based on that label in its forwarding table. An outgoing label is identified. That outgoing label is swapped for the incoming label, and the packet is switched out the appropriate port. This process or procedure is repeated hop by hop until you get to the edge of the network. At that point, the exit eLSR does a lookup based on the label, finds a match, determines that there is no outgoing label, strips off the incoming label, and forwards the packet on its way as a normal IP packet.

The things you need to notice here are that labels have only local significance, meaning that they are meaningful in the link only and not network-wide. That is normally the difference between an identifier, which is locally significant, and an address, which is globally significant.

A key point to remember is that MPLS provides a clean separation between routing (in the control plane) and forwarding (in the data plane), as opposed to normal IP routing, in which both routing and forwarding are almost indivisible.

A frame-based MPLS implementation and a cell-based MPLS implementation have some important differences:

- **Labels**—On frame-based interfaces, the MPLS label is carried in the shim header, at Layer 2 1/2, between the Layer 2 and Layer 3 headers. On ATM MPLS, VPI/VCI fields are used as MPLS labels, and in Frame Relay, DLCIs are used as labels.

- **Label distribution**—With frame-based MPLS, label distribution is Downstream Unsolicited. With cell-based MPLS, the label distribution is Downstream on Demand. Downstream on Demand is also called upstream-controlled because the VPI and VCI used as the label are scarce resources.

- **Label spaces**—LDP on frame-based interfaces uses a per-LSR label space. On the other hand, an ATM-LSR uses per-interface label spaces because two different label-controlled ATM interfaces can use the same VPI/VCI pair (the same label).

- **LSPs**—On ATM MPLS, LSPs are called label virtual circuits (LVCs).

- **Penultimate Hop Popping (PHP)**—On frame-based MPLS, the exit eLSR can ask the penultimate hop to pop the top label. This functionality does not exist in ATM MPLS because cells cannot be sent without VPI/VCI acting as the label. In LDPv1, it is assumed that all interfaces except LC-ATM and LC-Frame Relay are PHP-capable.

ATM MPLS Network Components

The following components are involved in an ATM MPLS network:

- **ATM-LSR**—The LSR has both router and switch functionality. A multiservice switching ATM-LSR is made up of a Label Switch Controller (LSC) and a controlled switch.

- **eLSR**—The eLSR provides IP-to-MPLS (at ingress into the network) and MPLS-to-IP (at egress) translations.

Conceptually, MPLS is based on separate routing and forwarding paradigms. Naturally, a one-to-one relationship or mapping with the multiservice switching architecture separates the controller and the controlled switch (as discussed in Chapter 1, "What Are Multiservice Switching Networks?"). Therefore, it is like a complete match, and it makes perfect sense to directly relate the two. That is why a multiservice switching ATM-LSR is made up of an LSC that takes care of the routing portion and a controlled virtual switch with a subswitching fabric that is responsible for the forwarding.

Label Switch Controller

The LSC is in essence a router that has two types of interfaces associated with MPLS:

- **Control interface**—This runs the Virtual Switch Interface (VSI), which is used to control and communicate with the ATM switch.

- **Extended tag ATM interfaces**—These are used to peer with other LSRs in the MPLS network.

VSI

Using VSI, an ATM switch can be controlled independently by multiple controllers, such as PNNI and MPLS, each having a different and independent control plane. This is possible because different control planes use different partitions of the switch resources. When a virtual interface is activated on an entity such as a port, trunk, or virtual trunk for use by the LSC, the resources of the virtual interface associated with that entity are made available to the controller. VSI requires a dedicated interface on the LSC and ATM switch for its exclusive use. Hence, it cannot be shared with ATM Forum connections. All related VSI information was covered in Chapter 2, "SCI: Virtual Switch Interface," and Chapter 3, "Implementations and Platforms."

LC-ATM

An LC-ATM interface is an ATM interface controlled by the LSC. When a packet traversing such an interface is received, it is treated as a labeled packet. The packet's top label is inferred from either the contents of the VCI field or the combined contents of the VPI and VCI fields. Any two LDP peers that are connected via an LC-ATM interface use LDP capability negotiations to determine which of these fields is applicable to that interface.

XTagATM

From the LSC's point of view, associated label-switched ports (LC-ATMs) on the ATM switch are represented as a Cisco IOS interface type called extended label ATM (XTagATM). LC-ATM ports on the ATM switch are essentially any interfaces that carry packets as ATM cells on LVCs controlled by MPLS. This would include a virtual path connection (VPC) whose opposite endpoint is an edge LSR or another ATM switch (which in turn has eLSR connections). The XTagATM interfaces could be viewed as virtual ATM interfaces or ATM tunnels, but they can carry only MPLS-related traffic, and they represent interfaces external to the LSC. No ATM Forum connections can be defined on them. It is worth noting that the XTagATM interface on the LSC does not actually carry customer data cells—the switch does.

XTagATM interfaces are used to transport control information such as the LDP and the interior routing protocol between participating label switch routers (LSR or eLSR types).

Each XTagATM interface maps to a physical interface or virtual trunk endpoint on the switch. The signaling VCs from that interface are cross-connected to the LSC control interface and terminate on the XTagATM logical interface. In this way, the LSC can participate in the MPLS signaling for each physical or virtual trunk interface. Data traffic from each physical interface is generally cross-connected through the switch without passing through the LSC. The exception to this general rule is traffic whose source or destination is the LSC itself, as well as routing updates or label distribution information. The adjacent node on an LSC XTagATM interface is any router or switch acting as an LSR or eLSR. On every XTagATM interface, VCI 32 within the defined tunnel is used by default for routing and signaling. This topic is covered extensively in Chapter 3.

ATM-LSR

ATM-LSR is an LSR with a number of LC-ATM interfaces. It forwards cells between these interfaces using labels carried in the VCI or VPI/VCI field and without reassembling the cells into frames before forwarding. When two ATM-LSRs are connected via an LC-ATM interface, a non-MPLS connection, responsible for carrying unlabeled IP packets, must be available. This non-MPLS connection is used to carry LDP packets between the two peers and also is used for other unlabeled packets (such as interior routing packets). The non-MPLS connection uses the LLC/SNAP encapsulation, defined in RFC 1483 and RFC 2684.

VCI values 0 through 32 are reserved for control traffic and should not be used by an LC-ATM interface as a label. These unlabeled packets use standard AAL5 encapsulation, as defined in RFC 1483. The default VPI/VCI value for an unlabeled packet is VPI 0 VCI 32. With the exception of these reserved values, the VPI/VCI values used can be treated as independent spaces. The allowable ranges of VPI and VCIs are communicated through LDP.

You can also connect two LSRs via an ATM VP, as in the case where an ATM cloud connects the two LSRs via an ATM virtual path. In this case, the VPI field is unavailable to MPLS, and the label must be encoded entirely within the VCI field. Again, VCI ranges 0 to 32 cannot be used for LVCs. The VPI value must be configured for a predefined VPI set by a virtual path over ATM.

In all cases, LVCs use ATM null encapsulation, as defined in RFC 1483 and RFC 2684. The control-vc, using VPI/VCI 0/32 by default, uses AAL5SNAP encapsulation.

Frame-Based LSR

Frame-based LSR is an LSR that forwards complete frames between its interfaces. Note that such an LSR can have zero, one, or more LC-ATM interfaces. Sometimes a single box might behave as an ATM-LSR with respect to certain pairs of interfaces but might behave as a frame-based LSR with respect to other pairs. For example, an ATM switch with an Ethernet interface might function as an ATM-LSR when forwarding cells between its LC-ATM interfaces but might function as a frame-based LSR when forwarding frames from its Ethernet to one of its LC-ATM interfaces.

ATM-Edge LSR

An ATM-edge LSR is an LSR that is connected to an ATM-LSR by LC-ATM interfaces and that resides at the edge of the MPLS network. An eLSR does not switch labels; it switches IP packets to labels and labels to IP packets. A packet-based LSR is a device that manipulates whole packets rather than cells. A router running packet-based MPLS is a packet-based LSR. An ATM-edge LSR is also a type of packet-based LSR.

A traditional ATM switch can support ATM MPLS within a permanent virtual circuit (PVC), which acts as a point-to-point interface. In this case, the traditional ATM switches do not actually perform multiprotocol label switching. They merely support "tunnels" through which MPLS packets are carried.

All the MPLS network components are shown in Figure 4-6.

Figure 4-6   *MPLS Network Components*

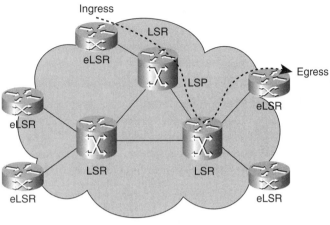

MPLS Domain

Label Distribution Protocol (LDP): Overview

LDP, defined in RFC 3035, contains a set of procedures and messages by which LSRs establish LSPs through a network by mapping Layer 3 routing information directly to Layer 2 1/2 switched paths. LDP associates a Forwarding Equivalence Class (FEC) with each LSP it creates. The FEC associated with an LSP specifies which packets are mapped to that LSP.

In essence, LDP is used to distribute <label, prefix> bindings. It runs in parallel with routing protocols. Other label-distribution mechanisms, such as TDP, Resource Reservation Protocol (RSVP), Protocol-Independent Multicast version 2 (PIMv2), and Border Gateway Protocol (BGP—see RFC 3107), can run in parallel with LDP. LDP descends from Cisco-proprietary TDP in such a way that LDP is a superset of TDP, using the same methods to achieve the same functions.

The four categories of LDP messages are as follows:

- **Peer discovery messages**—These LDP hello messages are used to advertise and maintain the presence of an LSR in a network. LDP neighbor discovery uses UDP port 646 to send periodic hello packets. The neighbor responds with an LDP hello if it is an LSR or eLSR. The LSR with the highest IP address becomes active and establishes TCP connections for LDP on port 646. The LSR with the lowest IP address becomes passive and waits on the establishment of the TCP connection for the LDP session.

- **Session management messages**—These establish, maintain, and terminate sessions between LDP peers. LDP session establishment allows negotiation of options over the TCP session established on port 646.

- **Advertisement or label distribution messages**—These create, change, and delete label mappings for FECs. These messages deal with label binding advertisements, requests, withdrawal, and release.

- **Notification messages**—These provide reporting information and signal error information and cause codes.

LDP Message Overview

LDP runs over TCP port 646 to provide reliable delivery of messages. The only exception is the discovery message, which uses UDP port 646.

LDP Message Structure

All LDP messages have a common structure that uses a type-length-value (TLV) encoding scheme to provide expandability. The value part of a TLV-encoded object (or TLV for short) can itself contain one or more TLVs. TLV encoding is very easy to deal with when it comes to adding a new capability because a new type is defined.

As described in RFC 3036, the LDP specification, all LDP messages have an LDP header followed by one or more TLVs. The LDP header format is shown in Figure 4-7.

Figure 4-7 *LDP Message Encoding—LDP Header*

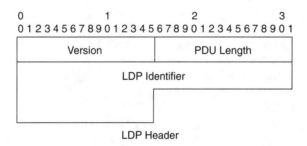

LDP Header

The fields in the LDP header are as follows:

- **Version**—This is the current LDP version is 1.

- **PDU Length**—This is the length of the protocol data unit (PDU) in octets, excluding the Version and PDU Length fields.

- **LDP Identifier**—This uniquely identifies the label space network-wide.

Figure 4-8 shows the LDP TLV format.

Figure 4-8   *LDP Message Encoding—TLV Format*

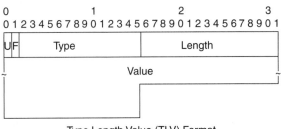

Type Length Value (TLV) Format

Figure 4-8 contains the following fields:

- **U bit**—Unknown message bit. This specifies the procedures to follow upon receipt of an unknown TLV. If it is cleared, a notification must be sent. If it is set, the unknown message must be silently ignored.

- **F bit**—Forward unknown TLV bit. Applies only when the U bit is set.

- **Type**—The type of message.

- **Length**—The length of the message in octets.

- **Value**—The value of the message. For LDP messages, the value is composed of a 32-bit message ID that identifies the LDP message, a set of mandatory parameters, and a set of optional parameters.

LDP Identifier

The LDP ID is a 32-bit magnitude that uniquely identifies the label space network-wide. Figure 4-9 shows the format of the LDP ID.

Figure 4-9   *LDP Identifier*

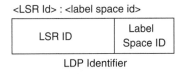

LDP Identifier

Figure 4-9 has two fields:

- **LSR ID**—The first four octets identify the LSR. It must be a globally unique value such as the router ID. It is generally derived from a router interface.

- **Label Space ID**—The last two octets identify the label space within the LSR. For platform-wide label spaces, the label space ID must be 0.

The combination of these two fields forms an LDP identifier that uniquely identifies the label space network-wide.

Each LDP identifier has it own LDP session, and each LDP session uses a separate TCP connection, as discussed in the upcoming section "Establishing LDP Sessions."

Label Spaces

Frame-based MPLS interfaces use a per-platform label space. In other words, labels are assigned from a unique pool of labels in the LSR. However, given the nature of labels on LC-ATM interfaces, per-platform label space cannot be used for ATM interfaces. Two LC-ATM interfaces can use the same VPI/VCI pair as a label. Therefore, LC-ATM interfaces use per-interface label spaces and a non-null label space ID.

In summary, we have the following:

- **Interface label space**—The label space is specific to an interface. It is used with LC-ATM and LC-Frame Relay. The label forwarding information base (LFIB), which is used to forward labeled packets, contains an incoming interface. This is because the same numerical label (with a different meaning) can be assigned on multiple interfaces.

- **Platform label space**—The label space is router-wide. It is generally used in frame-based MPLS—that is, POS, Ethernet, and Fast Ethernet interfaces. The label can be used on any interface. The LFIB does not contain an incoming interface.

From another angle, one label space is used per LDP session. When two LSRs are connected through three frame-based interfaces, as shown in Figure 4-10, you have only one LDP session and one label space.

Figure 4-10 *Link Topology and LDP Sessions, Part I*

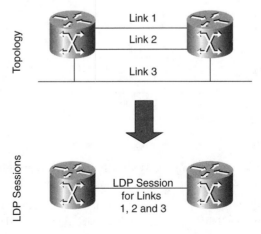

However, as shown in Figure 4-11, each LC-ATM interface has its own LDP session and label space.

Figure 4-11 *Link Topology and LDP Sessions, Part II*

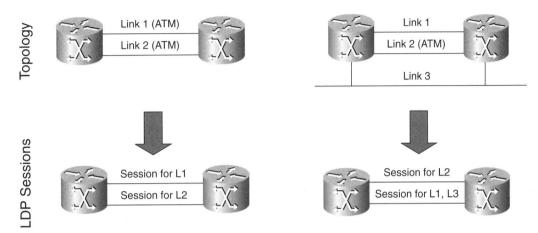

So an LSR has essentially two scenarios in which it needs to announce multiple label spaces to a peer and therefore use different LDP identifiers. One is when the LSR has two LC-ATM links to the peer, and the other is when the LSR has one LC-ATM link to the peer and one or more Ethernet links to the peer using per-platform label spaces.

In summary, the number of LDP sessions is determined by the number of different label spaces.

LDP Discovery

LSRs learn of LSR neighbors using LDP discovery and LDP link hellos. Figure 4-12 shows an example of LDP hello messages.

Figure 4-12 *LDP Discovery Using Hellos*

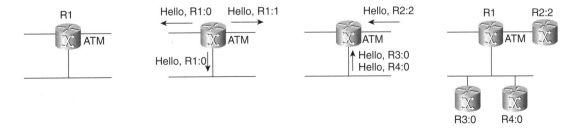

LDP discovery simplifies operational aspects of explicitly configuring peer LSRs.

LDP link hellos are sent out of MPLS-enabled interfaces as UDP packets to the well-known UDP port 646 for LDP discovery and for an "all routers on this subnet" group multicast address (they are sent to 224.0.0.2:646 UDP). The willingness to label-switch on a specific link is signaled by the LDP link hello messages.

Figure 4-13 shows a complete example of LDP discovery.

Figure 4-13 *LDP Discovery Example*

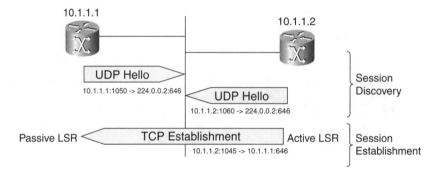

LDP sessions for nondirectly connected LSRs use an extended discovery process and require explicit configuration and the use of targeted hellos. There are two main differences between targeted hellos (or directed hellos) and normal hellos:

- Targeted hellos are sent to a specific IP address.

- The initiator solicits targeted hellos from the target (it is asymmetric).

Establishing LDP Sessions

LDP discovery triggers LDP session establishment. Session establishment is a two-step process:

- **Transport connection establishment**—The active LSR that has the highest transport address initiates the establishment of the LDP TCP connection by connecting to the well-known TCP port 646 at the neighbor LSR address. LDP uses TCP as a reliable transport for sessions. Each TCP connection has only one LDP session.

- **Session initialization**—After establishing the TCP transport connection, the active LSR sends the Init message to negotiate LDP session parameters exchanging LDP messages. The parameters negotiated include LDP protocol version, label distribution method, timer values, loop detection status, and maximum PDU length. Optional TLVs are also used to negotiate VPI/VCI ranges for label-controlled ATM, DLCI ranges for label-controlled Frame Relay, VC-merge capability, and VC directionality. Successful negotiation establishes an LDP session between LSRs. The session is ready to exchange label mappings after receiving the first keepalive.

An example of LDP session establishment is shown in Figure 4-14.

Figure 4-14 *LDP Session Establishment*

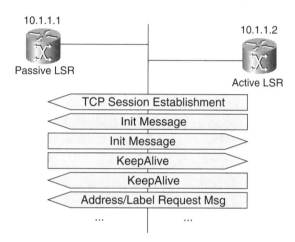

If multiple links connect two LSRs and use the same label space and LDP ID, there will be multiple link or hello adjacencies but only one transport adjacency.

Maintaining LDP Sessions

Two methods are used to maintain LDP sessions:

- Continued transmission of discovery hello messages as link keepalives, to monitor the willingness to label-switch on a specific interface. LSR keeps hello adjacency holddown timers.

- LDP session keepalives are used to monitor the health and integrity of the sessions and underlying TCP connection and to restart a session when keepalives are lost. An LSR maintains individual TCP session holdtime and keepalive intervals.

Label Distribution Methods

To help you understand the label distribution methods, I will first define the terms *downstream* and *upstream,* which are relative to an IP destination or FEC in general. Figure 4-15 shows upstream and downstream routers.

Figure 4-15 *Upstream and Downstream Routers*

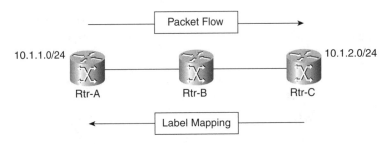

In Figure 4-15, Router C is the downstream neighbor of Router B for destination 10.1.2/24, and Router B is the downstream neighbor of Router A for destination 10.1.2/24. Likewise, Router A is the upstream neighbor of Router B for destination 10.1.2/24.

Every interface on an LSR uses one of two methods for label distribution:

- **Downstream Unsolicited**—The label distribution is downstream-controlled. The LSR downstream initiates and advertises the label mapping. Per-platform label space is distributed. Using this method, an LSR can distribute the same label to different neighbors.

- **Downstream on Demand**—The label distribution is upstream-controlled. The LSR upstream initiates the mapping request. It distributes per-interface (such as LC-ATM) labels. In a Downstream on Demand label assignment, the LVC is built from the headend platform to the tailend platform.

The advertisement mode is exchanged between peer LSRs during the initialization phase. Neighbor LSRs must agree on the distribution method used.

The label-mapping LDP message uses two TLVs to perform the mapping—the FEC TLV and the Label TLV. Different types of labels use different Label TLVs. In particular, you can have a Generic Label TLV (type 0x0200), an ATM Label TLV (type 0x0201), and a Frame Relay Label TLV (type 0x0202).

Two methods of label distribution control can be used:

- **Independent label control**—An LSR can assign a label for a FEC even if it does not have a downstream LSR. Upon receiving a label request message, the LSR can respond with a binding without waiting for a label mapping from the next-hop LSR.

- **Ordered label control**—An LSR can assign a label for a FEC only if it has already received a label from the downstream LSR. Upon receiving a label request message, the LSR must perform its own label request downstream.

One last concept has to do with whether an LSR maintains or requests a label binding from or to a neighbor that is not the next hop according to the routing table. This is called label *retention mode*, and it has two alternatives:

- **Conservative label retention mode**—An LSR operating in Downstream on Demand order requests label mappings only from the next-hop LSR. This saves valuable labels and therefore is preferred for LC-ATM interfaces.

- **Liberal label retention mode**—An LSR operating in Downstream Unsolicited order can accept label-to-FEC mappings from all LDP peers. An LSR operating in Downstream on Demand order might request label-to-FEC mappings from all LDP peers. This mode is useful only in Downstream Unsolicited mode because labels will already be assigned upon a next-hop change in a prefix's routing information base.

As a summary, Table 4-1 lists the different recommendations for label distribution modes.

Table 4-1   *Recommended Modes for Label Distribution*

	Per-Platform Labels	**Per-Interface Labels**
Order	Downstream Unsolicited	Downstream on Demand
Control	Independent	Ordered
Retention	Liberal	Conservative

Loop Detection

LDP implements three mechanisms to prevent looping LSPs beyond the loop detection mechanisms built into the IGP:

- **Time to Live**—The TTL field in the shim header is used to prevent looping labeled packets. Its use is equivalent to IP TTL, decrementing the TTL field at every hop as a packet travels through the network. This method can be used only in data link layers that use the shim header for label switching. It cannot be used for LC-ATM interfaces because the ATM cell header does not contain a TTL field. MPLS considers ATM a non-TTL segment.

- **Hop count TLV**—This additional TLV is used to count the number of hops in an LVC during LVC setup. The TTL field in the IP header is decremented by this hop count at ingress, and the IP packet is dropped if the new calculated TTL value is equal to or smaller than 0. A maximum TTL value can be configured. This method is required in RFC 3035, but it might be insufficient to prevent loops in all cases.

- **Path vector TLV**—This additional TLV was designed specifically to prevent loops. During an LVC setup, the path vector TLV records all the LSR IDs (the first 4 bytes of the Label ID) in the path, when the LSR adds its own router ID before sending the label request. If an LSR receives an LDP message containing its own router ID in the path vector TLV, the message is ignored, and a notification message is transmitted to signal the loop detected. LDP also supports a maximum path vector TLV length.

Figure 4-16 shows an example of the usage of the path vector TLV.

Figure 4-16 *Loop Prevention with the Path Vector TLV*

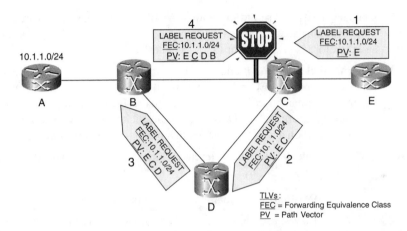

If loop detection is desired in an MPLS domain, it should be turned on in all LSRs and eLSRs.

Security Considerations

LDP runs on top of TCP, and it provides a TCP MD5 configurable signature option based on the one specified in RFC 2385 and used by BGP. The MD5 hash function is specified in RFC 1321. A shared secret is configured on the LDP peers, and the MD5 digest is computed from the shared password and the TCP segment.

ATM MPLS provides extra security capabilities. With per-interface label space, an LSR using Downstream on Demand label allocation mode discards labeled packets from upstream neighbors if the label was not previously advertised to that neighbor. This inherent capability prevents label spoofing.

Quality of Service and Multi-vc

MPLS support for CoS is inherited from IP CoS support. The MPLS shim header contains the 3-bit Exp field that allows the network to give special treatment to different labeled packets based on the Exp bits. That is why the Exp bits are also called CoS bits. In fact, they are modeled after the IP precedence bits. At the edge, the IP 3-bit precedence header field (or more precisely, the first three bits of the diff-serv code point [DSCP] called class selector) is copied to the Exp bits on label imposition. If the CoS is inferred from the Exp bits, the LSPs are called Exp-inferred PSC LSPs (E-LSPs). PSC stands for per-hop behavior Scheduling Class and refers to a core behavior.

The IPv4 header TOS octet (refer back to Figure 4-1) has come a long way from its original definition in RFC 791 to the current definition (at the writing of this book) in RFC 3168. Please refer to Figure 4-17 for the TOS byte historical evolution.

Figure 4-17 *Type of Service IPv4 Header Octet*

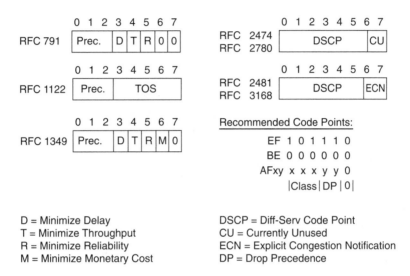

However, ATM MPLS encapsulation does not expose the Exp bits. The label-swapping paradigm is performed based solely on the label, which is the VCI field or VPI/VCI combined fields in the ATM cell header. A different method needs to be provided to support QoS on ATM MPLS. This method is called multi-vc, and it is shown in Figure 4-18.

Figure 4-18 *Multi-vc and ATM MPLS QoS Support*

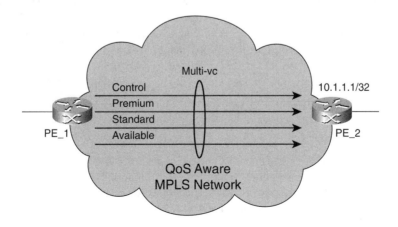

As you can see in Figure 4-18, up to four parallel LVCs are set up per FEC to support up to eight different CoSs, the same as with the Exp bits on frame-based MPLS. How? The Cell Loss Priority (CLP) bit is used in the ATM cell header to differentiate between two different classes per LVC, giving a loss priority. Because the CoS is inferred strictly from the LSP, as there are no Exp bits to use, these LSPs are called Label-only inferred-PSC LSPs (L-LSPs).

ATM MPLS QoS follows the DiffServ model, whose architecture is defined in RFC 2475. DiffServ clearly separates and distributes edge behaviors (classification, marking, policing, shaping, metering, and complex per-user and per-application tasks) from core functions (queuing, dropping, and simple tasks). Modeled after the ATM CoS scheme that includes ABR, CBR, VBR-RT, VBR-NRT, and UBR classes, it divides IP traffic into a small number of classes and allocates resources on a per-class basis. At the edge, the classification information is summarized in the DSCP, which gives a new interpretation to the TOS IPv4 header octet and IPv6 traffic class octet as defined in RFC 2474. RFC 2474 recommends the use of eight code points called Class Selector Code Points (any DSCP in the range 'xxx000' where 'x' can either be '0' or '1') to provide a degree of backward compatibility (see Figure 4-17). In accordance, both shim-based and multi-vc-based MPLS QoS approaches provide eight classes of service. Generally speaking, eight different QoS network treatments are enough for almost every network type. You can learn from ATM that the CBR, VBR-NRT, VBR-RT, UBR, and ABR classes have successfully been deployed and are sufficient for all multiservice types of applications.

MPLS edge devices can be application-aware and therefore give higher priority to a flow that matches certain applications. This guarantees a QoS in a more dynamic fashion. ATM MPLS QoS is covered in Chapter 7, "Practical Applications of MPLS."

VC-Merge-Capable ATM-LSRs

ATM MPLS was developed to control very large networks. The protocol's scalability was considered from the ground up. Very large networks with many LSRs, eLSRs, and destinations and that provide CoS support have an extra scalability tool: VC-merge reduces the number of labels required per link. A VC-merge MPLS link allocates at most one label per FEC. On multi-vc networks, a FEC is defined by a destination and a CoS; therefore, the VC merging occurs only for the same CoSs without jeopardizing multiple-class support.

VC-merge is the capability by which an LSR receives cells on several incoming LVCs (that is, on different VPI/VCI pairs) and transmits them on a single outgoing LVC without causing the cells of different AAL5 PDUs to become interleaved.

The primary difference is that a VC-merge-capable ATM-LSR needs only one outgoing label per FEC, even if multiple requests for label mapping to that FEC are received from upstream neighbors. In other words, VC-merge allows a multipoint-to-point connection to be implemented by queuing complete AAL5 frames in input buffers until the end of a frame has been received. This requires buffering because the cells from the same AAL5 frame are all transmitted before cells from any other frames.

Figure 4-19 shows an example of VC-merge.

Figure 4-19 *VC-Merge-Capable ATM-LSRs—Multipoint-to-Point*

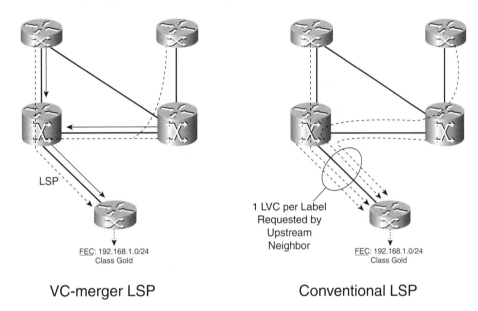

Figure 4-20 shows the network-wide LVC reduction effect in VC-merge MPLS networks.

Figure 4-20 *Network-Wide Effect of VC-Merge*

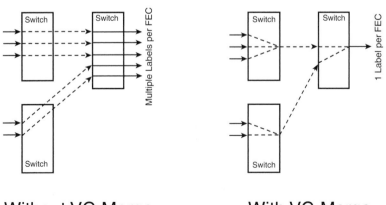

On Cisco multiservice switching platforms, VC-merge capability is implemented in the egress line cards. The primary endpoints are configured as with nonmerging endpoints, but the secondary endpoints in the egress line card are divided into leaves and root. The leaf receives cells from the switch fabric and translates the VC to the root. This is shown in Figure 4-21.

Figure 4-21 *VC-Merge Implementation in Multiservice Switches*

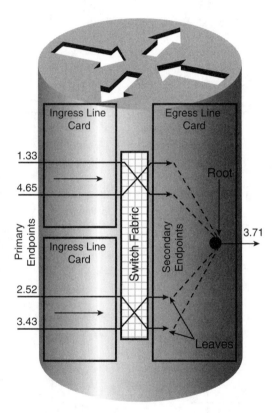

From a design perspective, VC-merge might be required only in core links.

There are differences between a conventional ATM-LSR and a VC-merge-enabled ATM-LSR with respect to label mapping. The messaging that takes place in a VC-merge-enabled LSR is shown in Figure 4-22.

Figure 4-22 *VC-Merge Label Mapping*

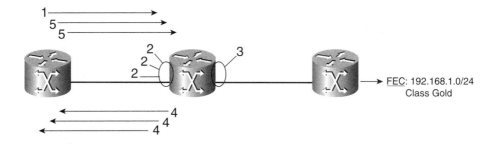

From Figure 4-22, we can identify the following steps:

Step 1   The LSR receives a label request from the upstream LSR.

Step 2   The LSR allocates an incoming label.

Step 3   Assuming that an outgoing mapping for the FEC has been set up by a previous request, no further downstream request is required.

Step 4   The LSR returns a label mapping to the upstream requester, regarding the label request from Step 1.

Step 5   Repeat for all further requests for that FEC.

In a non-VC-merge ATM-LSR, the procedure diverges in Step 3. A non-VC-merge environment has the following Step 3:

Step 3   The LSR requests a label from the downstream device. The downstream LSR returns a label mapping.

This non-merging scenario is shown in Figure 4-23.

Figure 4-23 *Non-VC-Merge Label Mapping*

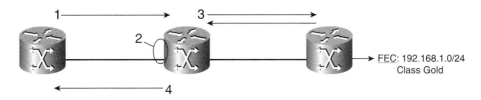

In summary, VC-merge improves signaling performance as well as label space.

IP Virtual Private Network Services

A VPN is a set of sites that is allowed to communicate with each other as a closed user group (CUG). There are many different VPN architectures. This section covers the benefits of the MPLS VPN architecture, as well as its operation.

I'll start by defining some terms used in an MPLS VPN environment:

- **Provider (P) router**—A router that resides in the provider network. It is an LSR in pure MPLS terminology.

- **Provider edge (PE) router**—A router that sits at the edge of the provider network and interfaces with customer edge routers. It is an eLSR using pure MPLS terminology.

- **Customer (C) router**—A router that resides in the customer network.

- **Customer edge (CE) router**—A router that resides at the edge of the customer network and interfaces with a PE router.

Figure 4-24 shows the MPLS VPN definitions.

Figure 4-24 *MPLS VPN Definitions*

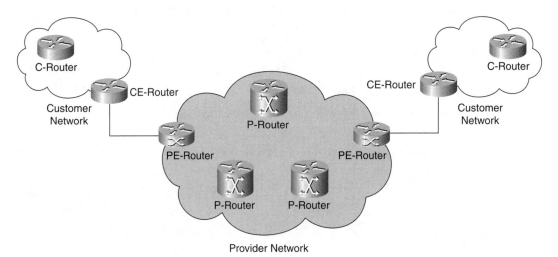

The major goal of an MPLS VPN solution is to overcome the limitations of an overlay VPN model, while maintaining its strengths. An overlay VPN model presents scalability limitations because the CE routers peer with each other, and the number of Layer 2 connections in the provider network increases with the square of the number of CE routers.

The MPLS VPN model is a peer model in which all the customer sites peer with the PE devices, guaranteeing optimum routing between sites and simplifying the provisioning of additional VPNs. This peer model lets the service provider support very large-scale VPN service offerings—up to millions of VPNs in a single network. Purchasing VPN services allows a VPN customer to rely on the service provider to deal with routing, scalability, QoS, and performance issues. Service providers can support customers with different needs using VPNs.

Figure 4-25 shows how a peer model works.

Figure 4-25 *MPLS VPN Route Distribution*

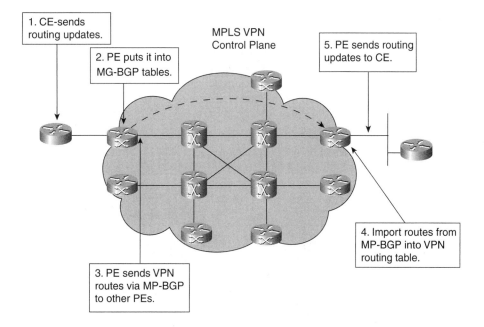

In summary, the MPLS VPN model combines the strengths of the overlay and peer-to-peer VPN models:

- **Peer-to-peer model**—Simplifies customer routing, as well as eliminates the requirement of maintaining full IP routing in the MPLS core.

- **Overlay model**—Provides isolation between customers, privacy, and security.

Service providers implementing MPLS VPNs distribute customer routes using the following steps (which are illustrated in Figure 4-25):

Step 1   The CE router for customer A sends routing updates to the ingress PE in the MPLS network using an IGP, such as OSPF, RIP, or eBGP. Note that no routing updates are sent if static routes are used.

Step 2 At the ingress PE, these routes are inserted into a separate routing table for this VPN and then are exported into the provider's multiprotocol BGP.

Step 3 These routes are then advertised within the provider's network among all the PEs using the MP-BGP extensions.

Step 4 At the egress PE, all the routing information is imported to customer A's VPN routing table from the provider's iBGP.

Step 5 The routing information is sent to the destination CE. As in Step 1, no routing updates are sent if static routes are used.

A new concept can be inferred from the MPLS VPN peer model functionality: CE devices have point-to-network connections, as opposed to the point-to-point connections in the overlay VPN model. In the MPLS VPN model, sites are configured, whereas in the overlay VPN model, links are configured. This is shown in Figure 4-26.

Figure 4-26 *Point-to-Cloud Connections*

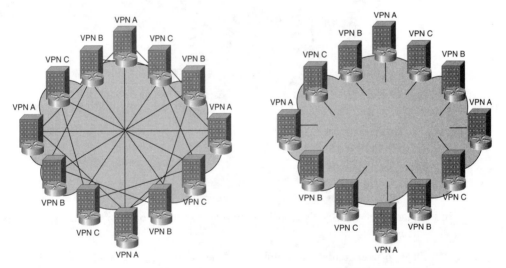

Overlay VPN Model   MPLS VPN Model

VPN Route Distribution and Filtering

VPN route distribution and filtering happen in the application plane on top of MPLS. VPN routes need to be distributed and VPN labels assigned to VPNv4 routes using multiprotocol BGP (MP-BGP) before user traffic can traverse the MPLS VPN network and MPLS is used to switch labeled packets through the provider network. You can control the routing information distribution using route filtering based on the BGP extended community attributes.

You can apply the filters in Steps 2 and 4 from the preceding section. Route filtering is performed against the route target (RT), which is a 64-bit value attached to MP-BGP VPNv4 routes.

This kind of operation is also used to ensure a secure VPN for each customer. Each PE has multiple VPN routing and forwarding (VRF) tables, one for each VPN customer. This is shown in Figure 4-27.

Figure 4-27 *MPLS VPN with Two Customers*

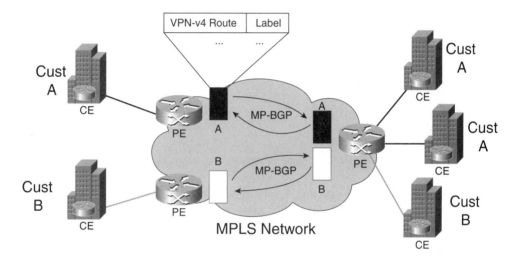

Each VRF is populated with routes received from directly connected CE routers, as well as routes received from other PEs via BGP filtering based on BGP extended community attributes.

Customer packets traversing a provider VPN MPLS network carry two labels. An inner VPN label called bottom label distributed by MP-BGP indicates VPN membership. MPLS uses an outer label called top label distributed by the IGP and LDP to switch the packet from ingress PE to egress PE. These two labels have the following characteristics:

- **Top label**—Distributed by LDP and derived from an IGP route. Corresponds to a PE address, which in turn is the MP-BGP next hop of VPNv4 routes.

- **Bottom label**—Distributed by MP-BGP. It corresponds to a VPNv4 route and identifies the outgoing interface or VRF to be used to reach the VPN destination.

This can be seen in Figure 4-28.

As a side note regarding Figure 4-28, in a frame-based MPLS environment, provider LSR P2 would perform PHP to remove the top label. PE LSR PE2 using LDP would request this action. In this case, PE2 eLSR would do only a single lookup and would forward the IP packet to CE2.

Figure 4-28 *MPLS VPN Label Stack*

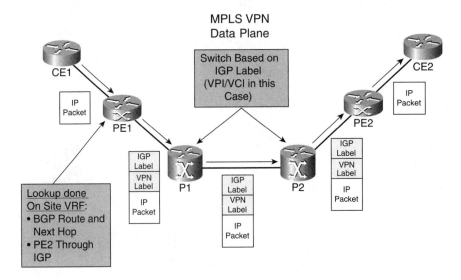

One of the reasons for the great scalability of an MPLS VPN solution is that provider routers do not have MP-BGP or VPN knowledge.

VPN IP Addressing

In an MPLS VPN network, different customers can use the same IPv4 address space, as well as private IP addresses (see RFC 1918). VPN-IPv4 addresses make each customer's IPv4 address unique within the provider's network. A VPN-IPv4 or VPNv4 address has a 64-bit field called a route distinguisher (RD) that is prepended to the 32-bit IPv4 address to make a unique 96-bit VPNv4 address. This is shown in Figure 4-29.

Figure 4-29 *VPNv4 Address*

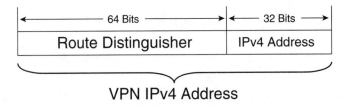

The RD is never carried in packets—only in label tables. PE routers perform the conversion between customer IPv4 addresses and provider VPNv4 addresses. This happens only in the control plane before the routes are exported into MP-BGP.

The RD can be seen as a VRF identifier that solves the overlapping address space problem.

MP-BGP lets BGP handle routes for multiple VPNv4 addresses. The general process is no different from handling traditional IPv4 addresses. The different addresses, such as IPv4, IPv6, NSAP, IPv4 multicast, and VPNv4, are called address families. Multiprotocol extensions for BGP-4 are defined in RFC 2283 and RFC 2858, using address families from RFC 1700, "Assigned Numbers." The VPN-IPv4 address family is defined in Section 4.1 of RFC 2547, "BGP/MPLS VPNs."

BGP tables can have a mixture of both VPN-IPv4 routes and normal IPv4 routes. CE routers have no knowledge of VPN-IPv4 addressing. CEs send and receive regular IPv4 routing updates.

The presence of RDs and independent RTs gives the MPLS VPN model great flexibility in implementing complex VPN scenarios.

Summary

MPLS provides benefits that service providers desperately need in their networks, such as predictability, scalability, and manageability, all built into one network. The key to MPLS labels is that they tell a device not just where to send packets, but also how to send them. All the information needed is encoded in the label, including the destination prefix, the service class, the QoS, the level of privacy, the VPN, and so on. The network does not have to make a decision at every device. It's all precomputed in the LSP by means of the label. MPLS is transparent to both routers and switches because it is largely a switching intelligence. This is why MPLS is so scalable. It is how you deliver services across very large networks. You make service decisions once, at the edge, and the core automatically supports services without reprocessing packets at every stop. Because hop-by-hop routing decisions no longer have to be made, switched networks, such as carrier ATM backbones, can implement end-to-end Layer 3-type intelligence.

In addition, MPLS provides QoS capabilities that service providers need in order to offer differentiated services. Although MPLS might require modifications to existing equipment, it does not require extensive equipment replacement (no forklift upgrades). You can add this functionality by adding a new card, controller, and/or software to an existing ATM switch. Overall, MPLS defines an evolutionary networking paradigm that combines the operating principles of Layer 2 and Layer 3 technologies while preserving service providers' investment in IP routing technology at the network's edge and switching technology in the network's core.

MPLS VPN technology is one of the most important drivers for large MPLS deployments.

In addition, all the RAS features developed for carrier-class ATM switches are leveraged to provide high-availability LSRs.

CHAPTER 5

MPLS Design in MSS

This chapter covers the important points and steps when you're designing an MPLS network based on IP + ATM. MPLS extends the capability of an ATM switch to all devices in a network using IP + ATM. With IP routing integrated into ATM switches, you can provide support for scalable VPN services that can guarantee a Service-Level Agreement (SLA) for all traffic types. ATM switches have a very high degree of reliability and availability, which is fundamental to delivering QoS and high-availability networks. A key to successful deployment of MPLS in a multiservice switching network is to be able to guarantee SLAs for applications such as voice and video that require high uptime and that have strict latency and jitter bounds.

Reliability, availability, and serviceability (RAS) features are significantly more robust in switches that have parallel and redundant operating devices. Although redundancy is critical, high reliability, availability, and serviceability relate not only to minimizing failures and downtime but also to switch recovery time in the event of a failure, failure reporting, diagnostics, and carefully planned upgrades. Cisco ATM multiservice switches have been around for years and have proven RAS records.

This chapter concentrates on the fundamentals of ATM MPLS network design.

The main design steps for a robust working network are as follows:

- Designing and dimensioning ATM MPLS points of presence
- Dimensioning edge and backbone links in the network
- IP routing in ATM MPLS networks
- Label Switch Controller (LSC) redundancy
- Refining the design as soon as the network is operational

These steps are discussed in the following sections.

Designing and Dimensioning ATM MPLS Points of Presence

There are a number of ways to connect a customer edge (CE) router to an ATM MPLS network. These functions can be implemented on various types of equipment and can be combined with access equipment in various ways. They are shown in Figure 5-1.

Figure 5-1 *Connecting a CE to an ATM MPLS Network*

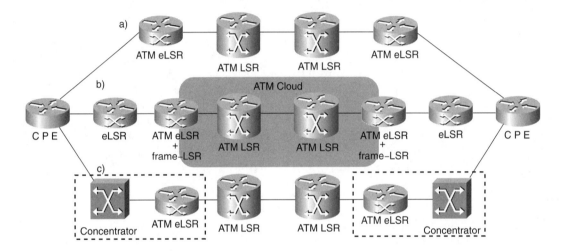

The three ways of connecting a CE to an ATM MPLS network are explained in the following sections.

ATM MPLS Network with Cell-Based Edge LSRs

The simplest ATM MPLS network structure is shown in part a) of Figure 5-1. CEs are connected directly to router-based ATM eLSRs, typically a Cisco 7200 or 7500 series. The edge Label Switch Routers (LSRs) are connected by ATM links to the core devices, which are ATM LSRs. The ATM LSRs may be BPX-8650 IP + ATM switches, MGX-8850 with PXM-45, MGX-8950, LS-1010, and other ATM switches.

Mixed Cell and Frame-Based MPLS Edge LSRs

You can have a network with a mixture of ATM MPLS and frame-based MPLS. A simple example of this is shown in part b) of Figure 5-1. In a network such as this, some links run packet-based MPLS, and some links run ATM MPLS. A cell-based backbone and a frame-based backbone may also exist. The devices that interface between packet-based MPLS and

ATM MPLS are the same routers that act as ATM MPLS edge LSRs as well as packet-based LSRs—anything from a Cisco 3600 up to a Cisco 12000.

MPLS Edge LSRs with Access Devices

ATM MPLS networks with router-based edge LSRs may also use access devices, as shown in part c) of Figure 5-1. The access devices can either be separate pieces of equipment or MGX-8230 or MGX-8250 access concentrators. This happens when access is required through a device that does not support MPLS services. There are some common situations in which this is required:

- Access is required to IP, Frame Relay, and ATM services through a single access device. The most common example when the access device doesn't support MPLS services is through the use of the MGX 8220. The access concentrator can support MPLS services such as the MGX-8230 or MGX-8250 access concentrator through the use of route processor module (RPM) cards. In this case, the access device can support legacy services in two different ways:

    - **Normal Frame Relay or ATM services**—This refers to Frame Relay or ATM point-to-point PVC mesh services.

    - **IP-enabled Frame Relay or ATM services**—MGX-8230 and MGX-8250 multiservice switching concentrators can IP-enable legacy services through the use of an RPM card in a point-to-cloud or point-to-network fashion. In this case, the CEs are connected through T1/E1 or fractional T1/E1 links to the MGX multiservice shelf, and the MGX cross-connects those PVCs to an RPM card to support connectionless IP VPN services.

- Higher densities of low-bandwidth access lines can be better supported by way of an access device than by simply using edge LSRs.

- Both point-to-point PVCs and MPLS VPN services are required by the customer network (C network).

ATM Label Switch Routers

There are in essence six main considerations when choosing ATM LSRs:

- LSR processing power and switching capacity
- Types and speeds of links
- Number of links
- Number of connections to be supported
- Whether VC-merge is required
- Redundancy and reliability requirements

Label Virtual Circuit (LVC) Resources

One logical connection number (LCN) is used for each first-level label allocated in an LVC in the core network. A first-level label is assigned to each Forwarding Equivalence Class (FEC) in the MPLS network (that is, for each entry in the global routing table that has a specific class of service). The global routing table in the network normally consists of the loopback addresses of each participating MPLS router, both LSR and eLSR. Therefore, the number of LCNs needed is directly related to the number of addresses in the global routing table. For example, it would not be possible to introduce the full Internet table (currently over 110,000 routers) into the global routing table because there would not be enough logical channels. An MPLS network should be designed to be simple in the core, with the objective of allowing edge LSR reachability through shortcut LVCs.

NOTE   It is important to remember that the LCN assignment is arranged into different port groups in multiservice switching line cards (LCs). The arrangement is either Broadband Switch Module (BXM) line cards for BPX switches, Universal Switch Module (UXM) line cards for IGX switches, or ATM Switch Service Module (AXSM) line cards for MGX-8850 and MGX-8950 switches.

As a good general design practice, when you interconnect LC-ATM links in the same line card, the busier links should not be configured in the same port group. Not taking this into consideration increases the possibility of exhausting the LCN pool.

Multiservice switching software contains an algorithm for allocating LCNs to service groups (partitions) within the port group on the line cards according to their minimum and maximum requirements. A number of LCNs can be held in a common pool, depending on the ratio of the configured maximum to the sum of the minimum. The formula that governs the LCN allocation and VSI partitions and common pools is covered in Chapter 2, "SCI: Virtual Switch Interface." Refer to the section "Hard, Soft, and Dynamic Resource Partitioning" in that chapter for formulas and further details.

Dimensioning MPLS LVC Space

Many of the issues in designing MPLS networks are similar to those in designing ordinary IP networks. One important exception to this resemblance is the dimensioning of MPLS LVC requirements on each link. A sufficient number of virtual circuits (VCs) must be reserved for use as LVCs on each link, based on a worst-case scenario. VCs are precious resources that need to be controlled. This is particularly important if multiple controllers, such as MPLS, PNNI, and redundant MPLS LSCs, are sharing the links' VSI VC resources. The design dilemma is determining how many LVCs are required on a per-link basis and in a worst-case scenario.

The required number of LVCs depends on the following:

- The number of IP destinations in the network, not counting blocked destinations
- The number of classes of service in a multi-vc environment, in which multiple LVCs are requested per destination
- The number of nodes with eLSR functionality
- Whether VC-merge is used
- The paths chosen by IP routing

Destinations

The number of LVCs used in a particular area of a network depends on the number of IP destination prefixes advertised in that area. These prefixes include the following:

- Loopback address of all routers in the area.
- The subnet address-prefix of any numbered point-to-point link. Because of this, it is best to use unnumbered links in MPLS networks or to configure the router not to request LVCs for numbered link destinations.
- Any other address prefixes advertised into the area. If addresses are summarized into a single address prefix at the area border router (or autonomous system boundary router), this counts as a single destination prefix.

It is strongly recommended that you not share eLSR and LSC functionality in a single router. The LSC should be a dedicated router not terminating customer traffic. It is also recommended that you disable headend and tailend edge functionality in the LSC, as covered in the section "Running Out of LVCs" in Chapter 6, "MPLS Implementation and Configuration."

Calculating the Number of LVCs Required

To ensure proper functioning of the network, you need to allocate enough resources to support a worst-case scenario for the number of LVCs required. The problem we are trying to solve is shown in Figure 5-2.

It is important to emphasize one of the concepts mentioned at the beginning of the section "Label Virtual Circuit Resources" earlier in this chapter: The connections or channel resources in the controlled switches are called logical connection numbers (LCNs), and different LCs support different numbers of LCNs. One LCN is used for every first-level label allotted in the core network, for every forwarding equivalence class (FEC) defined by a global routing entry and class of service. The number of LCNs needed is directly related to the number of FECs in the MPLS network. VSI provides a clean separation between the control and forwarding planes. Remember that LVC is a controller concept, and an LCN is a switch entity.

Figure 5-2 *LVC Usage in an MPLS MSS Network*

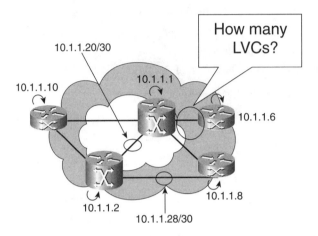

To calculate the number of LCNs required, you must take into account the number of edge LSRs behaving as MPLS devices in the MPLS network.

Equation 5-1 shows the formula used to calculate the number of LCNs required in the core network in a worst-case scenario of link failures.

Equation 5-1 *Worst-Case LVC Usage in an MPLS Link*

$$L \leq \frac{c \times (n+d)^2}{8} \qquad \text{if } 3 < d < 3 \times n - 4$$

$$L \leq c \times (n-1)(d-n+2) \qquad \text{if } d \geq 3 \times n - 4$$

The symbols used in Equation 5-1 are as follows:

- L = Number of active LVCs in an MPLS link.
- c = Number of classes of service.
- n = Number of eLSRs.
- d = Number of destinations. Destinations blocked for LVC requests are not counted.

During the calculation of Equation 5-1, the following simplifications take place and need to be true in order to use the formula:

- All eLSRs contribute with at least one destination.
- The number of eLSRs is greater than 2.
- The number of destinations is greater than 3.

If those conditions are not met, Equation 5-2 must be used.

Equation 5-2 *LVC Usage in an LSR Link Without VC-Merge*

$$L \leq c \times d \times (n - 1)$$

In the most general case, you can calculate the LCN requirement in a specific link by counting the sources (routers) and destinations (IP prefixes) at both ends of a link and using Equation 5-3.

Equation 5-3 *General-Case LVC Usage Calculation*

$$L \leq c \times (n_1 \times d_2 + n_2 \times d_1)$$

In Equation 5-3, subindex 1 indicates one side of the link, and subindex 2 indicates the other side of the link. The calculation is based on the fact that each eLSR (n_i) from one side requests LVCs for all destinations (d_j) on the other side, and vice versa.

At the edge of the network, Equation 5-3 can be simplified by assuming that subindex 1 corresponds to the eLSR side, so the following are true:

- $n_1 = 1$ —There is only one eLSR.
- $d_1 = 1$ —The eLSR is advertising only one prefix by summarization.

Based on that, the formula shown in Equation 5-4 can be used for eLSRs.

Equation 5-4 *LVC Usage Calculation in eLSR Links Without VC-Merge*

$$L = c \times (d_2 + n_2)$$

It is critical to take into account that by default, LSRs and LSCs have edge functionality. By default, they have eLSR behavior in the sense that LSRs and LSCs request headend LVCs for the destinations they learn and respond to tailend LVC requests for their own destinations. This behavior can be modified, as described in Chapter 6 in the section "Running Out of LVCs."

LVC Usage Per Link and VC-Merge

Each MPLS device behaving as an ATM edge LSR requests LVCs for the destinations it learns through the IGP. If an MPLS multi-vc class of service is used, it may ask for up to four LVCs for each destination prefix.

The requests for label prefix mappings or bindings flow through the network according to the paths chosen by the IGP IP routing.

With VC-merge, the LVCs to each FEC are merged at each ATM LSR. This means that each link has at most one LVC per FEC, where each FEC defines a destination in the area and a class of service.

Figure 5-3 shows the effect of VC-merge in an MPLS network.

Figure 5-3 *VC-Merge Effect in MPLS Networks*

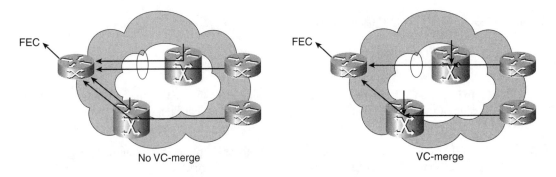

ATM Edge LSRs and VC-Merge

The existence of VC-merge can simplify some of the LVC calculations for links connecting an eLSR to an LSR. A VC-merge link has at most one LVC per FEC. Therefore, the LVC requirement can be calculated as shown in Equation 5-5.

Equation 5-5 *LVC Requirement on VC-Merge eLSRs*

$$L \leq c \times d$$

ATM LSRs with VC-Merge

The formula shown in Equation 5-5 for merge LVCs also applies to ATM LSRs with VC-merge. However, another important issue with LSRs needs to be taken into account: the number of merging LVCs.

As discussed in Chapter 4, "Introduction to Multiprotocol Label Switching," multiservice switches implement VC-merge in the egress line card. On a multiservice switching ATM LSR, the number of LCNs for each FEC equals one LCN per incoming merging LVC plus one LCN for the merged root LVC. For example, if 15 LVCs merge into one LVC, the outgoing line card uses 16 LCNs.

The formula shown in Equation 5-6 also needs to be considered.

Equation 5-6 *Number of LVCs to Be Merged in an LSR*

$$M \leq c \times d \times (k-1)$$

Equation 5-6 uses the following definitions:

- M = Number of merging LVCs
- k = Number of links in the ATM LSR

VP-Tunnels and LVC Usage

Some changes apply when you calculate the number of LVCs on VP-Tunnel interfaces: When multiple VP-Tunnels exist on an interface, the LVCs on all VP-Tunnels must be taken into account.

Either equation applicable to eLSR links or LSR links needs to be multiplied by the number of VP-Tunnels in a physical interface.

To include the presence of VP-Tunnel interfaces in the formulas discussed, Equations 5-1 through 5-5 need to be multiplied for the number of tunnel interfaces in the MPLS LSR. On the other hand, in order to update Equation 5-6, we need to redefine the variable k to include not only physical interfaces but also VP-Tunnel interfaces.

Figure 5-4 gives a summary view of LVCs with VC-merge and VP-Tunnels. VC-merge benefits normally increase when you avoid VP-Tunnels and ATM LSRs are used.

Figure 5-4 *VC-Merge and VP-Tunnels*

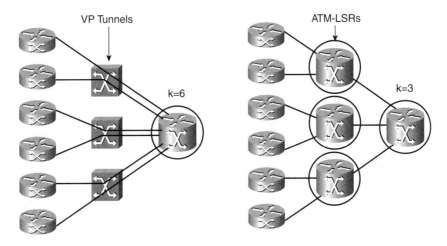

Tips on Saving LCN Resources When Designing ATM-MPLS Networks

Three design tips reduce the number of required LVCs in an MPLS network:

- Disable eLSR functionality in the LSCs. This includes disabling headend LVCs and tailend LVCs.

- Use unnumbered links whenever possible, or configure all eLSRs and LSRs to not request LVCs for the numbered links.

- Monitor the IGP, ensuring that unneeded prefixes are not injected into the global routing table.

These steps are covered in detail with configuration examples in Chapter 6 in the section "Reducing the Number of LVCs."

Summary of LVC Dimensioning

Table 5-1 summarizes the LVC calculations based on the equations presented in this chapter.

Table 5-1   *LVC Calculation Summary*

	Without VC-Merge	With VC-Merge
LSR	Equation 5-1 if d >= n, n > 2, d > 3 Equation 5-2 otherwise	Equations 5-5 and 5-6
eLSR	Equation 5-4	Equation 5-5

Two independent scenarios multiply the number of LVCs used in an MPLS network: LSC hot redundancy and multi-vc class of service.

IP Routing Protocols

This section touches on the subject of IP routing in an MPLS context. The MPLS paradigm builds on existing IP routing technologies, so there are few new IP routing concepts. In-depth IP routing is outside the scope of this book. The books *OSPF Network Design Solutions* and *IS-IS Network Design Solutions* from Cisco Press offer in-depth analysis of the two interior routing protocols used the most in MPLS environments.

This section covers interior routing protocol concepts in an MPLS domain, as well as some BGP route reflector and access routing protocol ideas needed in MPLS VPN environments.

Interior Gateway Protocol

Theoretically, any IGP can be used in an MPLS environment. Link-state protocols such as Open Shortest Path First (OSPF) and Intermediate System-to-Intermediate System (IS-IS) are needed for MPLS traffic engineering (TE) applications. They are also the choice of most ISPs. Today, from a Cisco perspective, features available for OSPF and IS-IS are very similar, if not identical. Link-state technology ensures the fastest convergence that is loop-free in terms of route calculation.

From an operational point of view, link-state protocols are easier to troubleshoot because all routers have the same link-state database. In particular, the advantage of IS-IS over OSPF is that a router inserts all the prefixes it announces on a single protocol packet. Therefore, it is easier to find all the routing information announced by a particular router.

IS-IS also has scalability and reliability advantages over OSPF:

- **Scalability**—IS-IS allows you to build larger areas than OSPF. Multiple OSPF areas might be necessary in the core to support the expected number of routers (core and PE routers) especially in the presence of unstable links. IS-IS has a different routing hierarchy than OSPF. The backbone concept still exists and has the same functionality for connecting areas. However, the backbone is implemented differently. It allows for more flexibility, particularly when the backbone must be extended. In IS-IS, the backbone is not an area, but a contiguous collection of area border routers. It should also be noted that a flat topology is desirable for traffic engineering applications.

- **Fast convergence**—IS-IS has a faster convergence time, with features such as incremental SPF. Like OSPF, IS-IS uses Dijkstra's algorithm to compute the topology tree. However, IS-IS uses fewer packet types than OSPF to propagate routing information. Therefore, the time taken in determining how to react to a given packet decreases, speeding up the convergence process. Also, flooding (especially on broadcast media) is more optimal with IS-IS.

- **Less resource usage**—IS-IS databases contain one link-state packet (LSP) per router in the routing domain or the area (depending on the routing hierarchy). All prefixes announced by a router (local prefixes, redistributed from other protocols) are part of the unique LSP that the router floods on the network. IS-IS has four different packet types in two forms, level 1 and level 2. The packet type depends on the router type and not on the nature of the prefix announced on the packet type. Therefore, the computation of the SPF tree is facilitated by the fact that all the routing information is on a limited number of LSPs for each router.

iBGP and Route Reflectors

In an MPLS VPN environment, iBGP sessions need to be set up among all Provider Edge (PE) routers. A full iBGP mesh is required among all PEs that need the same VPN information. Route reflectors (RRs) provide scalability and ease of management in that respect. RRs have the following advantages:

- They remove the requirement for a full mesh of iBGP sessions between all PE nodes. RRs relax this requirement by having the PEs peer with the RRs and then reflecting the routing information.

- They require fewer configurations as well as increase the network manageability by eliminating the full mesh requirement.

- They are more suitable for an iBGP network.

- They are better able to meet future service requirements.

- More RRs can be added as needed to increase performance.

- New RRs can be added to support special services.

RRs are recommended to be dedicated routers similar to PE routers, such as Cisco 72xx routers, as shown in Figure 5-5.

Figure 5-5 *Route Reflectors Connected to the Core*

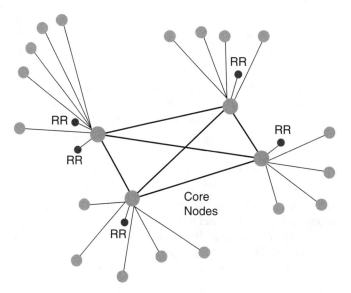

In Figure 5-5, peering is configured between a PE to each of the four RRs. This means that each PE has four neighbor statements. If not for the RRs, you would need to configure neighbor statements to all the PEs in the network. A router, or pair of redundant routers, dedicated for route reflector functionality and not in the forwarding path allows for faster convergence by saving CPU and memory.

You can improve BGP convergence and scalability by grouping neighbors with the same update policies into peer-groups, making update calculations more efficient (by reducing the number of times the BGP table needs to be walked) and also lowering CPU and memory requirements. Increasing the input hold-queue and the maximum segment size (MSS) can also improve the BGP performance. This last case will be covered in Chapter 7, "Practical Applications of MPLS."

Access Routing Protocols

As shown in Figure 5-6, access routing defines the routing protocol between the edge LSR (PE router) and a CE router. The following VPN-aware access routing protocols are supported: static, eBGP, OSPF, RIP2, and EIGRP.

Figure 5-6   *Access Routing Between the PE and CE Routers*

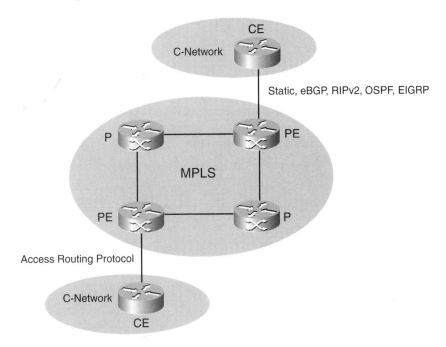

When connecting PE to CE devices, BGP, RIPv2, and static routing use separate routing contexts for each VRF, and OSPF uses separate routing processes per VRF.

Most CPE routers are connected to the core through static routes. This is true of small VPN sites where the routing table has few routes. Using static routes would provide a significant advantage for security, but it would require manual intervention to configure the static routes.

Dynamic routing could be offered as an additional service option, allowing the customer to dynamically change his network addressing. If dynamic routing were implemented, eBGP would be the preferred routing protocol when customers are dual homing to multiple PEs or when a large number of routes is present.

ATM MPLS Convergence

The previous section discussed routing protocol concepts including IGP convergence. However, ATM MPLS networks need a two-stage convergence:

- **IGP Convergence**—Core IP routing convergence.
- **MPLS Convergence**—Re-establishment of label mappings.

After the IGP converges, MPLS ATM eLSRs must resignal for label mappings. This is because the label distribution is downstream on demand, and the label retention is conservative. So in the event of a change in the next-hop, a new label must be requested by signaling downstream.

LSC Redundancy Options

Some of the most appealing features of multiservice switches are all the carrier class RAS and redundancy features, such as controller card redundancy, switching fabric redundancy, hitless switchovers, line card y-redundancy, 1:N redundancy, and so on.

However, the LSC and its control interface present a single point of failure, as shown in Figure 5-7.

Figure 5-7 *A Nonredundant ATM-MPLS Switch*

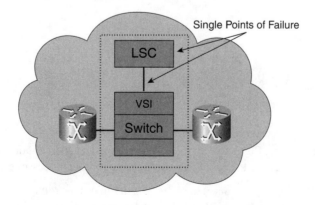

If any one of the LSC components shown in Figure 5-7 failed, it could bring down the BPX-based LSR IP switching functionality.

To achieve full system redundancy, each individual component of the LSR should be redundant. Apart from redundant line cards, it is desirable to implement LSC redundancy.

The VSI protocol detailed in Chapter 2 includes some characteristics that make LSC redundancy possible:

- VSI allows multiple, independent control planes to control a switch.
- VSI ensures that the master control processes (MPLS, PNNI) can act independently of each other.
- A VSI slave process manages the switch's resources among the different control planes.

All these characteristics are also true when you use two different LSCs to control the ATM switch. Two different independent control planes can actually be two different LSC instances.

To make an LSC redundant, you perform these basic steps:

- Partition the slave ATM switch's resources.
- Assign one resource partition to one LSC and another resource partition on the same physical interface to the redundant LSC.
- Both LSCs independently control the controlled switch. They use a different VSI controller-id.
- The LSCs need to have different nonoverlapping VPI ranges assigned on the same physical interface.
- The XTagATM interfaces in different LSCs pointing to partitions of the same physical interface must have different control-vcs.

To achieve full redundancy, you use a parallel model, as shown in Figure 5-8.

In Figure 5-8, ATM switch physical interfaces 1 to N are mapped to VSI slaves 1 to N from LSC-1. Similarly, the same physical interface is mapped to VSI slave N+1 to 2N from LSC-N-1. This mapping also results in independent masters and parallel network topology.

In this parallel redundant LSC model, each LC-ATM interface is extended to all its redundant LSCs. If two LSCs are present for redundancy, as shown in Figure 5-9, the four switch ATM physical interfaces are mapped as four XTagATM interfaces at LSC-1 and another four XTagATM interfaces at LSC-2. The mapping is independent and local to the LSC. One LSC mapping is not visible from the other LSC.

Figure 5-8 *LSC Redundancy Model*

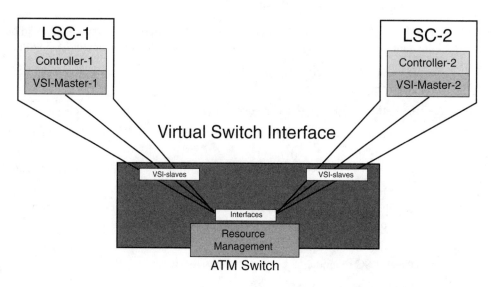

Figure 5-9 *Redundant LSC Model for an ATM-LSR*

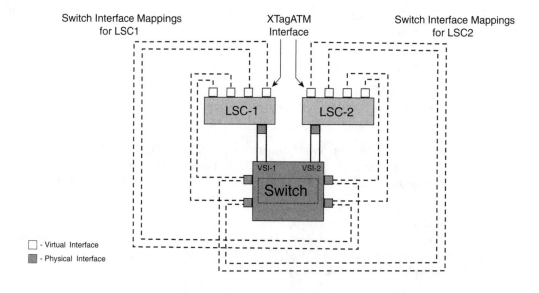

Both LSCs could be active all the time. The same sets of VSI slave protocol instances should be running in the controlled switch. In each LSC, only one VSI master runs all the time.

In the case of LSC redundancy, there is a slight change in the configuration of edge LSRs. This is shown in Figure 5-10. Instead of two ATM MPLS subinterfaces, you have to create four ATM MPLS subinterfaces at the edge LSRs. Two of the four interfaces are connected to LSC-1, and the other two are connected to LSC-2, as shown in the figure. LSC redundancy eliminates the single points of failure.

Figure 5-10 *Redundant LSC Model Attaching an ATM eLSR*

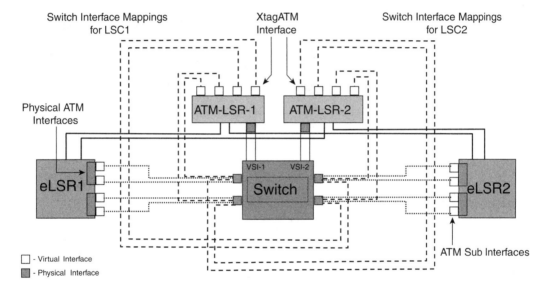

From a logical standpoint, the LSC redundancy model creates a parallel route topology, as shown in Figure 5-11. This provides redundancy in the route tables at the eLSRs.

Figure 5-11 *Virtual View When Using LSC Redundancy*

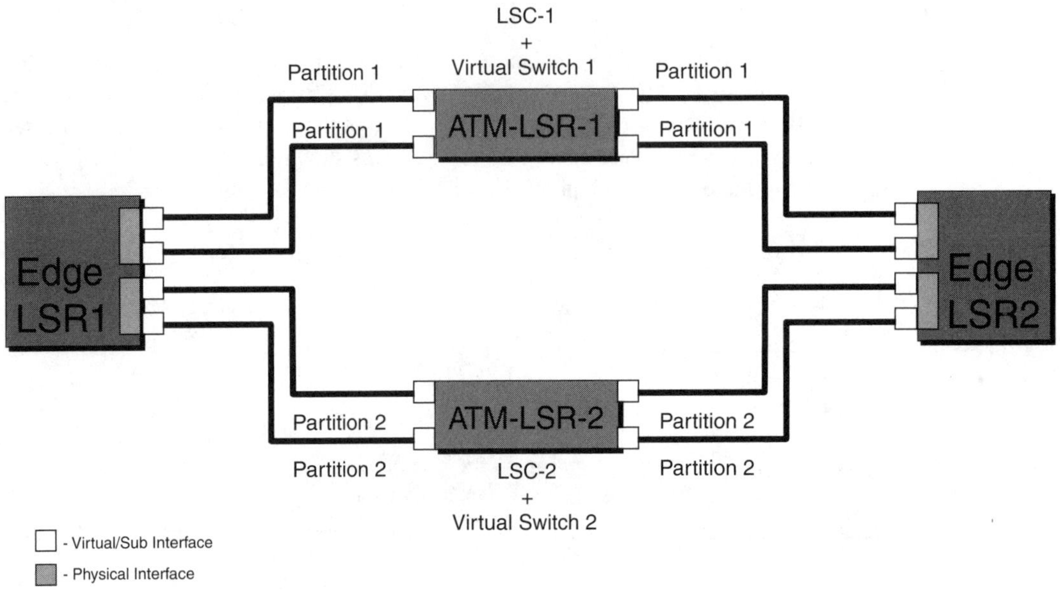

Levels of LSC Redundancy

There are two levels of LSC redundancy:

- **Hot redundancy**—Uses both redundant paths to route traffic. Both paths are set up to use the same routing cost so that traffic is load-balanced between the two paths. Hot redundancy uses twice the number of LVCs as warm redundancy but offers a higher level of resiliency. Either logical path is redundant to the other, and there's no traffic interruption in the event of failure.

- **Warm redundancy**—Uses only one path (one LSC) at a time. The paths are set up so that one path has a higher cost than the other. Traffic uses only the lower-cost path. The other path is a backup path. In this case, the failover time equals the sum of the reroute time and the LDP mapping request time.

NOTE   Although it was just mentioned, the fact that hot LSC redundancy mode uses twice as many LCN resources as a non-LSC redundant MPLS network is important enough that it deserves its own note.

The logical and physical views of both redundancy levels are shown in Figure 5-12.

Figure 5-12 *LSC Redundancy Levels*

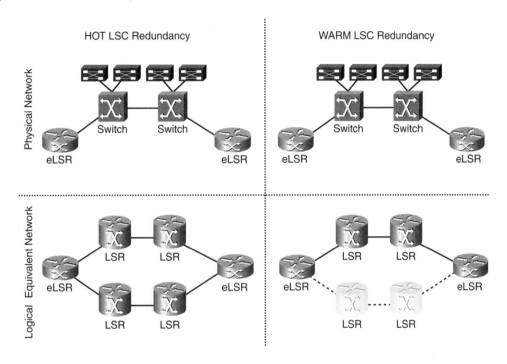

LSC redundancy not only enhances network reliability but also provides increased flexibility. For example, the following operational steps can be performed in one logical network without affecting the other:

- Changing the configuration in a live network
- Upgrading the software image without rebooting the entire system
- Upgrading LSC gracefully
- Running different or experimental configurations or software images
- Switching from warm to hot redundancy on-the-fly

Having several LSCs active on a single BPX switch turns the switch into several separate logical ATM-LSRs, as shown in Figure 5-13. Each logical ATM-LSR acts independently. The ATM-LSRs share the same trunks by way of VSI partitioning, creating separate logical networks.

Figure 5-13 *Parallel Redundant LSCs Controlling an ATM Switch*

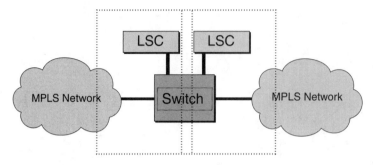

Figure 5-14 shows the resulting logical network topology when two ATM switches are connected in serial.

Figure 5-14 *LSC Redundancy in a Two-Switch Network*

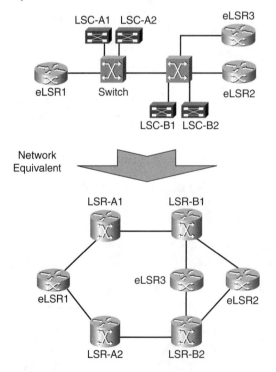

Figure 5-15 shows a more complex physical and logical MPLS redundant network.

Figure 5-15 *LSC Redundancy in a Three-Switch Network*

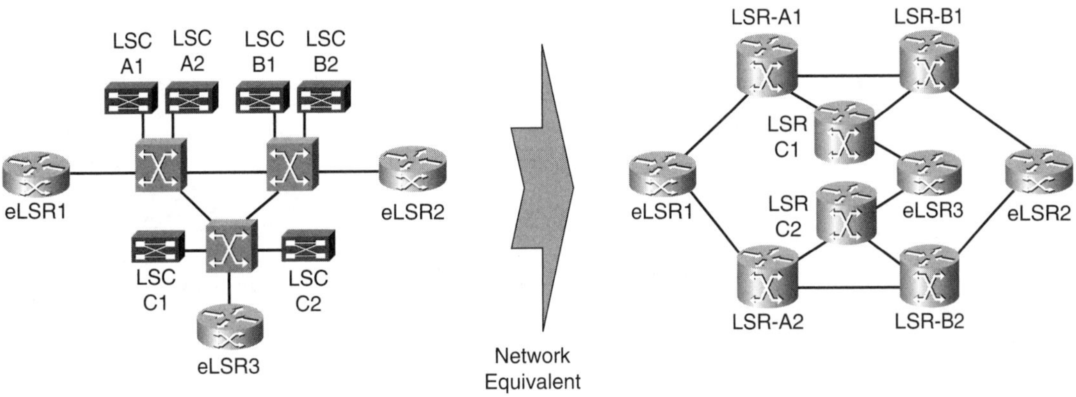

Link Bandwidth Considerations with LSC Redundancy

The LSC redundancy model requires that two LSCs manage resources in the control switch. Three different ways exist for multiple LSCs to control interface resources in the switch:

- **Virtual trunks or VNNI**—This mode provides better separation of bandwidth resources between the LSCs. Its disadvantage is that bandwidth cannot be shared between different virtual trunks. This is because different virtual interfaces such as virtual trunks actually use different sets of queues and shaping devices on the line-card hardware. Therefore, bandwidth cannot be used upon LSC failure, and bandwidth assigned to the failed LSC is locked up. An example is LC-ATM 1.2.1 partition 1 for LSC1 and LC-ATM 1.2.2 partition 2 for LSC2. In this mode, no control-vc collision is possible because the control-vc's VPI must be within the resource partition.

- **Physical interface with multiple partitions**—This mode can share bandwidth between partitions belonging to different LSCs. An example is LC-ATM 1.2 partition 1 for LSC1 and LC-ATM 1.2 partition 2 for LSC2. In this mode, control-vcs can collide and need to be manually configured, as they default to use VPI=0 in both LSCs.

- **Different physical interfaces for different LSCs**—An example is LC-ATM 1.2 partition 1 for LSC1 and LC-ATM 1.3 partition 2 for LSC2. No control-vc collision is possible because the two LSCs control different physical interfaces, but resources can be underused.

Figure 5-16 shows the difference between using multiple VSI partitions and multiple virtual trunks in a physical interface for LSC redundancy.

Figure 5-16   *Interface Bandwidth and LSC Redundancy*

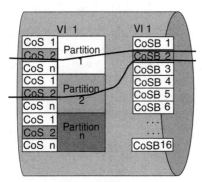
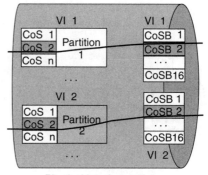

In Figure 5-16, you can see that a virtual interface (VI) is assigned to each interface in the controlled switch. This is done for physical interfaces as well as for virtual trunks and VNNI/VUNI interfaces. Each VI has in turn 16 class of service buffers (CoSBs) and a scheduler to serve bandwidth among them depending on the load and priority level of the CoSBs. There is a two-level bandwidth distribution in each CoSB based on guaranteed and excess bandwidth. A VI is defined as a group of CoSBs treated as a logical interface.

In a physical interface with multiple partitions, there is only one VI with 16 CoSBs, so different partitions can share the bandwidth in a CoSB for a given class. The formulas comparing the sum of the minimum bandwidth values and the largest maximum bandwidth values studied in the Chapter 2 section titled "Hard, Soft, and Dynamic Resource Partitioning" govern the bandwidth sharing. On the other hand, in a physical interface with virtual trunks or VNNI/VUNI interfaces, every virtual trunk has its own VI with 16 CoSBs each. Therefore, connections from different virtual trunks are serviced in different CoSBs, even for the same class of service.

Because of the bandwidth considerations just covered, it is recommended that you use multiple partitions whenever possible. You should use virtual trunks only when they are necessary for other reasons, such as connecting multiple RPM edge LSR devices in an MGX-8230 or MGX-8250 multiservice concentrator to an ATM MPLS LSR.

Summary

This chapter covered ATM MPLS design concepts. Different ATM MPLS architectures were presented, and detailed explanations of LCN requirements and LVC calculations were given.

This chapter ended with a discussion of LSC redundancy options and benefits. You are now ready to delve into MPLS configuration in multiservice switching networks.

Network design is an ongoing process. Network status and values need to be calculated and measured periodically, providing a feedback loop that refines the design.

CHAPTER 6

MPLS Implementation and Configuration

In this chapter and Chapter 7, "Practical Applications of MPLS", we will implement an Multiprotocol Label Switching (MPLS) network detailing the configuration steps and linking them to all the concepts you already know. The approach of these two chapters is the one a service provider or Enterprise network engineer would take. In this chapter we will implement the MPLS network that will serve as the foundation for services built on top of it. Different implementations of ATM MPLS LSRs and eLSR will build this MPLS network. In Chapter 7, we will configure a basic MPLS VPN. Finally, we will enable quality of service following the DiffServ model.

An important part of this chapter is that in addition to covering BPX-8600, IGX-8400, MGX-8850, and other platforms, a generic configuration model is introduced. Along with that common configuration model, we will present a summary of commands. The final part of MPLS implementation and configuration will be spent covering some troubleshooting scenarios—notably, Label VC (LVC) starvation.

MPLS Configuration in MSS

The goal of the following sections is to bring up a multiservice switching MPLS network that can be deployed in a service provider or enterprise customer. This network has three Points of Presence (PoPs), each with one Label Switch Router (LSR) and two edge Label Switch Routers (eLSRs), also called Label Edge Routers (LERs). Following MPLS VPN terminology, the LSRs are marked P (Provider), and the eLSRs are labeled PE (Provider Edge).

Each LSR is composed of a controlled switch and a Label Switch Controller (LSC). Each PoP has a different LSR platform implementation but the same MPLS controller software (which is the heart of the multiservice switching architecture). A full mesh interconnects the three PoPs, forming a triangle. In summary, the objective is to configure the network shown in Figure 6-1.

Figure 6-1 *MPLS Network*

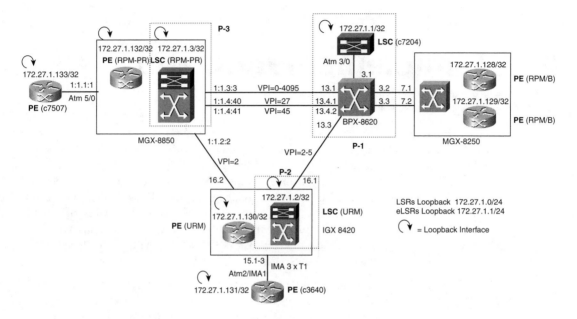

This MPLS network will be the foundation for the MPLS VPN and MPLS QoS configuration in the next chapter.

Generic Configuration Model

Before jumping into specific configurations, we will start by building a generic configuration model, removing the specifics and concentrating on the shared structure. The tasks required for LSR and eLSR bringup and setup are the same in all platform implementations, but they have different command sets. The key here is to understand the ideas and interconnect them with the previous chapter's concepts. All the specific configurations will come up naturally as a result of recognizing these general ideas.

After all the configuration cases in the following sections, a summary of all the commands and their contexts following this model is included in the later section "Summary of MPLS Configuration Commands."

Generic LSR Bringup

The common LSR bringup can be broken into four major steps. Steps 1 and 2 are performed in the controlled switch, and Steps 3 and 4 are applied to the LSC. Figure 6-2 shows these steps.

Figure 6-2   *Generic LSR Bringup Model*

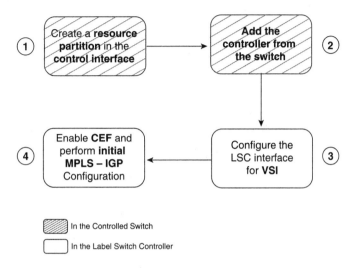

In BPX-8600 and IGX-8400 implementations, the first step includes a set of configuration tasks:

- Up the trunk in the BPX.
- Up a line. Add and then up the corresponding port in the IGX.
- Configure the resource partition in both BPX and IGX.

In MGX-8850 platforms, the partition configuration is performed in the RPM-PR card and is automatically sent to the PXM-45 card where the VSI slave lives.

In the BPX-8600 and IGX-8400 implementations with an external controller, the controller cannot be added (Step 2 in Figure 6-2) if the control interface is in alarm (such as a Loss of signal [LOS] Red Alarm). MGX-8850 implementations, as well as URM LSC in IGX-8400, do not have this check because the control interface is internal (such as the switch interface in MGX-8850) and cannot be shut down.

Steps 2 and 3 can be swapped, such that Step 3 is performed before Step 2. The order presented in this section was chosen so that the LSC configuration would be separated from the controlled switch configuration.

Generic eLSR Bringup

We can also build a diagram for the eLSR configuration. It starts with the LSR configured and involves two different paths: the LSR side and the eLSR side. Figure 6-3 shows this process. The LSR portion is on the left, and the eLSR part is on the right.

Figure 6-3 *Generic eLSR Bringup Model*

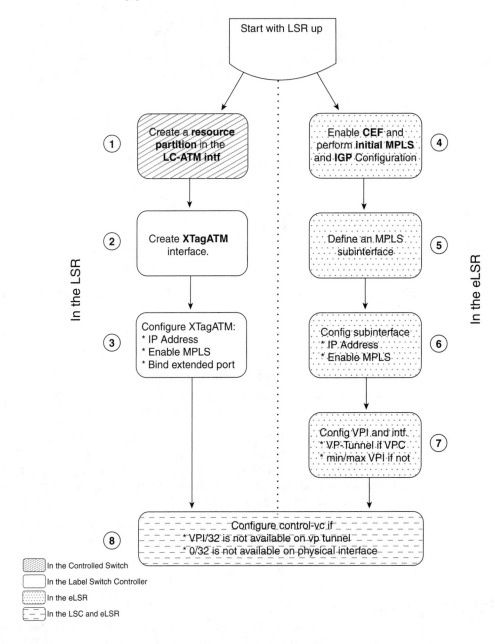

Figure 6-3 is straightforward. However, we can outline some details.

Step 1 can also involve bringing up and configuring the physical and ATM layers of the line in the controlled switch. It also includes configuring Class of Service Buffer (CoSB) resources by mapping a Service Class Template (SCT). In an AXSM platform, a Card SCT also must be configured with all the ports administratively down.

In Step 3, the IP address is normally unnumbered to a loopback interface to save IP address and LCN resources.

Step 7 does not have an equivalent step in the LSR. This is because the VSI interface messages in the LSR learn the logical interface's VPI range and VP tunnel characteristics. On the other hand, in the eLSR, there is no partitioning of resources or VSI protocol, so we need to explicitly configure those parameters. Both configuration commands (**vp-tunnel** and **vpi**) have an extra optional parameter that limits the VCI range at the control plane. The parameter *vci-range* is extremely important in cases where the slave cannot partition VCI resources.

Finally, the optional **control-vc** VPI/VCI configuration must match in both ends.

Tag Switching or MPLS?

We can configure tag switching using TDP (Tag Distribution Protocol) or MPLS using standards-based LDP (Label Distribution Protocol) on LSRs and eLSRs. As you know, tag switching is a precursor of MPLS. Functionally, LDP is a superset of TDP, and the protocol exchange patterns are almost identical between the two protocols. They have similar messages with different names (Open PIE and Request Bind PIE in TDP correspond to Initialization Message and Label Request Message in LDP) and different UDP ports (for session discovery) and TCP ports (for session establishment).

From an operations standpoint, Cisco IOS Parser has tag-switching and MPLS options for most commands. (There are no tag-switching commands for MPLS-specific functions such as a TCP MD5 signature of an LDP session or Path Vector type-length-value [TLV] loop detection.) You can configure either distribution protocol (TDP or standard LDP) with either form of the commands (**tag-switching** or **mpls**). One global and interface configuration command configures the use of TDP or LDP. That command is **mpls label protocol** *{ldp|tdp}*. The **[sub]interface** form of the command also has a **both** option to simultaneously run TDP and LDP. That command has only the **mpls** form.

In this chapter, we will configure standards-based MPLS running LDP using the **mpls** form of the commands. By default, the commands are saved to configuration in their **tag-switching** form. This is done for backward compatibility, in case you want to use that configuration on a router without **mpls** commands.

BPX-8600- and MGX-8250-Based PoP

The first PoP is made up of one LSR and two eLSRs. A BPX-8620 controlled switch and a Cisco 7204-based LSC (external controller) form the LSR. The two eLSRs are RPM/B cards in an MGX-8250 edge concentrator shelf. The PXM-1 controller card in the MGX edge concentrator connects to the BPX. The details of the setup are shown in Figure 6-4.

Figure 6-4 *BPX-8650- and MGX-8250-Based PoP*

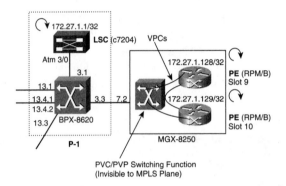

We will study this case in two main steps. First we'll look at the configuration steps to bring up the LSR, and then the steps to attach the eLSRs.

ATM LSR Bringup: BPX + LSC

We start the configuration by upping and configuring the 3.1 trunk that connects the BPX to the LSC (see Example 6-1). That interface is the control interface. On the trunk configuration, using **cnftrk**, we will leave the statistical reserve non-null. Even though there is no CC traffic to protect, we will protect VSI protocol traffic.

Example 6-1 *Upping the Control Interface in a BPX-Based LSR*

```
P1-b8620         TN    Cisco          BPX 8620  9.3.30    Nov. 23 2001 14:47 GMT

TRK      Type    Current Line Alarm Status              Other End
3.1      OC3     Clear - OK                             -
```

Example 6-1 *Upping the Control Interface in a BPX-Based LSR (Continued)*

```
Last Command: uptrk 3.1

256 PVCs allocated. Use 'cnfrsrc' to configure PVCs
Next Command:
```

We define a partition on the control interface so that VSI has resources to control. In this resource partition, we leave no resources (LCN and bandwidth) for PVCs because we will have only VSI traffic and VSI-controlled MPLS traffic. From the three VSI partitions that the BXM card supports, we enable partition ID 1 for MPLS. See Example 6-2.

Example 6-2 *Creating a VSI Resource Partition*

```
P1-b8620        TN    Cisco         BPX 8620  9.3.30    Nov. 23 2001 14:50 GMT

Trunk : 3.1
                                    Full Port Bandwidth: 353208
Maximum PVC LCNS:              0    Maximum PVC Bandwidth: 0
                                    (Statistical Reserve: 5000)

PVC VPI RANGE [1]:   -1   /-1       PVC VPI RANGE [2]: -1    /-1
PVC VPI RANGE [3]:   -1   /-1       PVC VPI RANGE [4]: -1    /-1

           Partition :  1            2              3
     Partition State :  Enabled      Disabled       Disabled
   VSI LCNS (min/max):  500    /1000  0      /0      0      /0
  VSI VPI (start/end):  1      /4     0      /0      0      /0
    VSI BW (min/max):   348207 /348207 0     /0      0      /0
      VSI ILMI Config:  CLR          CLR            CLR


Last Command: cnfrsrc 3.1 0 348207 y 1 e 500 1000 1 4 348207 348207


Next Command:
```

NOTE   Strictly speaking, a resource partition on the control interface is not needed to add the controller because VSI communication occurs on the VSI VCs. However, without a resource partition, VSI can't set up a cross-connect with a leg in the control interface (such as the private legs of the control-vcs) because it has no resources to control.

At this point, interface 3.1 is in LOS alarm. We need to clear that alarm to proceed with adding the controller (see Example 6-3) so that we can move on to the LSC configuration.

Example 6-3 *Adding a Label Switch Controller*

```
P1_LSC_c7204#
P1_LSC_c7204#conf t
Enter configuration commands, one per line.  End with CNTL/Z.
P1_LSC_c7204(config)#interface ATM 3/0
P1_LSC_c7204(config-if)#label-control-protocol vsi id 1 base-vc 0 40
P1_LSC_c7204(config-if)#no shutdown
P1_LSC_c7204(config-if)#
18:02:14: %LINK-3-UPDOWN: Interface ATM3/0, changed state to up
18:02:15: %LINEPROTO-5-UPDOWN: Line protocol on Interface ATM3/0, changed state to up
P1_LSC_c7204(config-if)#end
P1_LSC_c7204#
```

We use the interface-level command **label-control-protocol** to specify that this is an LSC and that VSI will be used to manage a controlled switch. In this case, we specify a controller ID of 1 and a base-vc for VSI communication of VPI/VCI = 0/40. The base-vc is the VPI/VCI value for slave 0. These are the LSC default values, and they need to match in the controlled switch.

We are ready now to start VSI communication by adding the VSI controller. To do so, we go to the BPX and use the command **addshelf** with the **VSI** option, as shown in Example 6-4. Adding a VSI shelf means adding a VSI controller.

Example 6-4 *Starting VSI Communication from the BPX-8600*

```
P1-b8620        TN    Cisco           BPX 8620  9.3.30    Nov. 23 2001 15:01 GMT

                     BPX 8620 Interface Shelf Information

Trunk    Name      Type     Part Id    Ctrl Id      Control_VC        Alarm
                                                    VPI   VCIRange
 3.1     VSI       VSI         1           1         0    40-54       OK

















Last Command: addshelf 3.1 VSI 1 1 0 40

Shelf has been added
Next Command:
```

The **addshelf** command triggers the creation of the VSI virtual circuits (VCs), both master-slave and slave-to-slave VCs.

The **addshelf** command also assigns resources in the LC-ATM interfaces to the specific controller. Mapping the partition ID to the controller ID does this. In this case, the controller with an ID of 1, which is the LSC, controls resource partition 1 in all interfaces.

NOTE	We could have selected any partition ID (1 to 3 in the BPX) and any controller ID. However, if we want to also run PNNI in the BPX, the service expansion shelf (SES) PNNI controller has a fixed controller ID of 2.

We can check the controller configured in a BPX switch by using the command **dspctrlrs**.

From the LSC, we can check the status of VSI communication using the command **show controllers vsi**, as shown in Example 6-5.

Example 6-5 *Showing the VSI Sessions*

```
P1_LSC_c7204#show controllers vsi session
Interface    Session   VCD    VPI/VCI     Switch/Slave Ids    Session State
ATM3/0       0         1      0/40        0/0                 UNKNOWN
ATM3/0       1         2      0/41        0/0                 UNKNOWN
ATM3/0       2         3      0/42        0/3                 ESTABLISHED
ATM3/0       3         4      0/43        0/0                 UNKNOWN
ATM3/0       4         5      0/44        0/0                 UNKNOWN
ATM3/0       5         6      0/45        0/0                 UNKNOWN
ATM3/0       6         7      0/46        0/0                 UNKNOWN
ATM3/0       7         8      0/47        0/0                 UNKNOWN
ATM3/0       8         9      0/48        0/0                 UNKNOWN
ATM3/0       9         10     0/49        0/0                 UNKNOWN
ATM3/0       10        11     0/50        0/0                 UNKNOWN
ATM3/0       11        12     0/51        0/0                 UNKNOWN
ATM3/0       12        13     0/52        0/0                 UNKNOWN
ATM3/0       13        14     0/53        0/0                 UNKNOWN
P1_LSC_c7204#
```

The parameter **session** displays all the VSI sessions that the VSI master maintains. In the state machine, ESTABLISHED is the last state for a discovered and active VSI slave. Because we know there's one VSI session per slave, slaves are also called sessions.

We can also display all the session details by specifying the session number. Refer to Example 6-6.

Example 6-6 *Displaying the VSI Session Details*

```
P1_LSC_c7204#show controllers vsi session 2
Interface:            ATM3/0    Session number:       2
VCD:                  3         VPI/VCI:              0/42
Switch type:          BPX       Switch id:            0
Controller id:        1         Slave id:             3
Keepalive timer:      15        Powerup session id:   0x00000001
Cfg/act retry timer:  8/8       Active session id:    0x00000001
Max retries:          10        Ctrl port log intf:   0x00030100   ! LIN for 3.1
Trap window:          50        Max/actual cmd wndw:  21/21
Trap filter:          all       Max checksums:        79
Current VSI version:  2         Min/max VSI version:  2/2
Messages sent:        16        Inter-slave timer:    0.004
Messages received:    11        Messages outstanding: 0
P1_LSC_c7204#
```

Session 2 is the only active session at this point. The field Switch/Slave Id is equal to 0/3. The switch ID is 0 and is meaningful only in multishelf slaves because VSI supports multishelf. The slave ID equals the slot number.

NOTE   The session state is ESTABLISHED for all active BXM cards or VSI slaves in general, even if they do not have resource partitions.

The parameter **descriptor** shows the VSI logical interface details, including the Logical Interface Number (LIN), as shown in Example 6-7. The descriptor is an ASCII field in VSI interface messages that can be thought of as the interface's "name." Even though we can work with the 32-bit LIN, it's easier for humans to remember the ASCII descriptor.

Example 6-7 *Showing VSI Interface Partition Details by Descriptor*

```
P1_LSC_c7204#show controllers vsi descriptor

Phys desc: 0.3.1.0
Log intf:  0x00030100    ! The slave resides in BXM Card in slot 3.
Interface: switch control port
IF status: n/a                 IFC state: ACTIVE
Min VPI:   1                   Maximum cell rate:  348207
Max VPI:   4                   Available channels: 1000
Min VCI:   32                  Available cell rate (forward):  348207
Max VCI:   65535               Available cell rate (backward): 348207

P1_LSC_c7204#
```

Finally, we can also enable a debug to see the VSI messages, as shown in Example 6-8.

Example 6-8 *Debugging VSI Packets on a VSI Slave*

```
P1_LSC_c7204#debug vsi packets interface ATM 3/0 slave 2
VSI Master packet debugging is on
Displaying packets on interface ATM3/0, selected slave(s) only
P1_LSC_c7204#
00:35:12: VSI Master (session 2 on ATM3/0): sent msg SW GET CNFG CMD on 0/42
00:35:12: VSI Master (session 2 on ATM3/0): rcvd msg SW GET CNFG RSP on 0/42
00:35:27: VSI Master (session 2 on ATM3/0): sent msg SW GET CNFG CMD on 0/42
00:35:27: VSI Master (session 2 on ATM3/0): rcvd msg SW GET CNFG RSP on 0/42
00:35:42: VSI Master (session 2 on ATM3/0): sent msg SW GET CNFG CMD on 0/42
00:35:42: VSI Master (session 2 on ATM3/0): rcvd msg SW GET CNFG RSP on 0/42
P1_LSC_c7204#
```

The VSI messages **SW GET CNFG** (switch get configuration) command and response are used as keepalives. As displayed in the VSI **session details show** command, the keepalive timer defaults to 15 seconds.

NOTE   The VSI channels in the controlled switch can be seen in the BPX-8650 and IGX-8400 platforms using the command **dspvsich** *{Slot}*. Both **control-port msvc** and **interslav** channels are displayed, with the local and remote slots.

The next step is the basic MPLS configuration at the LSR (effectively at the LSC).

In the initial MPLS configuration in the LSC, we enable Cisco Express Forwarding (CEF), we configure the label protocol to be LDP (as opposed to TDP), we give the LSR a loopback IP address, and we assign that loopback address to be the LDP router ID to provide LDP stability (see Example 6-9).

Example 6-9 *Initial MPLS LSC Configuration*

```
P1_LSC_c7204#conf t
Enter configuration commands, one per line.  End with CNTL/Z.
P1_LSC_c7204(config)#ip cef
P1_LSC_c7204(config)#mpls label protocol ?
  ldp  Use LDP
  tdp  Use TDP (default)

P1_LSC_c7204(config)#mpls label protocol ldp   mpls label protocol ldp
P1_LSC_c7204(config)#interface loopback 0
P1_LSC_c7204(config-if)#ip address 172.27.1.1 255.255.255.255
P1_LSC_c7204(config-if)#exit
P1_LSC_c7204(config)#mpls ldp router-id loopback 0
P1_LSC_c7204(config)#^Z
P1_LSC_c7204#
```

NOTE   The global configuration command **mpls label protocol ldp** enables LDP in all the interfaces in the LSR. This command has an interface configuration mode as well, to enable TDP, LDP, or both on a per-interface basis.

MGX-8250 eLSR Configuration

Now that we have our MPLS LSR, we will attach the two eLSRs. We will first configure the LSR end of the links to the eLSRs. Then we will bring up the RPM/Bs as eLSRs.

We are ready to configure the interfaces between LSR and eLSRs. As covered in Chapter 3, "Implementations and Platforms," we cannot partition resources in the PXM-1 controller card. Therefore, we will add Virtual Path Connections (VPCs) from the PXM-1 to the RPM/Bs and use VP-Tunnel LC-ATM interfaces.

From the BPX side, we start by upping a virtual trunk, as shown in Example 6-10.

Example 6-10 *Upping a VP-Tunnel LC-ATM Interface*

```
P1-b8620        TN   Cisco         BPX 8620  9.3.30    Nov. 23 2001 15:30 GMT

TRK       Type    Current Line Alarm Status            Other End
3.1       OC3     Clear - OK                           VSI(VSI)
3.3.1     OC3     Clear - OK                           -













Last Command: uptrk 3.3.1

256 PVCs allocated. Use 'cnfrsrc' to configure PVCs
Next Command:
```

We configure the trunk so that we set the statistical reserve to 0 and do not misuse bandwidth resources, as well as configure the virtual trunk's VPI number. For simplicity, we will use VPI=9 for the RPM/B in slot 9 and VPI=10 for the RPM/B in slot 10. See Example 6-11.

Example 6-11 *Configuring the VP-Tunnel from the BPX-8600 Controlled Switch*

```
P1-b8620        TN    Cisco          BPX 8620  9.3.30    Nov. 23 2001 15:32 GMT

TRK  3.3.1 Config   OC3    [2867 cps]   BXM slot:      3
Transmit Rate:          3000            VPC Conns disabled:    --
Protocol By The Card:   --              Line framing:          STS-3C
VC Shaping:             No                 coding:             --
Hdr Type NNI:          No                 recv impedance:     --
Statistical Reserve:    0       cps        cable type:         --
Idle code:             7F hex                    length:       --
Connection Channels:   256             Pass sync:             No
Traffic:V,TS,NTS,FR,FST,CBR,N&RT-VBR,ABR Loop clock:          No
Restrict CC traffic:   No              HCS Masking:           Yes
Link type:             Terrestrial     Payload Scramble:      Yes
Routing Cost:          10              Frame Scramble:        Yes
F4 AIS Detection:      No              Vtrk Type / VPI:       CBR / 9
                                       Incremental CDV:       0
                                       Deroute delay time:    0 seconds

Last Command: cnftrk 3.3.1 3000 N 0 7F V,TS,NTS,FR,FST,CBR,NRT-VBR,ABR,RT-VBR N
  TERRESTRIAL 10 N 0 N N Y Y Y CBR 9 0

Next Command:
```

Finally, we partition resources in the virtual trunk, setting the PVC LCNs to 0 (so as not to misuse LCN resources). We use partition ID = 1 because that's the one controlled by the LSC. Refer to Example 6-12.

Example 6-12 *Creating a Resource Partition in the BPX-8600*

```
P1-b8620        TN    Cisco          BPX 8620  9.3.30    Nov. 23 2001 15:33 GMT

Virtual Trunk : 3.3.1

                                  Full Port Bandwidth: 3000
Maximum PVC LCNS:            0    Maximum PVC Bandwidth: 0
                                  (Statistical Reserve: 0)

PVC VPI RANGE [1]:   -1    /-1    PVC VPI RANGE [2]: -1    /-1
PVC VPI RANGE [3]:   -1    /-1    PVC VPI RANGE [4]: -1    /-1

          Partition :  1            2            3
       Partition State :  Enabled     Disabled     Disabled
     VSI LCNS (min/max):  256   /512   0     /0    0     /0
     VSI VPI (start/end): 9     /9     9     /9    9     /9
       VSI BW (min/max):  1400  /2867  0     /0    0     /0
       VSI ILMI Config:   CLR          CLR          CLR
```

continues

Example 6-12 *Creating a Resource Partition in the BPX-8600 (Continued)*

```
Last Command: cnfrsrc 3.3.1 0 0 y 1 e 256 512 9 9 1400 2867


Next Command:
```

We can also check that we are using SCT 1 in this virtual trunk, using the command
dspvsiif (see Example 6-13). SCT 1 is the default SCT and is an MPLS-only SCT.

Example 6-13 *Checking the Service Class Template in the BPX-8600 VSI Slaves*

```
P1-b8620        TN    Cisco          BPX 8620  9.3.30   Nov. 23 2001 15:34 GMT

    Virtual Trunk : 3.3.1

    Service Class Template ID: 1

    VSI Partitions :
                  channels          bw              vpi
    Part  E/D   min   max    min     max     start  end      ilmi
    1     E     256   512    1400    2867    9      9        D
    2     D     0     0      0       0       9      9        D
    3     D     0     0      0       0       9      9        D




Last Command: dspvsiif 3.3.1


Next Command:
```

From the LSC, we can see the just-added partition that is not yet bound to any extended
MPLS interface, as shown in Example 6-14.

Example 6-14 *Showing the VSI Partition Details from the LSC by Descriptor*

```
P1_LSC_c7204#show controllers vsi descriptor

Phys desc: 0.3.1.0
Log intf:  0x00030100
Interface: switch control port
IF status: n/a              IFC state: ACTIVE
Min VPI:   1                Maximum cell rate:  348207
Max VPI:   4                Available channels: 1000
Min VCI:   32               Available cell rate (forward):  348207
Max VCI:   65535            Available cell rate (backward): 348207
```

Example 6-14 *Showing the VSI Partition Details from the LSC by Descriptor (Continued)*

```
Phys desc: 0.3.1.0
Log intf:  0x00030301
Interface: n/a
IF status: n/a                   IFC state: ACTIVE
Min VPI:   9                     Maximum cell rate:  2867
Max VPI:   9                     Available channels: 512
Min VCI:   32                    Available cell rate (forward):  2867
Max VCI:   65535                 Available cell rate (backward): 2867

P1_LSC_c7204#
```

The LSC knows this because the VSI slave in the BXM card sent a VSI interface trap to the VSI master when we configured the resource partition. This is done so that the LSC is aware of the new virtual interface and its resources that it has to control.

For the LSC to control those resources, we create an extended MPLS ATM interface (XTagATM), which is an extension of the physical interface's resources in the BPX switch. See Example 6-15.

Example 6-15 *Creating and Configuring an Extended MPLS Interface from the LSC*

```
P1_LSC_c7204#conf t
Enter configuration commands, one per line.  End with CNTL/Z.
P1_LSC_c7204(config)#int XTagATM 331
P1_LSC_c7204(config-if)#mpls ip
P1_LSC_c7204(config-if)#ip unnumbered loopback 0
P1_LSC_c7204(config-if)#extended-port ATM 3/0 bpx 3.3.1

P1_LSC_c7204(config-if)#end
18:38:56: %LINK-3-UPDOWN: Interface XTagATM331, changed state to up
18:38:57: %LINEPROTO-5-UPDOWN: Line protocol on Interface XTagATM331, changed state
P1_LSC_c7204#
```

After creating the XTagATM interface, we enable MPLS in it with **mpls ip**, unnumber the interface's IP address to the loopback IP address, and tell the XTagATM interface that is bound to port 3.3.1 in the controlled switch.

NOTE   Using unnumbered interfaces saves IP addresses as well as precious LCN resources because by default, MPLS sets up LVCs to all IP addresses in the global routing table.

We can see that the VSI interface is bound to the XTagATM interface. See Example 6-16.

Example 6-16 *Displaying the VSI Controller Status*

```
P1_LSC_c7204#show controllers vsi status
Interface Name            IF Status    IFC State  Physical Descriptor
switch control port             n/a    ACTIVE     0.3.1.0
XTagATM331                       up    ACTIVE     0.3.3.1
P1_LSC_c7204#
```

Finally, we need to configure an Interior Gateway Protocol (IGP) in our MPLS network. We choose OSPF with only the backbone area, as shown in Example 6-17.

Example 6-17 *Configuring the IGP*

```
P1_LSC_c7204(config)#router ospf 100
P1_LSC_c7204(config-router)#network 172.27.1.1 0.0.0.0 area 0
P1_LSC_c7204(config-router)#^Z
P1_LSC_c7204#
```

At this point we can begin the PE router side of the configuration. Portable AutoRoute (PAR) reigns in the MGX-8250 switch. PAR is the only control plane that manages the MGX-8250 switch resources. So we will assign the RPM resource partition to PAR and add the VPCs using PAR. This is the reason why we need a VP-Tunnel LC-ATM interface.

As soon as PAR has learned the VPC from the RPMs to the PXM-1 uplink port and MPLS is configured, the MPLS control plane sets up LVCs (even though MPLS will not directly manage switch resources). That's the edge behavior in the architecture.

We will start by adding the 7.2 line and port (so that the PAR controller manages the PAR If), as shown in Example 6-18. The line 7.1 will be used later, in the section "Adding an AutoRoute Control Plane."

Example 6-18 *Adding a PAR Interface to a Feeder Shelf*

```
m8250-7a.1.7.PXM.a > addln -sonet 7.2

m8250-7a.1.7.PXM.a > addport 2 2 100 0 4095 100

m8250-7a.1.7.PXM.a > dspports

   Port  Status  Line  PctBw  minVpi  maxVpi  maxRatePct
   ------------------------------------------------------
     2    ON       2    100      0     4095     100

m8250-7a.1.7.PXM.a >

m8250-7a.1.7.PXM.a > dspparifs
```

Example 6-18 *Adding a PAR Interface to a Feeder Shelf (Continued)*

```
slot.port   type        status    vpi           vci           txRate    rxRate
----------------------------------------------------------------------------
      7.2   UNI_IF        UP    0 to 4095    0 to 65535     353208    353208
     7.33   UNI_IF        UP    0 to  255    0 to 65535     176604    176604
     7.34   UNI_IF        UP    0 to  255    0 to 65535     176604    176604
     7.35   CLK_IF      FAILED  0 to    0    0 to     0          0         0

m8250-7a.1.7.PXM.a >
```

Then we add the slave end of the VPC, as shown in Example 6-19.

Example 6-19 *Adding the VPC Slave Endpoint*

```
m8250-7a.1.7.PXM.a > addcon 2 1 9 0 1 1 2
    Connection ID: m8250-7a.0.2.9.0

m8250-7a.1.7.PXM.a > cnfupccbr 2.9.0 4 2867 10000 100 2867 100

m8250-7a.1.7.PXM.a > dspchans

Chan Stat Intf locVpi locVci conTyp srvTyp PCR[0+1] Mst rmtVpi rmtVci State
-------------------------------------------------------------------------
  18 MOD    2      9      0   VPC    CBR      2867  Slv   N/A    N/A normal

m8250-7a.1.7.PXM.a >
```

NOTE   In an MGX-8250 feeder shelf, the command **dspcons** shows node-wide cross-connects that have a master endpoint, whereas the command **dspchans** shows per-card connection endpoint information for both master and slave endpoints.

The rest of the configuration takes place on the RPM card.

As mentioned, we start with the initial MPLS, CEF, and loopback configuration (see Example 6-20).

Example 6-20 *Initial MPLS Configuration*

```
PE_m8250_RPMB_9#conf t
Enter configuration commands, one per line.  End with CNTL/Z.
PE_m8250_RPMB_9(config)#ip cef
PE_m8250_RPMB_9(config)#interface loopback 0
PE_m8250_RPMB_9(config-if)#ip address 172.27.1.128 255.255.255.255
PE_m8250_RPMB_9(config-if)#exit
PE_m8250_RPMB_9(config)#mpls ldp router-id loopback 0
PE_m8250_RPMB_9(config)#mpls label protocol ldp
```

We add the PAR resource partition in the RPM to create the VP-Tunnel, as shown in Example 6-21. We assign all resources to PAR because it is the only control plane that directly controls MGX-8250 switch resources (and this is why the MGX-8250 supports only edge MPLS functionality).

Example 6-21 *Configuring the RPM PAR Resource Partition*

```
PE_m8250_RPMB_9(config)#rpmrscprtn par 100 100 0 255 0 3840 4080
```

At this point, the RPM switch interface partition is visible to the PAR control plane in the MGX-8250. See Example 6-22.

Example 6-22 *Using the Command dspparifs*

```
m8250-7a.1.7.PXM.a > dspparifs
slot.port   type      status     vpi            vci          txRate      rxRate
-----------------------------------------------------------------------------
     7.1    FTRK_IF    ADDED    0 to 4095    0 to 65535      353208      353208
     7.2    UNI_IF       UP     0 to 4095    0 to 65535      353208      353208
     7.33   UNI_IF       UP     0 to  255    0 to 65535      176604      176604
     7.34   UNI_IF       UP     0 to  255    0 to 65535      176604      176604
     7.35   CLK_IF     FAILED   0 to    0    0 to     0           0           0
     9.1    UNI_IF       UP     0 to  255    0 to  3840      353208      353208

m8250-7a.1.7.PXM.a >
```

Subsequently, we configure an MPLS subinterface, where we enable MPLS, unnumber the IP address to the loopback, and configure the interface to be a VP-Tunnel with VPI=9 (see Example 6-23).

Example 6-23 *LC-ATM VP-Tunnel Configuration*

```
PE_m8250_RPMB_9(config)#interface switch 1.1 mpls
PE_m8250_RPMB_9(config-subif)#mpls ip
PE_m8250_RPMB_9(config-subif)#ip unnumbered loopback 0
PE_m8250_RPMB_9(config-subif)#atm pvc 9 9 0 aal5snap
PE_m8250_RPMB_9(config-subif)#mpls atm vp-tunnel 9
PE_m8250_RPMB_9(config-subif)#exit
```

At this stage, a valid question is, "Why do we need to configure the interface to be a VP-Tunnel in the PE and not in the LSC?"

The answer would be, "Because the LSC runs VSI with the controlled switch." And when we configure a virtual trunk in the BPX switch, the LSC learns through VSI that the interface is VP-Tunnel. In the PE, we don't have MPLS partitions or VSI. Therefore, there's no mechanism for the PE to learn that the interface is a VP-Tunnel.

In essence, a VP-Tunnel interface has two implications:

- The control-vc will be within the VPI (Tunnel_VPI/32 by default and not in 0/32).

- Only the VCI space (and not also the VPI space) will be the label space. This implies that the VPIs can be different at both ends.

These ideas are expanded on in the section "A Note on LC-ATM Interfaces."

Finally, as shown in Example 6-24, we add the master end of the PAR VPC and configure the IGP.

Example 6-24 *Adding the VPC Master Endpoint*

```
PE_m8250_RPMB_9(config)#addcon vpc switch 1.1 9 rslot 0 2 9 master local
PE_m8250_RPMB_9(config)#router ospf 100
PE_m8250_RPMB_9(config-router)#network 172.27.1.128 0.0.0.0 area 0
```

At this point, our P-PE LDP session should be up. We follow the same procedure to bring up the RPM in slot 10 as a second PE, and we also add a VPC and configure another MPLS subinterface as a VP-Tunnel between the two RPMs. The configuration of MPLS subinterface 1.2 connecting the two RPMs is equivalent to the configuration covered already. The only difference resides in the **addcon vpc switch** command, in which the remote slot is the other RPM's slot number. In the **addcon vpc switch** configuration, one RPM is set up as **master local**, and the other RPM is configured as **master remote**.

This last step is a topology optimization. If that VPC is not there, traffic from one RPM to the other goes through the BPX. It's also a VPC-level redundancy implementation.

LDP Session, Bindings, and LVC show Commands

We can start by checking the MPLS interfaces. See Example 6-25.

Example 6-25 *Using the **show mpls interfaces** Command*

```
P1_LSC_c7204#show mpls interfaces
Interface            IP          Tunnel  Operational
XTagATM331           Yes (ldp)   No      Yes          (ATM labels)
XTagATM332           Yes (ldp)   No      Yes          (ATM labels)
P1_LSC_c7204#

PE_m8250_RPMB_9#show mpls interfaces
Interface            IP          Tunnel  Operational
Switch1.1            Yes (ldp)   No      Yes          (ATM labels)
Switch1.2            Yes (ldp)   No      Yes          (ATM labels)
PE_m8250_RPMB_9#
```

Note that interface Switch 1.1 is connected (via a VPC) to the BPX-8600-based LSR, and interface Switch 1.2 is connected to the RPM-based eLSR in slot 10.

We can verify the LDP discovery and the status of the LDP neighbors as well. See Example 6-26.

Example 6-26 *Showing the LDP Neighbor Discovery*

```
P1_LSC_c7204#show mpls ldp discovery
Local LDP Identifier:
    172.27.1.1:0
Discovery Sources:
    Interfaces:
        XTagATM331 (ldp): xmit/recv
            LDP Id: 172.27.1.128:1
        XTagATM332 (ldp): xmit/recv
            LDP Id: 172.27.1.129:1
P1_LSC_c7204#
```

The command **show mpls ldp discovery** should show that the interface is in transmit/ receive. It also displays the remote LDP ID. (The local LDP ID displayed is the per-router LDP ID.)

If the state is xmit only, we are transmitting LDP messages but are not receiving any. We should check the status of the control-vc at both ends, as well as the CoS buffers (QBins in BXM and UXM) and the routing configuration. If the status is xmit/recv but the neighbor relationship is not established, there might be a problem with the routing (such as no route to the destination), or CEF is not enabled.

The LDP identifier uniquely identifies the label space in the network. It is 6 bytes long. The format from RFC 3036 is to use the first 4 bytes (32 bits) as the router ID that uniquely identifies the router in the network and the remaining 2 bytes as the label space ID that identifies the label space in that router. This is shown in Figure 6-5.

Figure 6-5 *LDP Identifier from RFC 3036 (LDP Specification)*

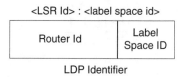

With frame-based MPLS, there is a per-router label space, and the label space ID is 0. For LC-ATM interfaces, because different interfaces can use the same label (VPI/VCI), there is a per-interface label space with a nonzero ID value.

If for any reason an LDP neighbor is not coming up, we need to use the **discovery** command. The output would be blank in the command in Example 6-27 (**neighbor**) because there's no LDP neighbor yet.

Example 6-27 *Showing the LDP Neighbor Details*

```
P1_LSC_c7204#show mpls ldp neighbor
Peer LDP Ident: 172.27.1.128:1; Local LDP Ident 172.27.1.1:1
        TCP connection: 172.27.1.128.11015 - 172.27.1.1.64(
        State: Oper; Msgs sent/rcvd: 81/81; Downstream on demand
        Up time: 01:04:44
        LDP discovery sources:
          XTagATM331, Src IP addr: 172.27.1.128
Peer LDP Ident: 172.27.1.129:1; Local LDP Ident 172.27.1.1:2
        TCP connection: 172.27.1.129.11003 - 172.27.1.1.646
        State: Oper; Msgs sent/rcvd: 8/8; Downstream on demand
        Up time: 00:02:00
        LDP discovery sources:
          XTagATM332, Src IP addr: 172.27.1.129
```

In an LDP peering relationship, there is an active LDP peer and a passive LDP peer. The active peer has the responsibility of connecting to the passive LDP peer in the LDP well-known port (646). The LSR with the largest LDP ID is the active LDP peer.

This can be seen in the **show mpls ldp neighbor** command:

```
TCP connection: 172.27.1.128.11015 - 172.27.1.64646
TCP connection: 172.27.1.129.11003 - 172.27.1.64646
```

172.27.1.128 is larger than 172.27.1.1. And 172.27.1.129 is also larger than 172.27.1.1.

The UDP port for LDP hello messages (discovery) is 646. The TCP port for establishing LDP session connections is 646. If the label protocol were TDP (instead of standard LDP), the well-known ports would be 711 for UDP and TCP:

```
TCP connection: 10.10.10.5.11138 - 10.10.10.3.711
TCP connection: 10.10.10.2.711 - 10.10.10.3.11001
```

Going back to our MPLS network running standards-based LDP, we can check the ATM MPLS local, remote, and negotiated capabilities, as shown in Example 6-28.

Example 6-28 *Showing the ATM LDP Capabilities*

```
P1_LSC_c7204#show mpls atm-ldp capability

              VPI         VCI          Alloc   Odd/Even VC Merge
XTagATM331    Range       Range        Scheme  Scheme   IN   OUT
   Negotiated [9 - 9]     [33 - 65518] BIDIR            -    -
   Local      [9 - 9]     [33 - 65535] BIDIR   EVEN     NO   NO
   Peer       [9 - 9]     [33 - 65518] UNIDIR  ODD      -    -
```

continues

Example 6-28 *Showing the ATM LDP Capabilities (Continued)*

```
               VPI          VCI             Alloc   Odd/Even VC Merge
XTagATM332     Range        Range           Scheme  Scheme   IN   OUT
  Negotiated   [10 - 10]    [33 - 65518]    BIDIR            -    -
  Local        [10 - 10]    [33 - 65535]    BIDIR   EVEN     NO   NO
  Peer         [10 - 10]    [33 - 65518]    UNIDIR  ODD      -    -
P1_LSC_c7204#
```

We see with this command the Local, Peer, and Negotiated ranges. One thing to mention is that on a BPX-8650, IGX-8400, or MGX-8850 switches, the Allocation Scheme is always bidirectional (because they don't support unidirectional cross-connects). This means that a VPI/VCI pair can appear in only one binding. It can be unidirectional in an eLSR (such as RPM cards) because it is not cross-connecting but terminating LVCs. The Negotiated Scheme is unidirectional only if both Local and Peer are unidirectional. In a bidirectional Allocation Scheme, either peer is bidirectional. One peer allocates even VCIs, and the other odd VCIs. The peer with the lower TDP ID uses even.

NOTE   The definition of unidirectional and bidirectional, as well as the MPLS capability command output, are consistent with industry-standard ATM usage and nomenclature. Nonetheless, they are exactly the reverse of the IETF LDP specification (RFC 3036) definition.

These capabilities are negotiated as part of the LDP initialization message, in the optional parameters, with the ATM session parameters (type 0x0501 and variable length), according to RFC 3036 (LDP specification).

They include merge capability, VC directionality, and one or more ATM label range components.

The label distribution mechanism used in ATM MPLS is Downstream on Demand. It distributes per-interface label space (label-controlled ATM).

The upstream LSR requests a label for a specific destination from the downstream LSR. Label VCs are built from headend platform to tailend platform.

It is important to note that "on demand" is not traffic-related. "On-demand" means "whenever the LSR learns a destination." It could be called "upstream-controlled" (as opposed to frame-based MPLS, which is downstream-controlled).

Figure 6-6 shows an example of the distribution for the destination D learned by R1.

Figure 6-6 *Downstream on Demand Label Allocation*

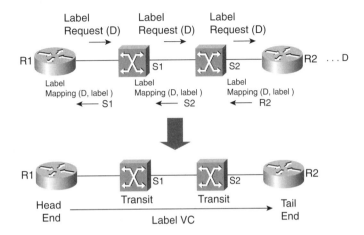

Figure 6-6 shows the headend and tailend platforms, as well as transit platforms. In summary, headend LVCs originate at this router, tailend LVCs terminate at this platform, and transit LVCs pass through this switch.

NOTE	We can use the global configuration command **mpls ldp atm control-mode** to change between **independent** (answering requests for label mappings immediately, without waiting for a label mapping from the next hop) and **ordered** (transmitting a label mapping only for a Forwarding Equivalence Class (FEC) for which it has a label mapping for the FEC next hop) Downstream on Demand LVC setup. Please refer to RFC 3036 for more details.

The bindings—or, more precisely, the LIB (Label Information Base)—can be displayed using three commands. First (see Example 6-29) is **show mpls ldp bindings**.

Example 6-29 *Showing the LDP Bindings*

```
P1_LSC_c7204#show mpls ldp bindings
  172.27.1.1/32, rev 2
      local binding:  label: imp-null
  172.27.1.128/32, rev 8
      local binding:  label: 16
  172.27.1.129/32, rev 10
      local binding:  label: 17
P1_LSC_c7204#
```

This command is not useful here because it displays only the generic (not LC-ATM) bindings. We included it to show that for an attached destination, the label advertised for frame-based is implicit-null.

The command that is important for this scenario is **show mpls atm-ldp bindings**. It displays the ATM label binding database, as shown in Example 6-30.

Example 6-30 *Showing the ATM LDP Bindings*

```
P1_LSC_c7204#show mpls atm-ldp bindings
 Destination: 172.27.1.1/32
    Tailend Switch XTagATM331 9/34 Active -> Terminating Active, VCD=3
    Tailend Switch XTagATM332 10/34 Active -> Terminating Active, VCD=2
 Destination: 172.27.1.128/32
    Headend Switch XTagATM331 (1 hop) 9/33  Active, VCD=4
    Transit XTagATM332 10/36 Active -> XTagATM331 9/35 Active
 Destination: 172.27.1.129/32
    Headend Switch XTagATM332 (1 hop) 10/35  Active, VCD=3
    Transit XTagATM331 9/38 Active -> XTagATM332 10/33 Active
 P1_LSC_c7204
```

The transit bindings are specifically the LSC functionality. In an eLSR, you see headend and tailend LVCs, but not transit. The LSC by default also has eLSR functionality, and that's why you also see headend and tailend bindings. 172.27.1.1/32 is a local destination, for which you see only tailend LVCs. The other two destinations are remote destinations, and that's why you see headend LVCs.

One further command (see Example 6-31) actually combines the output of the previous two: **show mpls ip binding**. The optional parameter **[atm]** displays only atm-ldp bindings.

Example 6-31 *Using the Command* **show mpls ip bindings**

```
P1_LSC_c7204#show mpls ip binding
  172.27.1.1/32
        in label:      imp-null
        in vc label:   9/34       lsr: 172.27.1.128:1    XTagATM331
                       Active     egress (vcd 3)
        in vc label:   10/34      lsr: 172.27.1.129:1    XTagATM332
                       Active     egress (vcd 2)
  172.27.1.128/32
        in label:      16
        out vc label:  9/33       lsr: 172.27.1.128:1    XTagATM331
                       Active     ingress 1 hop (vcd 4)
        in vc label:   10/36      lsr: 172.27.1.129:1    XTagATM332
                       Active     transit
        out vc label:  9/35       lsr: 172.27.1.128:1    XTagATM331
                       Active     transit
  172.27.1.129/32
        in label:      17
```

Example 6-31 *Using the Command* **show mpls ip bindings** *(Continued)*

```
             out vc label: 10/35      lsr: 172.27.1.129:1     XTagATM332
                           Active      ingress 1 hop (vcd 3)
             in vc label:  9/38       lsr: 172.27.1.128:1     XTagATM331
                           Active      transit
             out vc label: 10/33      lsr: 172.27.1.129:1     XTagATM332
                           Active      transit
P1_LSC_c7204
```

NOTE   In non-TTL (Time to Live) decrementing media such as ATM, the hop count is calculated using the Hop Count TLV in the LDP label mapping message.

Finally (see Example 6-32), we can display the Label Forwarding Information Base (LFIB) in the eLSRs or the LSC using the command **show mpls forwarding-table**.

Example 6-32 *Showing the Label Forwarding Information Base*

```
P1_LSC_c7204#show mpls forwarding-table
Local  Outgoing    Prefix         Bytes tag  Outgoing    Next Hop
tag    tag or VC   or Tunnel Id   switched   interface
16     9/33        172.27.1.128/32  0          XT331      point2point
17     10/35       172.27.1.129/32  0          XT332      point2point
P1_LSC_c7204#
```

It is important to note that on the LSC, **show mpls forwarding-table** displays only the outgoing label for headend VCs. If we disable headend VCs or include an access list blocking binding requests to some destinations, the outgoing VC shows "Untagged."

In addition, specific to the LSC, the status of the VSI interfaces and the control interface can be displayed. Refer to Example 6-33.

Example 6-33 *Using the Command* **show controllers vsi**

```
P1_LSC_c7204#
P1_LSC_c7204#show controllers vsi control-interface
Interface:          ATM3/0        Connections:            8
P1_LSC_c7204#show controllers vsi status
Interface Name            IF Status   IFC State  Physical Descriptor
switch control port          n/a       ACTIVE    0.3.1.0
XTagATM331                   up        ACTIVE    0.3.3.1
XTagATM332                   up        ACTIVE    0.3.3.2
P1_LSC_c7204#
```

We can see that the XTagATM interfaces are up and that there are eight connections in the control interface.

Finally, we display the VCs and cross-connects (X-conns) present in the LSC.

The command **show atm vc** (see Example 6-34) displays the ATM view of the VCs in the LSC's interface connected to the BPX switch.

Example 6-34 *Showing the ATM Virtual Circuits*

```
P1_LSC_c7204#show atm vc
                VCD /                                 Peak  Avg/Min Burst
Interface   Name        VPI   VCI  Type   Encaps   SC  Kbps  Kbps    Cells   Sts
3/0         1             0    40  PVC    SNAP     UBR 155000                UP
3/0         2             0    41  PVC    SNAP     UBR 155000                UP
3/0         3             0    42  PVC    SNAP     UBR 155000                UP
3/0         4             0    43  PVC    SNAP     UBR 155000                UP
3/0         5             0    44  PVC    SNAP     UBR 155000                UP
3/0         6             0    45  PVC    SNAP     UBR 155000                UP
3/0         7             0    46  PVC    SNAP     UBR 155000                UP
3/0         8             0    47  PVC    SNAP     UBR 155000                UP
3/0         9             0    48  PVC    SNAP     UBR 155000                UP
3/0         10            0    49  PVC    SNAP     UBR 155000                UP
3/0         11            0    50  PVC    SNAP     UBR 155000                UP
3/0         12            0    51  PVC    SNAP     UBR 155000                UP
3/0         13            0    52  PVC    SNAP     UBR 155000                UP
3/0         14            0    53  PVC    SNAP     UBR 155000                UP
3/0         16            1    33  PVC    XTAGATM  UBR 155000                UP
3/0         17            1    34  TVC    XTAGATM  UBR 155000                UP
3/0         18            1    35  TVC    XTAGATM  UBR 155000                UP
3/0         19            1    36  PVC    XTAGATM  UBR 155000                UP
3/0         20            1    37  TVC    XTAGATM  UBR 155000                UP
3/0         21            1    38  TVC    XTAGATM  UBR 155000                UP
P1_LSC_c7204#
```

First we see a set of VCs with VPIs 0/40 to 0/53. Those are the VSI master-slave VCs for VSI communication. There's one master-slave VC per VSI slave. They are PVCs using AAL5Snap encapsulations, as detailed in Chapter 2, "SCI: Virtual Switch Interface," in the section "Controller Location Options."

At the end of the list are some VCs with XTagATM encapsulations. This means that these are private VCs, as explained in Chapter 3 in the section "Label Switch Controller." All of these use VPI=1 because that's the first VPI configured in the control interface's resource partition. Two of these are PVCs (1/33 and 1/36), and they are the private leg of the control-vcs (for untagged traffic). The rest are type TVC (and the **show** command here displays TVCs for both tag virtual circuits [TVCs] and label virtual circuits [LVCs]).

A more general look at this can be seen using the **show xtagatm vc** command, as shown in Example 6-35. It is more general because it also shows the "Transit" cross-connects not displayed with the previous command. It shows the Controlled Switch view of the LVCs, which is only the cross-connect from one port to another port, as shown in Figure 6-7.

Figure 6-7   *LVCs and Cross-Connects*

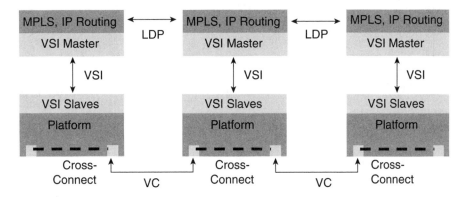

Example 6-35 *Showing the XTagATM Virtual Circuits*

```
P1_LSC_c7204#show xtagatm vc
                              AAL /           Control Interface
Interface    VCD    VPI    VCI Type  Encapsulation   VCD   VPI   VCI   Status
XTagATM331     2      9     32 PVC   AAL5-SNAP        16     1    33   ACTIVE
XTagATM331     3      9     34 TVC   AAL5-MUX         17     1    34   ACTIVE
XTagATM331     4      9     33 TVC   AAL5-MUX         18     1    35   ACTIVE
XTagATM332     1     10     32 PVC   AAL5-SNAP        19     1    36   ACTIVE
XTagATM332     2     10     34 TVC   AAL5-MUX         20     1    37   ACTIVE
XTagATM332     3     10     35 TVC   AAL5-MUX         21     1    38   ACTIVE
P1_LSC_c7204#
```

Here we explicitly see that PVCs 1/33 and 1/36 are the private VCs of the control-vcs 9/32 and 10/32, respectively. We also differentiate between AAL5-SNAP encapsulation for control-vcs and AAL5-MUX encapsulation for LVCs.

A last viewpoint be seen by displaying the XTagATM cross-connects, as shown in Example 6-36.

Example 6-36 *Showing the XTagATM Cross-Connects*

```
P1_LSC_c7204#show xtagatm cross-connect
Phys Desc    VPI/VCI    Type   X-Phys Desc   X-VPI/VCI   State
0.3.1.0      1/38       ->     0.3.3.2       10/35       UP
0.3.1.0      1/36       <->    0.3.3.2       10/32       UP
0.3.1.0      1/35       ->     0.3.3.1       9/33        UP
0.3.1.0      1/33       <->    0.3.3.1       9/32        UP
```

continues

Example 6-36 *Showing the XTagATM Cross-Connects (Continued)*

```
0.3.1.0       1/37      <-      0.3.3.2     10/34    UP
0.3.1.0       1/34      <-      0.3.3.1     9/34     UP
0.3.3.1       9/38      ->      0.3.3.2     10/33    UP    !   Type: -> :
0.3.3.1       9/34      ->      0.3.1.0     1/34     UP    !   Even LVCs in 3.3.1
0.3.3.1       9/35      <-      0.3.3.2     10/36    UP    !    Type <- :
0.3.3.1       9/33      <-      0.3.1.0     1/35     UP    !    Odd LVCs in 3.3.1
0.3.3.1       9/32      <->     0.3.1.0     1/33     UP    ! Control-vc in 3.3.1
0.3.3.2       10/36     ->      0.3.3.1     9/35     UP    !   Type: -> :
0.3.3.2       10/34     ->      0.3.1.0     1/37     UP    !   Even LVCs in 3.3.2
0.3.3.2       10/35     <-      0.3.1.0     1/38     UP    !    Type: <- :
0.3.3.2       10/33     <-      0.3.3.1     9/38     UP    !    Odd LVCs in 3.3.2
0.3.3.2       10/32     <->     0.3.1.0     1/36     UP    ! Control-vc in 3.3.2
P1_LSC_c7204#
```

It is important to note that the control-vcs are bidirectional (<->) and the LVCs are uni-directional (-> or <-). There's an LVC to reach a destination. Also, in a bidirectional allocation scheme, we see that for a given LC-ATM interface facing an LSR or eLSR (such as 3.3.1 or 3.3.2), the VCIs of LVCs going in one direction (->) are even, whereas in the other direction (<-), VCIs are odd.

IGX-8400-Based PoP

The second PoP in our MPLS network is also made of an ATM LSR and two ATM eLSRs, as shown in Figure 6-8.

Figure 6-8 *IGX-8400-Based PoP*

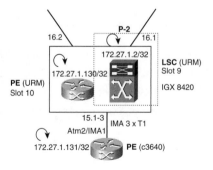

The LSR (P-2) is composed of an IGX-8420-controlled switch (16 slots) and a URM card acting as LSC. The URM-LSC is in slot 9.

One of the eLSRs is another URM card in slot 10 in the IGX-8420, and the other eLSR is a Cisco c3640 router connected via a 3xT1 IMA link.

The IGX-8400-based POP configuration is explored more fully in the following sections. The organization is the same as with the BPX-8600-based POP. The upcoming sections highlight the LSR and eLSR bringup, as well as an IMA-connected eLSR configuration.

ATM LSR Bringup: IGX + LSC

The first step is to put together our LSR (P-2). As I mentioned, we will be using a URM card as a cocontroller card LSC.

If we were to use an external router acting as LSC, only one preliminary step would be performed, and the rest would be exactly the same as with the URM: We would need to up the physical line to the external router using **upln** from the IGX and **no shutdown** from the router. The URM internal ATM interface cannot be shut down.

We will start by adding (**addport**) and upping (**upport**) the URM port. See Example 6-37.

Example 6-37 *Adding and Upping a Port*

```
P2-i8420         TN    Cisco          IGX 8420  9.3.30    Nov. 23 2001 17:49 GMT

Port:        9.1     [ACTIVE  ]
Interface:           INTERNAL         CAC Override:        Enabled
Type:                UNI              %Util Use:           Disabled
Speed:               353208 (cps)     GW LCNs:             200
SIG Queue Depth:     640              Reserved BW:         0 (cps)
Alloc Bandwidth:     353208 (cps)     VC Shaping:          Disabled
Protocol:            NONE




Last Command: upport 9.1

14 UBUs allocated to slot 9. Allocation can be modified using cnfbusbw
Next Command:
```

We will also create a resource partition in the URM port (for VSI to control) using partition ID = 1. Refer to Example 6-38.

Example 6-38 *Creating a VSI Resource Partition*

```
P2-i8420        TN    Cisco          IGX 8420   9.3.30    Nov. 23 2001 17:50 GMT

Line : 9.1
Maximum PVC LCNS: 0              Maximum PVC Bandwidth: 0



                 State   MinLCN   MaxLCN   StartVPI   EndVPI   MinBW     MaxBW
Partition 1:     E       256      512      100        110      352898    352898
Partition 2:     D
Partition 3:     D




Last Command: cnfrsrc 9.1 0 0 y 1 e 256 512 100 110 352898 352898

Cnfrsrc successful.
Next Command:
```

Note that we did not use all the bandwidth for the VSI resource partition. We set aside 310 cells per second (CPS). This was done on purpose to reserve some bandwidth for the VSI protocol in the control interface (VSI master-slave VCs). We set aside those 310 CPS using the command **cnfport**, as shown in Example 6-39. The parameter Reserved BW supports a range of 0 CPS to 310 CPS.

Example 6-39 *Configuring the Port Reserved Bandwidth*

```
P2-i8420        TN    Cisco          IGX 8420   9.3.30    Nov. 23 2001 17:51 GMT

Port:        9.1     [ACTIVE  ]
Interface:           INTERNAL         CAC Override:         Enabled
Type:                UNI              %Util Use:            Disabled
Speed:               353208 (cps)     GW LCNs:              0
SIG Queue Depth:     640              Reserved BW:          310 (cps)
Alloc Bandwidth:     0 (cps)          VC Shaping:           Disabled
Protocol:            NONE
```

Example 6-39 *Configuring the Port Reserved Bandwidth (Continued)*

```
Last Command: cnfport 9.1 N N N N 0 310 N


Next Command:
```

Finally, we add the controller using the command **addctrlr**. See Example 6-40.

Example 6-40 *Adding a VSI Controller*

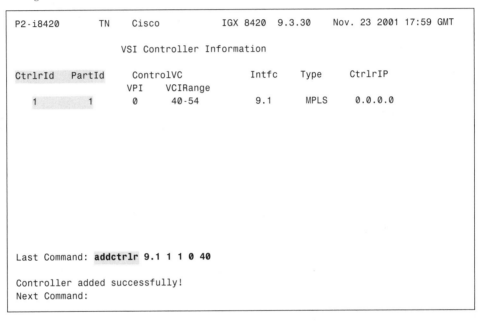

```
P2-i8420        TN    Cisco          IGX 8420  9.3.30    Nov. 23 2001 17:59 GMT


                  VSI Controller Information


CtrlrId   PartId    ControlVC          Intfc   Type    CtrlrIP
                    VPI    VCIRange
   1         1       0      40-54        9.1    MPLS    0.0.0.0
```

```
Last Command: addctrlr 9.1 1 1 0 40

Controller added successfully!
Next Command:
```

Now we can go to the LSC and finish the configuration. See Example 6-41.

Example 6-41 *Enabling VSI in the Label Switch Controller*

```
P2_LSC_URM_9#conf t
Enter configuration commands, one per line.  End with CNTL/Z.
P2_LSC_URM_9(config)#interface ATM 0/0
P2_LSC_URM_9(config-if)#label-control-protocol vsi id 1 base-vc 0 40 slaves 15
P2_LSC_URM_9(config-if)#^Z
P2_LSC_URM_9#
```

As before, we do not have to specify **id** and **base-vc** because we are using defaults. We also specify the **slaves** parameter. The **slaves** parameter indicates the number of VSI sessions that the LSC will use. In the case of an IGX-8430 with 32 slots, we need to specify **slaves 31**, or there will not be a VSI session and VSI master-slave VC for the higher-number slots (in the lower shelf).

NOTE   The number of slaves defaults to 14, so we do not need to change it for a BPX switch. For IGX switches, we specify a number of slaves equal to the number of slots in the IGX minus 1. Slots 1 and 2 (reserved for NPM controller cards) share one VSI VC because they are a redundant pair and only one is active at a time. On an IGX-8420, we configure 15 slaves— session 0 for redundant slots 1 and 2 and sessions 1 to 14 for slots 3 to 16. Currently, NPM cards do not house a VSI slave.

VSI communication is established, as shown in Example 6-42.

Example 6-42 *Showing the VSI Sessions*

```
P2_LSC_URM_9#show controllers vsi session
Interface    Session  VCD   VPI/VCI     Switch/Slave Ids    Session State
ATM0/0       0        2     0/40        0/0                 UNKNOWN
ATM0/0       1        3     0/41        0/0                 UNKNOWN
ATM0/0       2        4     0/42        0/0                 UNKNOWN
ATM0/0       3        5     0/43        0/0                 UNKNOWN
ATM0/0       4        6     0/44        0/0                 UNKNOWN
ATM0/0       5        7     0/45        0/0                 UNKNOWN
ATM0/0       6        8     0/46        0/0                 UNKNOWN
ATM0/0       7        9     0/47        0/9                 ESTABLISHED
ATM0/0       8        10    0/48        0/0                 UNKNOWN
ATM0/0       9        11    0/49        0/0                 UNKNOWN
ATM0/0       10       12    0/50        0/0                 UNKNOWN
ATM0/0       11       13    0/51        0/0                 UNKNOWN
ATM0/0       12       14    0/52        0/0                 UNKNOWN
ATM0/0       13       15    0/53        0/0                 UNKNOWN
ATM0/0       14       16    0/54        0/0                 UNKNOWN
P2_LSC_URM_9#show controllers vsi session 7
Interface:          ATM0/0         Session number:     7
VCD:                9              VPI/VCI:            0/47
Switch type:        IGX            Switch id:          0
Controller id:      1              Slave id:           9
Keepalive timer:    15             Powerup session id: 0x00000002
Cfg/act retry timer: 8/8           Active session id:  0x00000002
Max retries:        10             Ctrl port log intf: 0x00090100 ! LIN for 9.1
Trap window:        50             Max/actual cmd wndw: 10/10
Trap filter:        all            Max checksums:      19
Current VSI version: 2             Min/max VSI version: 2/2
Messages sent:      16             Inter-slave timer:  4.000
Messages received:  9              Messages outstanding: 0
P2_LSC_URM_9#
```

Finally, we will create a loopback interface and assign the IP address 172.27.1.2/32 to it. We will also put together the initial MPLS configuration (enable CEF, enable LDP using the loopback IP as the router-id, and advertise the loopback under OSPF area 0) as shown in Example 6-43.

Example 6-43 *Initial MPLS Configuration*

```
P2_LSC_URM_9#conf t
Enter configuration commands, one per line.  End with CNTL/Z.
P2_LSC_URM_9(config)#interface loopback 0
P2_LSC_URM_9(config-if)#ip address 172.27.1.2 255.255.255.255
P2_LSC_URM_9(config-if)#exit
P2_LSC_URM_9(config)#ip cef
P2_LSC_URM_9(config)#mpls ip
P2_LSC_URM_9(config)#mpls label protocol ldp
P2_LSC_URM_9(config)#mpls ldp router-id loopback 0
P2_LSC_URM_9(config)#router ospf 100
P2_LSC_URM_9(config-router)#network 172.27.1.2 0.0.0.0 area 0
```

To summarize, the BPX-8600 and the IGX-8400 configurations have some subtle differences between them. These differences are outlined in Table 6-1.

Table 6-1   *BPX and IGX Differences in LSR Bringup*

BPX-8600	IGX-8400
The LSC is attached to a trunk (**uptrk**).	The controller is attached to a line/port (**upln**, **addport**, and **upport**).
Protects VSI protocol traffic (master-slave VCs) with **Stats Reserve**.	Protects VSI protocol traffic (master-slave VCs) with **Reserved BW**.
The LSC is added using **addshelf VSI**.	The LSC is added using **addctrlr**.
Uses the default number of slaves in the LSC.	You might need to increase the number of slaves in the LSC.

URM as eLSR Configuration

The eLSR configuration does not differ from the previous eLSR configuration in general terms. For the URM acting as eLSR, we have an internal port that we need to add (**addport**) and then up (**upport**).

We configure a resource partition in URM port 10.1 using the same partition ID as the LSC. At this point (see Example 6-44), the LSC discovers this interface it has to control and all its resources.

Example 6-44 *Showing a VSI Resource Partition from the LSC*

```
P2_LSC_URM_9#show controllers vsi descriptor | begin 10.1

Phys desc: 0.10.1.0
Log intf:  0x000A0100    ! The slave resides in the URM card in slot 10.
Interface: XTagATM101
IF status: n/a            IFC state: ACTIVE
Min VPI:   50             Maximum cell rate:   253208
Max VPI:   52             Available channels: 512
Min VCI:   32             Available cell rate (forward):   253208
Max VCI:   65535          Available cell rate (backward): 253208

P2_LSC_URM_9#
```

From the LSC, we create an extended MPLS ATM interface, enable MPLS on it, configure the IGX port mapping, and unnumber its IP address to the loopback. Refer to Example 6-45.

Example 6-45 *Creating an Extended MPLS ATM Interface*

```
P2_LSC_URM_9#conf t
Enter configuration commands, one per line.  End with CNTL/Z.
P2_LSC_URM_9(config)#interface xTagATM 101
P2_LSC_URM_9(config-if)#ip unnumbered loopback 0
P2_LSC_URM_9(config-if)#mpls ip
P2_LSC_URM_9(config-if)#extended-port ATM 0/0 igx 10.1

P2_LSC_URM_9(config-if)#end
P2_LSC_URM_9#
```

We finish with the eLSR URM configuration in an MPLS subinterface of the internal ATM port adapter (PA ATM0/0), as shown in Example 6-46.

Example 6-46 *Configuring the PE Subinterfaces*

```
PE_URM_10#conf t
Enter configuration commands, one per line.  End with CNTL/Z.
PE_URM_10(config)#ip cef
PE_URM_10(config)#mpls ip
PE_URM_10(config)#mpls label protocol ldp
PE_URM_10(config)#interface loopback 0
PE_URM_10(config-if)#ip address 172.27.1.130 255.255.255.255
PE_URM_10(config-if)#exit
PE_URM_10(config)#mpls ldp router-id loopback 0
PE_URM_10(config)#interface atM 0/0.1 mpls
```

Example 6-46 *Configuring the PE Subinterfaces (Continued)*

```
PE_URM_10(config-subif)#ip unnumbered loopback 0
PE_URM_10(config-subif)#mpls ip
PE_URM_10(config-subif)#mpls atm vpi 50-52
PE_URM_10(config-subif)#exit
PE_URM_10(config)#router ospf 1
PE_URM_10(config-router)#network 172.27.1.130 0.0.0.0 area 0
PE_URM_10(config-router)#^Z
PE_URM_10#
```

It is important to note that we use the subinterface-level command **mpls atm vpi** to instruct the eLSR to use the VPI range configured in the resource partition, 50 to 52. As you know, the LSC learns those values from the VSI slave.

IMA-Connected eLSR Configuration

The configuration of the external eLSR that is connected via a 3xT1 IMA link is in essence the same as with the URM. The first difference, as shown in Example 6-47, is that we need to up the IMA lines from the IGX first.

Example 6-47 *Upping IMA Lines*

```
P2-i8420        TN    Cisco          IGX 8420   9.3.30    Nov. 23 2001 18:39 GMT

Line       Type   Current Line Alarm Status
15.1(3)    T1/71  Clear - OK


















Last Command: upln 15.1-3

256 LCNs allocated. Use 'cnfrsrc' to configure LCNs
Next Command:
```

NOTE   The 3 x T1 line has a usable bandwidth of 71 DS0s. This is because 1 DS0 is being used by the IMA protocol. In other words, 3 times 24 DS0s minus 1 DS0 equals 71 DS0s.

The LSR configuration after this is the same. We **upport** 15.1, we configure resources for the use of VSI (using VPI range 5 to 10), and we create the Extended MPLS ATM interface in the URM LSC.

Likewise, from the c3640 configuration, the only difference is that we need to create the IMA group first. See Example 6-48.

Example 6-48 *Configuring an IMA Interface*

```
PE_3640#conf t
Enter configuration commands, one per line.  End with CNTL/Z.
PE_3640(config)#interface ATM 2/0
PE_3640(config-if)#ima-group 1
PE_3640(config-if)#no shut
PE_3640(config-if)#interface ATM 2/1
PE_3640(config-if)#ima-group 1
PE_3640(config-if)#no shut
PE_3640(config-if)#interface ATM 2/2
PE_3640(config-if)#ima-group 1
PE_3640(config-if)#no shut
PE_3640(config-if)#exit
PE_3640(config)#interface ATM 2/IMA1
PE_3640(config-if)#^Z
PE_3640#
```

The rest of the configuration is the same as creating a subinterface of ATM2/IMA1, except that we need to specify the VPI range in the eLSR with the interface-level command **mpls atm vpi**. Refer to Example 6-49.

Example 6-49 *Specifying the LC-ATM VPI Range*

```
PE_3640(config)#interface ATM 2/IMA1.1
PE_3640(config-subif)#mpls atm vpi 5-10
```

Interconnecting the Two PoPs

This section shows how to configure the P-to-P link to interconnect the two PoPs, using the interfaces and VPI ranges shown in Figure 6-9.

Figure 6-9   *Connecting Two PoPs*

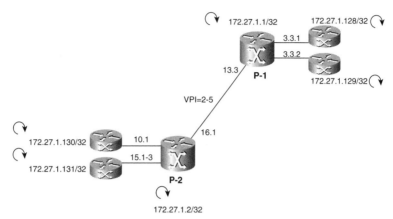

Figure 6-9 also shows the MPLS view of the network. The LSC-switch connection (the control interface) is invisible (as if it were an internal bus). In general, all the ATM VC and VP switching functions are invisible to the MPLS control plane.

The task is straightforward, and it is the same in both ends, so we will show only the IGX end.

In a nutshell, we need to up a trunk, configure resources, and assign those resources to the LSC. All these tasks take place in the controlled switch.

In the LSC, the configuration is the same as when configuring the LC-ATM interface toward an eLSR: create an XTagATM interface, link it to the port in the switch, and enable MPLS on it with an unnumbered IP.

The resource partition in the IGX is as shown in Example 6-50.

Example 6-50 *Creating a VSI Resource Partition*

```
P2-i8420        TN    Cisco            IGX 8420  9.3.30    Nov. 23 2001 20:44 GMT

Trunk : 16.1
Maximum PVC LCNS: 256             Maximum PVC Bandwidth: 41000
                                  (Statistical Reserve: 5000)


                State   MinLCN   MaxLCN   StartVPI   EndVPI   MinBW    MaxBW
   Partition 1:   E      512     1000      2          5       50000    50000
   Partition 2:   D
   Partition 3:   D




```

continues

Example 6-50 *Creating a VSI Resource Partition (Continued)*

```
Last Command: cnfrsrc 16.1 256 y 1 e 512 1000 2 5 50000 50000

Cnfrsrc successful.
Next Command:
```

Example 6-51 shows the LSC configuration in the URM LSC.

Example 6-51 *Label Switch Controller Configuration*

```
P2_LSC_URM_9#conf t
Enter configuration commands, one per line.  End with CNTL/Z.
P2_LSC_URM_9(config)#interface xTagATM 161
P2_LSC_URM_9(config-if)#ip unnumbered loopback 0
P2_LSC_URM_9(config-if)#mpls ip
P2_LSC_URM_9(config-if)#extended-port ATM 0/0 igx 16.1

P2_LSC_URM_9(config-if)#^Z
P2_LSC_URM_9#
```

Before we move on, I want to make an important point. The categorization of links in trunks and lines/ports belongs to the AutoRoute (AR) world. For a VSI master, an interface is a set of resources under its control. The VSI master, LSC in this case, does not know if the resource partition belongs to a trunk or a line/port in a BPX- or IGX-controlled switch. For the MPLS control plane, it makes no difference. So the configuration rule of thumb is to configure the link as a trunk only if AR will be enabled on it. Otherwise, configure it as a line. The special case is VP-Tunnel interfaces. They need to be configured as virtual trunks in the controlled switch.

Example 6-52 introduces an extremely helpful **show** command to see that everything is working: **show mpls atm-ldp summary**.

Example 6-52 *Showing the ATM MPLS LSR Summary*

```
P1_LSC_c7204#show mpls atm-ldp summary
Total number of destinations: 6

ATM label bindings summary
      interface  total  active  local  remote  Bwait  Rwait  IFwait
      XTagATM331     10      10      5       5      0      0       0
      XTagATM332     10      10      5       5      0      0       0
      XTagATM133     18      18      9       9      0      0       0
P1_LSC_c7204#
```

Example 6-52 *Showing the ATM MPLS LSR Summary (Continued)*

```
P2_LSC_URM_9#show mpls atm-ldp summary
Total number of destinations: 6

ATM label bindings summary
        interface   total  active   local  remote  Bwait  Rwait  IFwait
        XTagATM101      10      10       5       5      0      0       0
        XTagATM151      10      10       5       5      0      0       0
        XTagATM161      18      18       9       9      0      0       0
P2_LSC_URM_9#
```

We want the total number of bindings to be the same as the active number of bindings. In other words, no label bindings should be in a wait state (Bind Wait, Resource Wait, or Interface Wait). Table 6-2 describes these states.

Table 6-2 *Bindings in Wait State*

State	Description
Bwait	The number of bindings that are waiting for a label assignment from the neighbor LSR
Rwait	The number of bindings that are waiting for resources (VPI/VCI space) to be available on the downstream device
IFwait	The number of bindings that are waiting for learned labels to be installed for switching use

If there's any destination with binding in **bindwait**, it can be pinpointed with the command **show mpls atm-ldp bindwait**.

We can calculate the number of LVCs in the interfaces shown in the **show mpls atm-ldp summary** command output. Consider the interface XTagATM 133 in the P1 LSR. That interface divides the network into two sections, each with one LSR and two eLSRs. Those LSRs have eLSR functionality (they originate headend and terminate tailend LVCs), so they can be considered eLSRs for this calculation. Each eLSR on each side of the network sets up headend LVCs to all destinations on the other side. So the number of LVCs through a link that divides the network can be calculated like this:

$$LVCs = c * (ni * d2 + n2 * di)$$

where:

ni equals the number of eLSRs on side i

di equals the number of destinations on side i

c equals the number of Classes of Service

In our MPLS network at this time, each eLSR is advertising only one destination (its loopback IP address). So from the headend functionality, the number of LVCs equals the number of eLSRs on one side times the number of eLSRs on the other side. But each eLSR also terminates tailend LVCs from each eLSR on the other side, duplicating the number of LVCs. In summary, making the replacements n1 = d1, n2 = d2, and c = 1 (because there is only one Class of Service), the number of LVCs traversing the link is

LVCs = 2 * n1 * n2

where:

n1 equals the number of eLSRs on one side

n2 equals the number of eLSRs on the other side

Figure 6-10 shows how interfaces XTagATM133 and XTagATM331 divide the network.

Figure 6-10 *LC-ATM Dividing the Network*

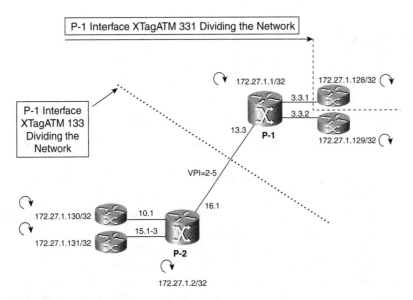

XTagATM133 has three eLSRs on each side, so the number of LVCs is 2 * 3 * 3 = 18.

XTagATM331 has one eLSR on one side and five eLSRs on the other side, so the number of LVCs is 2 * 1 * 5 = 10.

This line of reasoning can be generalized to quantifying and dimensioning the maximum number of LVCs in arbitrary networks, given that the worst case for a link with regards to the number of LVCs occurs when the network is divided in two.

MGX-8850-Based PoP

We will now configure the third PoP. In this case, the controlled switch is a PXM-45-based MGX-8850. An RPM-PR card in slot 9 acts as LSC. Another RPM-PR card functions as eLSR. There is also an external eLSR using a Cisco 7507 router. This topology is shown in Figure 6-11.

The upcoming sections delve into the MGX-8850-based POP configuration, following the same sequence as with the BPX-8600- and IGX-8400-based POPs.

Figure 6-11 *MGX-8850-Based PoP*

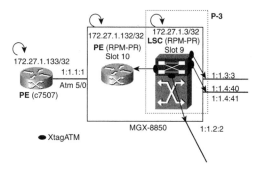

ATM LSR Bringup: PXM-45-Based MGX-8850 + LSC

On an MGX-8850, the different controllers have reserved controller IDs. Controller ID 1 is assigned to PAR, 2 to PNNI, 3 to LSC, and 4 through 20 to different LSC instances.

We can start this configuration by adding the controller in the PXM-45 controller card. See Example 6-53.

Example 6-53 *Adding the LSC from the Switch*

```
P3-m8850.8.PXM.a > addcontroller 3 i 3 9 LSC_RPM_PR_9

P3-m8850.8.PXM.a > dspcontroller
P3-m8850                        System Rev: 02.01   Nov. 24, 2001 12:33:31 GMT
MGX8850                                             Node Alarm: NONE
Number of Controllers:      1
Controller Name:            LSC_RPM_PR_9
Controller Id:              3
Controller Location:        Internal
Controller Type:            LSC
Controller Logical Slot:    9
Controller Bay Number:      0
Controller Line Number:     0
Controller VPI:             0
```

continues

Example 6-53 *Adding the LSC from the Switch (Continued)*

```
Controller VCI:              0
Controller In Alarm:         NO
Controller Error:



P3-m8850.8.PXM.a >
```

We create an MPLS resource partition in the switch interface in the RPM-PR acting as LSC
(see Example 6-54). The switch interface is the internal ATM interface facing the backplane. We will partition ingress and egress bandwidth resources and VPI and VCI ranges.
After the resource partition is configured for VSI to use, we will configure the RPM-PR as
LSC by enabling VSI as the label-control protocol.

Example 6-54 *Creating an MPLS Resource Partition in the RPM*

```
P3_LSC_m8850_RPM-PR_9(config)#interface switch 1
P3_LSC_m8850_RPM-PR_9(config-if)#switch partition vcc 2 3      ! <------------------+
P3_LSC_m8850_RPM-PR_9(config-if-swpart)#ingress-percentage-bandwidth 100 100 !   |
P3_LSC_m8850_RPM-PR_9(config-if-swpart)#egress-percentage-bandwidth 100 100  !   |
P3_LSC_m8850_RPM-PR_9(config-if-swpart)#vpi 0 0                              !   |
P3_LSC_m8850_RPM-PR_9(config-if-swpart)#vci 1 3808                           !   |
P3_LSC_m8850_RPM-PR_9(config-if-swpart)#exit                                 !   |
P3_LSC_m8850_RPM-PR_9(config-if)#                                            !   |
P3_LSC_m8850_RPM-PR_9(config-if)#label-control-protocol vsi id 3   ! <-----------+
```

We are using partition ID = 2 and controller ID = 3. This specifies the partition-to-controller
mapping. Controller ID 3 (as specified in the **addcontroller** command) manages resource
partitions with ID = 2.

Even though we apply the configuration command in the RPM-PR card, the resource partition resides in the PXM-45 controller card because that's the location of the VSI proxy
slave.

The controller ID needs to be 3, as specified in the resource partition. We can verify that the
LSC discovers the slaves using **show controller vsi session**, as shown in Example 6-55.

Example 6-55 *Checking the VSI Sessions*

```
P3_LSC_m8850_RPM-PR_9#show controller vsi session
Interface  Session  VCD  VPI/VCI   Switch/Slave Ids  Session State
Switch1    0        3    0/65507   0/1               ESTABLISHED
Switch1    1        4    0/65508   0/2               ESTABLISHED
Switch1    2        5    0/65509   0/0               UNKNOWN
Switch1    3        6    0/65510   0/0               UNKNOWN
Switch1    4        7    0/65511   0/0               UNKNOWN
Switch1    5        8    0/65512   0/0               UNKNOWN
Switch1    6        9    0/65513   0/7               ESTABLISHED ! Logical Slot 7
```

Example 6-55 *Checking the VSI Sessions (Continued)*

```
Switch1    7         10    0/65514    0/0          UNKNOWN
Switch1    8         11    0/65515    0/0          UNKNOWN
Switch1    9         12    0/65516    0/11         ESTABLISHED
Switch1    10        13    0/65517    0/12         ESTABLISHED
Switch1    11        14    0/65518    0/0          UNKNOWN
Switch1    12        15    0/65519    0/0          UNKNOWN
P3_LSC_m8850_RPM-PR_9#
P3_LSC_m8850_RPM-PR_9#show controller vsi session 6
Interface:           Switch1    Session number:       6
VCD:                 9          VPI/VCI:              0/65513
Switch type:         MGX        Switch id:            0
Controller id:       3          Slave id:             7
Keepalive timer:     15         Powerup session id:   0x3BFD9DED
Cfg/act retry timer: 2/4        Active session id:    0x3BFD9DED
Max retries:         4          Ctrl port log intf:   0x01074B01    ! Slave in PXM-45
Trap window:         50         Max/actual cmd wndw:  21/21
Trap filter:         all        Max checksums:        83
Current VSI version: 2          Min/max VSI version:  2/2
Messages sent:       14         Inter-slave timer:    2.000
Messages received:   14         Messages outstanding: 0
P3_LSC_m8850_RPM-PR_9#
```

From the output of the command **show controller vsi session**, you can see that logical slot 7, which spans redundant physical slots 7 and 8, uses session 6.

As mentioned in Chapter 3 in the section "MGX-8850 and MGX-8950 with an RPM-PR-Based LSC," the LSC implementation in an MGX-8850 switch has a fixed-base VC equal to 0/65507. There are also two well-known VCs (0/65526 and 0/65528) for IPC communication.

We now enable CEF and configure a loopback interface. Refer to Example 6-56.

Example 6-56 *Initial MPLS Configuration*

```
P3_LSC_m8850_RPM-PR_9#conf t
Enter configuration commands, one per line.  End with CNTL/Z.
P3_LSC_m8850_RPM-PR_9(config)#ip cef
P3_LSC_m8850_RPM-PR_9(config)#interface loopback 0
P3_LSC_m8850_RPM-PR_9(config-if)#ip address 172.27.1.3 255.255.255.0
P3_LSC_m8850_RPM-PR_9(config-if)#exit
P3_LSC_m8850_RPM-PR_9(config)#mpls ip
P3_LSC_m8850_RPM-PR_9(config)#mpls ldp router-id loopback 0
P3_LSC_m8850_RPM-PR_9(config)#mpls label protocol ldp
P3_LSC_m8850_RPM-PR_9(config)#end
P3_LSC_m8850_RPM-PR_9#
```

We finish the LSC configuration by setting up the initial MPLS configuration: LDP router-id, label protocol, and IGP routing.

RPM-PR as an eLSR Configuration

To start the RPM-PR configuration as an eLSR, we perform the initial configuration and create a resource partition of the switch interface using partition ID 2 and controller ID 3 so that it is managed from the LSC. Once more, the partition command is executed in the RPM-PR (even for the PE) but is kept in the PXM-45 proxy slave. These steps are shown in Example 6-57.

Example 6-57 *RPM eLSR Configuration*

```
PE_m8850_RPM-PR_10#
PE_m8850_RPM-PR_10#conf t
Enter configuration commands, one per line.  End with CNTL/Z.
PE_m8850_RPM-PR_10(config)#ip cef
PE_m8850_RPM-PR_10(config)#interface loopback 0
PE_m8850_RPM-PR_10(config-if)#ip address 172.27.1.132 255.255.255.255
PE_m8850_RPM-PR_10(config-if)#exit
PE_m8850_RPM-PR_10(config)#mpls label protocol ldp
PE_m8850_RPM-PR_10(config)#mpls ldp router-id loopback 0
PE_m8850_RPM-PR_10(config)#interface switch 1
PE_m8850_RPM-PR_10(config-if)#switch partition vcc 2 3
PE_m8850_RPM-PR_10(config-if-swpart)#ingress-percentage-bandwidth 100 100
PE_m8850_RPM-PR_10(config-if-swpart)#egress-percentage-bandwidth 100 100
PE_m8850_RPM-PR_10(config-if-swpart)#vpi 0 0
PE_m8850_RPM-PR_10(config-if-swpart)#vci 1 3808
PE_m8850_RPM-PR_10(config-if-swpart)#^Z
PE_m8850_RPM-PR_10#show switch partitions
Part Ctrlr  Guar    Max     Guar    Max
Id   Id     Ing%Bw  Ing%Bw  Egr%Bw  Egr%Bw  minVpi maxVpi minVci maxVci MaxCons
2    3      100     100     100     100     0      0      1      3808   3808

PE_m8850_RPM-PR_10#
```

Finally, on the eLSR, we configure an MPLS subinterface and routing. Refer to Example 6-58.

Example 6-58 *Configuring the PE MPLS Subinterface*

```
PE_m8850_RPM-PR_10#conf t
Enter configuration commands, one per line.  End with CNTL/Z.
PE_m8850_RPM-PR_10(config)#interface switch 1.27 mpls
PE_m8850_RPM-PR_10(config-subif)#ip unnumbered loopback 0
PE_m8850_RPM-PR_10(config-subif)#mpls ip
PE_m8850_RPM-PR_10(config-subif)#exit
PE_m8850_RPM-PR_10(config)#router ospf 1
PE_m8850_RPM-PR_10(config-router)#network 172.27.1.132 0.0.0.0 area 0
PE_m8850_RPM-PR_10(config-router)#^Z
PE_m8850_RPM-PR_10#
```

We now go to the LSC RPM-PR in slot 9. We can see in Example 6-59 that the PE interface is now visible to VSI (via a VSI interface trap from the PXM-45).

Example 6-59 *Showing the VSI Interfaces*

```
P3_LSC_m8850_RPM-PR_9#show controllers vsi descriptor

Phys desc: 9.1
Log intf:  0x01074B01   ! The slave resides in the PXM-45 card (proxy slave)
Interface: switch control port
IF status: n/a                 IFC state: ACTIVE
Min VPI:   0                   Maximum cell rate:  353208
Max VPI:   0                   Available channels: 3808
Min VCI:   1                   Available cell rate (forward):  353208
Max VCI:   3808                Available cell rate (backward): 353208

Phys desc: 10.1
Log intf:  0x01075301   ! The slave resides in the PXM-45 card (proxy slave)
Interface: n/a
IF status: n/a                 IFC state: ACTIVE
Min VPI:   0                   Maximum cell rate:  353208
Max VPI:   0                   Available channels: 3808
Min VCI:   1                   Available cell rate (forward):  353208
Max VCI:   3808                Available cell rate (backward): 353208

P3_LSC_m8850_RPM-PR_9#
```

As shown in Example 6-60, we need to create an XTagATM interface, map it to the controlled interface, enable MPLS on it, and advertise the loopback IP address in area 0 under an OSPF process.

Example 6-60 *Creating an Extended MPLS ATM Interface*

```
P3_LSC_m8850_RPM-PR_9#conf t
Enter configuration commands, one per line.  End with CNTL/Z.
P3_LSC_m8850_RPM-PR_9(config)#interface xTagATM 101
P3_LSC_m8850_RPM-PR_9(config-if)#extended-port switch 1 descriptor "10.1"

P3_LSC_m8850_RPM-PR_9(config-if)#ip unnumbered loopback 0
P3_LSC_m8850_RPM-PR_9(config-if)#mpls ip
P3_LSC_m8850_RPM-PR_9(config-if)#^Z
P3_LSC_m8850_RPM-PR_9#
```

In this example, we use the **extended-port** command with the parameter **descriptor**. As described earlier, this descriptor is an ASCII value that can be displayed using **show controllers vsi descriptor**. In the following section, we will use another method of identifying the controlled switch interface to be used with the **extended-port** command.

7507 as an eLSR Configuration

To configure an external eLSR, we start by configuring the LC-ATM in the controlled switch connecting to the eLSR. In an AXSM card, first we configure the card SCT, choosing SCT 5 (for PNNI plus MPLS support). Refer to Example 6-61. In the vast majority of cases (always in practical terms), card SCT 5 (without policing for ATMF service category) should be used.

Example 6-61 *Configuring the AXSM Card SCT*

```
P3-m8850.1.AXSM.a > dspcd
                        Front Card          Upper Card          Lower Card
                        ----------          ----------          -----------

Card Type:              AXSM-16-T3E3/B      SMB-8-T3            - - -
State:                  Active              Present             Absent
Serial Number:          SAG05274SE8         SBK045200R0         - - -
Boot FW Rev:            2.1(70.107)P2       - - -               - - -
SW Rev:                 2.1(70.107)P2       - - -               - - -
800-level Rev:          A0                  A0                  - - -
Orderable Part#:        800-07911-05        800-05029-02        - - -
PCA Part#:              73-5045-4           73-4223-2           - - -
CLEI Code:              BAA5YL6CAA          BA7IW8HAAA          - - -
Reset Reason:           Power ON Reset

SCT File Configured Version: 1

SCT File Operational Version: 1

Card SCT Id: 0 !DefaultSCT used!

#Lines #Ports #Partitions   #SPVC   #SPVP   #SVC
------ ------ -----------   -----   -----   -----

Type <CR> to continue, Q<CR> to stop:
     0      0            0       0       0       0

Port Group[1]:
#Chans supported:65280  Lines:1.1 1.2 1.3 1.4 1.5 1.6 1.7 1.8
Port Group[2]:
#Chans supported:65280  Lines:2.1 2.2 2.3 2.4 2.5 2.6 2.7 2.8
P3-m8850.1.AXSM.a > cnfcdsct 5

P3-m8850.1.AXSM.a >
```

We will now up the physical line to the eLSR, configure it to use PLCP to match the default in the router, create a logical port, and create a resource partition for MPLS. We need to use partition ID 2 and controller ID 3. These steps are shown in Example 6-62.

Example 6-62 *Configuring the AXSM VSI Slave*

```
P3-m8850.1.AXSM.a > upln 1.1

P3-m8850.1.AXSM.a > cnfln -ds3 1.1 -lt 2

P3-m8850.1.AXSM.a > addport 1 1.1 96000 96000 4 1

P3-m8850.1.AXSM.a > dspports
ifNum Line Admin Oper. Guaranteed Maximum      Port SCT Id    ifType  VPI
           State State Rate        Rate                               (VNNI only)
----- ---- ----- ----- ---------- --------- ----------------- ------ ----------
    1  1.1   Up  Down       96000     96000    4                 UNI    0

P3-m8850.1.AXSM.a > addpart 1 2 3 1000000 1000000 1000000 1000000 0 255 1 65535
   512 1024

P3-m8850.1.AXSM.a > dspparts
if   part Ctlr egr      egr      ingr     ingr     min max   min   max   min   max
Num  ID   ID   GuarBw   MaxBw    GuarBw   MaxBw    vpi vpi   vci   vci   conn  conn
          (.0001%)(.0001%)(.0001%)(.0001%)
-------------------------------------------------------------------------------
 1    2     3 1000000 1000000 1000000 1000000    0  255     1 65535    512  1024

P3-m8850.1.AXSM.a >
```

Port SCT 4 is specified with the **addport** command. SCT 4 supports policing for the ATMF service category. It was chosen because policing for SVCs and SPVCs is applied in this port facing Customer Premises Equipment (CPE).

A VSI Intf Trap is sent to the LSC. The partition is now visible from the LSC, as shown in Example 6-63.

Example 6-63 *Checking the VSI Resource Partition from the LSC*

```
P3_LSC_m8850_RPM-PR_9#show controllers vsi descriptor

Phys desc: 1:1.1:1
Log intf:  0x01011801    ! The slave resides in the AXSM card in slot 1.
Interface: n/a
IF status: n/a                IFC state: FAILED_EXT
Min VPI:   0                  Maximum cell rate:  96000
Max VPI:   255                Available channels: 1024
Min VCI:   1                  Available cell rate (forward):  96000
Max VCI:   65535              Available cell rate (backward): 96000
```

The LSC already has the initial MPLS configuration, so we only need to create an Extended MPLS interface and bind it to the partition. See Example 6-64.

Example 6-64 *Creating and Binding an LC-ATM from the LSC*

```
P3_LSC_m8850_RPM-PR_9#conf t
Enter configuration commands, one per line.  End with CNTL/Z.
P3_LSC_m8850_RPM-PR_9(config)#int xtag 1111
P3_LSC_m8850_RPM-PR_9(config-if)#ip unnumbered loopback 0
P3_LSC_m8850_RPM-PR_9(config-if)#mpls ip
P3_LSC_m8850_RPM-PR_9(config-if)#extended-port switch 1 descriptor "1:1.1:1"

P3_LSC_m8850_RPM-PR_9(config-if)#^Z
P3_LSC_m8850_RPM-PR_9#
```

The LSR portion is ready. We perform the initial configuration in the eLSR. See Example 6-65.

Example 6-65 *Performing the Initial MPLS Configuration in the eLSR*

```
PE_7507#conf t
Enter configuration commands, one per line.  End with CNTL/Z.
PE_7507(config)#ip cef distributed
PE_7507(config)#interface loopback 0
PE_7507(config-if)#ip address 172.27.1.133 255.255.255.255
PE_7507(config-if)#exit
PE_7507(config)#mpls ip
PE_7507(config)#mpls label protocol ldp
PE_7507(config)#mpls ldp router-id loopback 0
```

Next we configure an interface for MPLS, as shown in Example 6-66. In this step, we also set up the IGP.

Example 6-66 *Configuring an MPLS-Enabled Subinterface*

```
PE_7507(config)#interface ATM 5/0
PE_7507(config-if)#no shutdown
PE_7507(config-if)#exit
PE_7507(config)#interface ATM 5/0.27 mpls
PE_7507(config-subif)#ip unnumbered loopback 0
PE_7507(config-subif)#mpls ip
PE_7507(config-subif)#exit
PE_7507(config)#router ospf 1
PE_7507(config-router)#network 172.27.1.133 0.0.0.0 area 0
PE_7507(config-router)#^Z
PE_7507#
```

Displaying Cross-Connect Details

We can see the cross-connects in the AXSM card using the command **dspvsicons**. Refer to Example 6-67.

Example 6-67 *Using the Command* ***dspvsicons***

```
m8850-7a.1.AXSM.a > dspvsicons
  LCN    Type     lLin    lVpi  lVci    rLin    rVpi  rVci  cksmVal  pCref
  =========================================================================
  00434 s/svc    01011801 0000 000032 01074b01 0000 000052 00000000 0000
  00435 p/svc    01011801 0001 000038 01074b01 0000 000053 00000000 0000
  00436 s/svc    01011801 0001 000033 01074b01 0000 000054 00000000 0000
  00437 p/svc    01011801 0001 000048 01075301 0000 000049 00000000 0000
  00438 s/svc    01011801 0001 000035 01075301 0000 000102 00000000 0000

m8850-7a.1.AXSM.a >
```

We can also see the details of the cross-connects from the LSC, as shown in Example 6-68.

Example 6-68 *Checking the LSR Cross-Connects*

```
P3_LSC_m8850_RPM-PR_9#
P3_LSC_m8850_RPM-PR_9#show xtagatm cross-connect descriptor "1:1.1:1" 1 36
Phys desc:    1:1.1:1
Interface:    XTagATM1111
Intf type:    extended tag ATM
VPI/VCI:      1/36
X-Phys desc: 10.1
X-Interface: XTagATM101
X-Intf type: extended tag ATM
X-VPI/VCI:    0/35
Conn-state:   UP
Conn-type:    input
Cast-type:    point-to-point
Rx service type:   Tag COS 0
Rx cell rate:      n/a
Rx peak cell rate: 96000
Tx service type:   n/a
Tx cell rate:      n/a
Tx peak cell rate: n/a
P3_LSC_m8850_RPM-PR_9#

P3_LSC_m8850_RPM-PR_9#show xtagatm cross-connect descriptor "1:1.1:1" 0 32
Phys desc:    1:1.1:1
Interface:    XTagATM1111
Intf type:    extended tag ATM
VPI/VCI:      0/32
X-Phys desc: 9.1
X-Interface: n/a
X-Intf type: switch control port
```

continues

Example 6-68 *Checking the LSR Cross-Connects (Continued)*

```
X-VPI/VCI:     0/35
Conn-state:    UP
Conn-type:     input/output
Cast-type:     point-to-point
Rx service type:   Tag COS 7
Rx cell rate:      n/a
Rx peak cell rate: 96000
Tx service type:   Tag COS 7
Tx cell rate:      n/a
Tx peak cell rate: 96000
P3_LSC_m8850_RPM-PR_9#
```

It is important to note two things when comparing the first cross-connect (an LVC) to the second one (a control-vc). First, control-vcs are bidirectional, and LVCs are unidirectional; for LVCs, we only see either the Tx or the Rx directions. Second, LVCs by default use the Tag COS 0 service type, and control-vcs use the Tag COS 7 service type.

Connecting the MGX-Based PoP with the BPX-Based PoP

Finally, we will add an MPLS link between the MGX-8850 and the BPX-8600. We will add a virtual link over a VP to cover all the options.

On the BPX side, we up a virtual trunk (in this case using **uptrk 13.4.1**), configure the trunk for a null stats reserve, and configure resources on the virtual trunk. On the external LSC, we bind these resources to an XTagATM interface. See Example 6-69.

Example 6-69 *Configuring an LC-ATM for a Virtual Trunk*

```
P1_LSC_c7204#conf t
Enter configuration commands, one per line.  End with CNTL/Z.
P1_LSC_c7204(config)#interface xTagATM 1341
P1_LSC_c7204(config-if)#ip unnumbered loopback 0
P1_LSC_c7204(config-if)#mpls ip
P1_LSC_c7204(config-if)#extended-port atM 3/0 bpx 13.4.1

P1_LSC_c7204(config-if)#^Z
P1_LSC_c7204#
```

On the MGX side, we start by configuring the AXSM card. We up a line and up a port. That port is a VNNI port using VPI = 27 (see Example 6-70). It is important to match all line parameters such as framing, scrambling, and so on.

Example 6-70 *Configuring an AXSM Port*

```
P3-m8850.1.AXSM.a > addport 40 1.4 10000 10000 4 3 27

P3-m8850.1.AXSM.a > dspports
ifNum Line Admin Oper. Guaranteed Maximum       Port SCT Id    ifType VPI
            State State Rate       Rate                               (VNNI only)
----- ---- ----- ----- ---------- --------- ------------------ ------ ----------
    1  1.1   Up    Up      96000     96000   4                  UNI    0
   40  1.4   Up    Up      10000     10000   4                  VNNI   27

P3-m8850.1.AXSM.a >
```

We now add a resource partition (see Example 6-71).

Example 6-71 *Adding a VSI Resource Partition in an AXSM Port*

```
P3-m8850.1.AXSM.a > addpart 40 2 3 1000000 1000000 1000000 1000000 27 27 1 65535
   512 1024

P3-m8850.1.AXSM.a > dspparts
if   part Ctlr egr       egr      ingr      ingr     min max  min    max   min   max
Num  ID   ID   GuarBw    MaxBw    GuarBw    MaxBw    vpi vpi  vci    vci   conn  conn
               (.0001%)(.0001%)(.0001%)(.0001%)
------------------------------------------------------------------------------------
 1   2    3 1000000 1000000 1000000 1000000    0  255    1  65535    512  1024
40   2    3 1000000 1000000 1000000 1000000   27   27    1  65535    512  1024

P3-m8850.1.AXSM.a >
```

From the LSC, we now see the partition. Refer to Exampel 6-72.

Example 6-72 *Checking the VSI Resource Partition from the LSC*

```
P3_LSC_m8850_RPM-PR_9(config)#do show controllers vsi descriptor | begin 1.4

Phys desc: 1:1.4:40
Log intf: 0x01011828
Interface: n/a
IF status: n/a                    IFC state: ACTIVE
Min VPI:   27                     Maximum cell rate:  10000
Max VPI:   27                     Available channels: 1024
Min VCI:   1                      Available cell rate (forward):  10000
Max VCI:   65535                  Available cell rate (backward): 10000
```

Next we create the XTagATM interface. We use the **vsi** option of the **extended-port** command in this case, as shown in Example 6-73. The parameter of the **vsi** option is the 32-bit

LIN in hexadecimal form with a leading 0x. It uniquely identifies the Switch Logical Interface that the XTagATM extends.

Example 6-73 *Configuring an LC-ATM Using the LIN*

```
P3_LSC_m8850_RPM-PR_9(config)#interface xtagatm 1144
P3_LSC_m8850_RPM-PR_9(config-if)#mpls ip
P3_LSC_m8850_RPM-PR_9(config-if)#ip unnumbered loopback 0
P3_LSC_m8850_RPM-PR_9(config-if)#extended-port switch 1 vsi 0x01011828

P3_LSC_m8850_RPM-PR_9(config-if)#^Z
P3_LSC_m8850_RPM-PR_9#
```

More show Commands

It's time to reintroduce one of the most helpful **show** commands in the LSC (see Example 6-74) because it provides a global picture of the LSR state.

Example 6-74 *Using the Command show mpls atm-ldp summary*

```
P3_LSC_m8850_RPM-PR_9#show mpls atm-ldp summary
Total number of destinations: 9

ATM label bindings summary
      interface    total   active   local   remote   Bwait   Rwait   IFwait
   XTagATM101         16       16       8        8       0       0        0
   XTagATM1111        16       16       8        8       0       0        0
   XTagATM1144        36       36      18       18       0       0        0
P3_LSC_m8850_RPM-PR_9#
```

We can also calculate the number of LVCs that traverse each link following the previous reasoning.

XTagATM1144 has three eLSR-behaving routers on one side and six on the other side:

LVCs = 2 * 3 * 6 = 36

XTagATM101 and 1111 have one eLSR on one side and eight eLSR-behaving LSCs and eLSRs:

LVCs = 2 * 1 * 8 = 16

It is also interesting to observe the LDP negotiated capabilities (see Example 6-75) that were discussed in the section "LDP Session, Bindings, and LVC **show** Commands."

Example 6-75 *Showing the LDP ATM Capabilities*

```
P3_LSC_m8850_RPM-PR_9#show mpls atm-ldp capability

               VPI           VCI           Alloc   Odd/Even VC Merge
XTagATM101     Range         Range         Scheme  Scheme   IN   OUT
   Negotiated  [0 - 0]       [33 - 3808]   BIDIR            -    -
   Local       [0 - 0]       [33 - 3808]   BIDIR   EVEN     NO   NO
   Peer        [0 - 0]       [33 - 65518]  UNIDIR  ODD      -    -

               VPI           VCI           Alloc   Odd/Even VC Merge
XTagATM1111    Range         Range         Scheme  Scheme   IN   OUT
   Negotiated  [1 - 1]       [33 - 1018]   BIDIR            -    -
   Local       [0 - 255]     [33 - 65535]  BIDIR   EVEN     NO   NO
   Peer        [1 - 1]       [33 - 1018]   UNIDIR  ODD      -    -

               VPI           VCI           Alloc   Odd/Even VC Merge
XTagATM1144    Range         Range         Scheme  Scheme   IN   OUT
   Negotiated  [27 - 27]     [33 - 65535]  BIDIR            -    -
   Local       [27 - 27]     [33 - 65535]  BIDIR   ODD      NO   NO
   Peer        [27 - 27]     [33 - 65535]  BIDIR   EVEN     -    -
P3_LSC_m8850_RPM-PR_9#
```

The procedure to bring up the remaining P-P links is equivalent to what we already did, so we will not include the captures. At this point, our MPLS network is complete.

Summary of MPLS Configuration Commands

Now that we have configured all the different platforms, we can create a synopsis of all the configuration commands for them. We will follow the "Generic Configuration Model" section and identify all the required steps to complete each task.

Generic LSR Bringup

Table 6-3 enumerates all the required commands to finish each step in the LSR bringup. Refer back to Figure 6-2 to recall each task. The numbers in parentheses mean the following:

(1) In the BPX or IGX switch command-line interface (CLI)
(2) In the RPM-PR CLI
(3) In the PXM-45 CLI
(4) In the LSC CLI

Table 6-3   *Summary of ATM MPLS LSR Configuration*

Task Number	BPX-8650	IGX-8400 URM LSC	IGX-8400 External LSC	MGX-8850
1	(1) **uptrk** **cnfrsrc**	(1) **addport** **upport** **cnfrsrc**	(1) **upln** **addport** **upport** **cnfrsrc**	(2) (config-if)#**switch partition vcc**
2	(1) **addshelf**	(1) **addctrlr**		(3) **addcontroller**
3	(4) (config-if)#**label-control-protocol vsi**			
4	(4) (config)#**ip cef** (config)#**mpls label protocol ldp** (config)#**mpls ldp router-id loopback 0**			

In addition to the commands listed in Table 6-3, we still need to configure the routing portion in Step 4, which depends on the IGP used.

Generic eLSR Bringup

Table 6-4 is a review of the entire configuration command set needed to bring up an ATM MPLS eLSR in the different platforms. Refer to Figure 6-3 to recall each task. The numbers in parentheses mean the following:

(1) In the BPX or IGX switch CLI
(2) In the RPM-PR CLI
(3) In the AXSM CLI
(4) In the LSC CLI
(5) In the eLSR CLI

Table 6-4   *Summary of ATM MPLS eLSR Configuration*

Task Number	BPX-8650	IGX-8400 URM eLSR	IGX-8400 External eLSR	MGX-8850 RPM eLSR	MGX-8850 External eLSR
1	(1) **upln** **addport** **upport** **cnfrsrc**	(1) **addport** **upport** **cnfrsrc**	(1) **upln** **addport** **upport** **cnfrsrc**	(2) (config-if)#**switch partition vcc**	(3) **cnfcdsct** **upln** **addport** **addpart**
2	(4) (config)#**interface xtagatm**				
3	(4) (config-if)#**ip unnumbered loopback 0** (config-if)#**mpls ip** (config-if)#**extended-port**				
4	(5) (config)#**ip cef** (config)#**mpls label protocol ldp** (config)#**mpls ldp router-id loopback 0**				
5	(5) (config)#**interface <atm|switch> X.1 mpls**				
6	(5) (config-subif)#**ip unnumbered loopback 0** (config-subif)#**mpls ip**				
7	(5) (config-subif)#**mpls atm vp-tunnel <VPI> [vci-range<VCIRange>]** (config-subif)#**mpls atm vpi <minVPI>-<maxVPI> [vci-range<VCIRange>]**				
8	(4) and (5) (config-subif)#**mpls atm control-vc <VPI> <VCI>**				

Additionally, a routing protocol needs to be configured in Step 4.

Adding an AutoRoute Control Plane

An AutoRoute control plane can be added to the network. We can connect the BPX-8600 to the IGX-8420 switches with an AutoRoute routing trunk, as well as the BPX-8600 to the MGX-8250 using an AutoRoute feeder trunk.

We start by checking the trunks in the BPX switch, as shown in Example 6-76.

Example 6-76 *Displaying the Trunks in a BPX or IGX Switch*

```
P1-b8620         TN    Cisco          BPX 8620  9.3.30    Nov. 24 2001 09:21 GMT

TRK        Type    Current Line Alarm Status                Other End
 3.1       OC3     Clear - OK                               VSI(VSI)
 3.3.1     OC3     Clear - OK                               VSI only
 3.3.2     OC3     Clear - OK                               VSI only
13.3       T3      Clear - OK                               -
13.4.1     T3      Clear - OK                               VSI only




Last Command: dsptrks


Next Command:
```

You can see that the virtual trunks (VT) are **VSI only**. Virtual trunks can be used by either VSI (in this case, MPLS) or AutoRoute, but not both simultaneously. **VSI only** means that a VSI partition exists in the VT, making it unusable by AutoRoute.

On the other hand, for physical trunks, we have either **VSI(VSI)** or **-**. The first state means that VPI = 1, which is used by AutoRoute, is included in a VSI resource partition; therefore, it cannot be used by AutoRoute. A dash means that the trunk can be added as an AutoRoute trunk.

We'll now add trunk 13.3, which connects the BPX-8600 with the IGX-8420. Refer to Example 6-77.

Example 6-77 *Adding an AutoRoute Trunk*

```
P1-b8620         TN    Cisco          BPX 8620  9.3.30    Nov. 24 2001 09:21 GMT

TRK        Type    Current Line Alarm Status                Other End
 3.1       OC3     Clear - OK                               VSI(VSI)
```

Example 6-77 *Adding an AutoRoute Trunk (Continued)*

```
   3.3.1    OC3    Clear - OK                        VSI only
   3.3.2    OC3    Clear - OK                        VSI only
   13.3     T3     Clear - OK                        P2-i8420/16.1
   13.4.1   T3     Clear - OK                        VSI only




Last Command: addtrk 13.3


Next Command:
```

Finally, we attach the MGX-8250 to the BPX-8600 as an AutoRoute feeder shelf. We use trunk 3.2 in the BPX-8620 and trunk 7.1 in the MGX-8250.

Starting with the PXM-1 configuration, we add the line, up the port (assigning resources to PAR), and configure the line as a trunk. The trunk appears as a PAR interface, as shown in Example 6-78.

Example 6-78 *Configuring PAR in a PXM-1 Card*

```
m8250-7a.1.7.PXM.a > addln -sonet 7.1

m8250-7a.1.7.PXM.a > addport 1 1 100 0 4095 100

m8250-7a.1.7.PXM.a > dspparifs
slot.port   type       status    vpi          vci         txRate    rxRate
-----------------------------------------------------------------------------
    7.1     UNI_IF     FAILED    0 to 4095    0 to 65535   353208    353208
    7.2     UNI_IF     UP        0 to 4095    0 to 65535   353208    353208
    7.33    UNI_IF     UP        0 to  255    0 to 65535   176604    176604
    7.34    UNI_IF     UP        0 to  255    0 to 65535   176604    176604
    7.35    CLK_IF     FAILED    0 to    0    0 to     0        0         0
    9.1     UNI_IF     UP        0 to  255    0 to  3840   353208    353208
   10.1     UNI_IF     UP        0 to  255    0 to  3840   353208    353208

m8250-7a.1.7.PXM.a > cnfifastrk 7.1 ftrk

m8250-7a.1.7.PXM.a >
```

On the BPX side, we up trunk 3.2 with **uptrk** and add the shelf using **addshelf** as an AAL5 feeder shelf. See Example 6-79.

Example 6-79 *Adding an AutoRoute Feeder Shelf*

```
P1-b8620        TN    Cisco          BPX 8620  9.3.30   Nov. 24 2001 09:27 GMT

                  BPX 8620 Interface Shelf Information

Trunk   Name      Type    Part Id   Ctrl Id      Control_VC      Alarm
                                               VPI   VCIRange
 3.1    VSI       VSI       1         1          0    40-54       OK
 3.2    m8250-7a  AAL/5     -         -          -      -         MIN




Last Command: addshelf 3.2 X

Shelf has been added
Next Command:
```

As shown in Example 6-80, the shelf appears in the BPX-8600 display.

Example 6-80 *Checking the AR Trunks*

```
P1-b8620        TN    Cisco          BPX 8620  9.3.30   Nov. 24 2001 09:27 GMT

TRK       Type    Current Line Alarm Status          Other End
 3.1      OC3     Clear - OK                         VSI(VSI)
 3.2      OC3     Clear - OK                         m8250-7a(AAL/5)
 3.3.1    OC3     Clear - OK                         VSI only
 3.3.2    OC3     Clear - OK                         VSI only
13.3      T3      Clear - OK                         P2-i8420/16.1
13.4.1    T3      Clear - OK                         VSI only
```

Example 6-80 *Checking the AR Trunks (Continued)*

```
Last Command: dsptrks


Next Command:
```

It also appears in the MGX-8250 display. See Example 6-81.

Example 6-81 *Displaying the Feeder Shelf Trunk*

```
m8250-7a.1.7.PXM.a > dsptrks
TRK        Current Alarm Status        Other End

7.1        CLEAR                       P1-b8620

m8250-7a.1.7.PXM.a >
```

A Note on LC-ATM Interfaces

We can differentiate between two types of LC-ATM interfaces:

- **Directly connected LC-ATM interfaces**—These have one logical port per physical line.

- **Indirectly connected LC-ATM interfaces**—These are connected by means of a virtual path. They are also called VP-Tunnel interfaces. There can be multiple logical ports per physical line.

As we mentioned, the difference resides in the label space, as well as in the VPI used for the control-vc. Normally, both VPI/VCI fields are used as label space. In a VP-Tunnel interface, only the VCI field of the ATM cell header is used as label space.

We can use the command **show mpls interface detail** in the LSC or in an eLSR to differentiate between the two types of LC-ATM interfaces. See Example 6-82.

Example 6-82 *Directly and Indirectly Connected LC-ATM Interfaces*

```
P1_LSC_c7204#! Indirectly connected LC-ATM interface (VP-Tunnel)
P1_LSC_c7204#show mpls interfaces xTagATM 331 detail
Interface XTagATM331:
        IP labeling enabled (ldp)
        LSP Tunnel labeling not enabled
        Tag Frame Relay Transport tagging not enabled
        BGP tagging not enabled
        Tagging operational
        Optimum Switching Vectors:
          IP to MPLS Turbo Vector
          MPLS Punt Vector
```

continues

Example 6-82 *Directly and Indirectly Connected LC-ATM Interfaces (Continued)*

```
                    Fast Switching Vectors:
                      IP to MPLS Fast Switching Vector
                      MPLS Turbo Vector
                    MTU = 4470
                    ATM labels: Label VPI = 9 (VP Tunnel)
                            Label VCI range = 33 - 65535
                            Control VC = 9/32
P1_LSC_c7204#! Directly connected LC-ATM interface
P1_LSC_c7204#show mpls interfaces xTagATM 133 detail
Interface XTagATM133:
        IP labeling enabled (ldp)
        LSP Tunnel labeling not enabled
        Tag Frame Relay Transport tagging not enabled
        BGP tagging not enabled
        Tagging operational
        Optimum Switching Vectors:
          IP to MPLS Turbo Vector
          MPLS Punt Vector
        Fast Switching Vectors:
          IP to MPLS Fast Switching Vector
          MPLS Turbo Vector
        MTU = 4470
        ATM labels: Label VPI range = 2 - 10
                Label VCI range = 33 - 65535
                Control VC = 0/32
P1_LSC_c7204#
```

In the LSC only, we can also use the command **show controller xtagatm** to see the two LC-ATM interface types. See Example 6-83.

Example 6-83 *Directly and Indirectly Connected LC-ATM Interfaces*

```
P1_LSC_c7204#! Indirectly connected LC-ATM interface (VP_Tunnel)
P1_LSC_c7204#show controllers xTagATM 331
Interface XTagATM331 is up
  Hardware is Tag-Controlled ATM Port (on BPX switch BPX-VSI2)
  Control interface ATM3/0 is up
  Physical descriptor is 0.3.3.1
  Logical interface 0x00030301 (0.3.3.1)
  Oper state ACTIVE, admin state UP
  VPI range 9-9, VCI range 32-65535
  VPI may be translated at end of link
  Tag control VC must be strictly in VPI/VCI range
  Available channels: ingress 241, egress 241
  Maximum cell rate: ingress 2867, egress 2867
  Available cell rate: ingress 2867, egress 2867
  Endpoints in use: ingress 7, egress 7, ingress/egress 1
  Rx cells 182513
    rx cells discarded 0, rx header errors 0
    rx invalid addresses (per card): 364124
```

Example 6-83 *Directly and Indirectly Connected LC-ATM Interfaces (Continued)*

```
      last invalid address 0/0
   Tx cells 281966
      tx cells discarded: 0
 P1_LSC_c7204#! Directly connected LC-ATM interface
 P1_LSC_c7204#show controllers xTagATM 133
 Interface XTagATM133 is up
   Hardware is Tag-Controlled ATM Port (on BPX switch BPX-VSI2)
   Control interface ATM3/0 is up
   Physical descriptor is 0.13.3.0
   Logical interface 0x000D0300 (0.13.3.0)
   Oper state ACTIVE, admin state UP
   VPI range 2-10, VCI range 32-65535
   VPI is not translated at end of link
   Tag control VC need not be strictly in VPI/VCI range
   Available channels: ingress 475, egress 475
   Maximum cell rate: ingress 50000, egress 50000
   Available cell rate: ingress 50000, egress 50000
   Endpoints in use: ingress 18, egress 18, ingress/egress 1
   Rx cells 223934
     rx cells discarded 0, rx header errors 0
     rx invalid addresses (per card): 44
     last invalid address 0/0
   Tx cells 211581
     tx cells discarded: 0
 P1_LSC_c7204#
```

NOTE   Every time we use the command **show controller xtagatm** in the LSC, we are pulling the counters from the VSI slave (in this case, the BXM card). Those counters reside on the VSI slave and are retrieved for display using VSI commands. If the controlled switch is a BPX-8600 or IGX-8400, the counters can also be seen from the switch CLI using **dsptrkstats** or **dspportstats**.

If we use only one VPI in a non-VP-Tunnel LC-ATM interface, the VPI cannot be different at both ends of the link. This is because the VPI will be used as label space, and there could be a disjoint VPI range (null intersection). As a result, LDP will not come up. Consequently, if we use a PVP and the VPI is different at both ends, we need to define the XTagATM interfaces as VP-Tunnel.

In the LSC, a VP-Tunnel interface is auto-discovered with VSI messages as soon as we configure a virtual trunk or VNNI interface. On the eLSR, we need to specify that the interface is VP-Tunnel and also specify the VPI it will be using.

In RFC 3036, "LDP Specification," the field **V-bits** in the ATM Label TLV specifies whether VPI or VCI or both are significant.

The following is a related excerpt from RFC 3035, "MPLS Using LDP and ATM VC Switching":

7.2. Connections via an ATM VP

Sometimes it can be useful to treat two LSRs as adjacent (in some LSP) across an LC-ATM interface, even though the connection between them is made through an ATM "cloud" via an ATM Virtual Path. In this case, the VPI field is not available to MPLS, and the label MUST be encoded entirely within the VCI field.

In this case, the default VCI value of the non-MPLS connection between the LSRs is 32. Other values can be configured, as long as both parties are aware of the configured value. The VPI is set to whatever is required to make use of the Virtual Path.

Connecting MGX-8230 and MGX-8250 eLSRs

It is common practice to use RPM/B or RPM-PR cards in the MGX-8230 or MGX-8250 as PE routers. The Customer Edge (CE) routers connect to the PE using either Fast Ethernet (FE) interfaces in the RPM backcard or Frame Relay (FR), Point-to-Point Protocol (PPP), or ATM PVCs from FRSM or AUSM service modules.

The two designs to connect an MGX-8230 or MGX-8250 edge concentrator shelf to either a BPX-8600 or MGX-8850 routing switch, using AutoRoute and MPLS are as follows:

- Using separate physical interfaces for AutoRoute and MPLS (as we did in this chapter)
- Using only one physical link and a wraparound solution (a physical loop between two interfaces) in the BPX-8600 or MGX-8850 LSR

Using Separate Physical Interfaces

This design is straightforward. Basically, we use one physical port in the PXM-1 for the feeder trunk (running Annex.G) and separate physical ports for MPLS traffic. This design is shown in Figure 6-12.

This design is especially practical for colocated LSRs and eLSRs.

VPCs from the RPM cards to the PXM-1 port terminate in VNNI ports in the MGX-8850 or virtual trunks in the BPX-8600. The virtual path connections and VP-Tunnel LC-ATM interfaces cannot use the same interface because feeder trunks and VNNI/VTs cannot coexist in the same physical interface. This setup also provides the advantage of separating MPLS and other traffic. We can use only one feeder trunk but as many MPLS VP-Tunnel LC-ATM interfaces as needed.

Figure 6-12 *MGX-8230 or MGX-8250 Using Separate Uplinks*

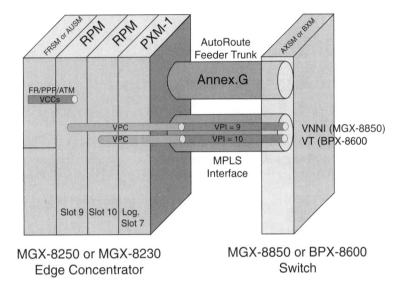

Using One Uplink and a Wraparound

In some cases, the design just outlined is impractical, cost-prohibitive, or impossible. It might be impractical in situations where MGX-8200s (except MGX-8220, which doesn't support RPM cards) are in remote locations. This is impossible with OC-12 interfaces because the PXM-1 backcard has only one port. In these cases, you can use the design shown in Figure 6-13.

In this setup, the VPCs in the MGX-8230 or MGX-8250 from RPM cards terminate on the feeder trunk in the PXM-1. On the MGX-8850 AXSM port or BPX-8600 BXM port, we cannot have VNNIs or VTs in a feeder trunk. So we create VPCs to another port. These VPCs are PVPs (Permanent Virtual Paths) in a BPX-8600 under the AutoRoute control plane, or SPVPs (Soft Permanent Virtual Paths) in an MGX-8850 under the PNNI control plane.

This port where the VPCs terminate is physically looped back to the port where the VNNI or VT is configured.

This design has one further benefit. As shown in Figure 6-14, only one physical loop is needed per LSR for multiple MGX-8230 and MGX-8250 edge concentrator shelves, as long as bandwidth and LCN permit.

Figure 6-13 *MGX-8230 or MGX-8250 Using a Wraparound*

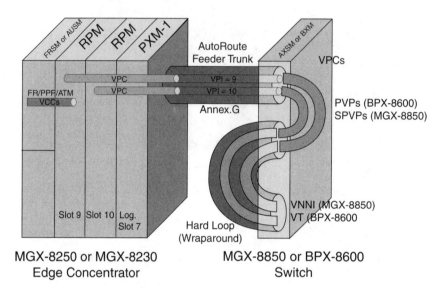

Figure 6-14 *Multiple MGX-8200 Feeders Using a Wraparound*

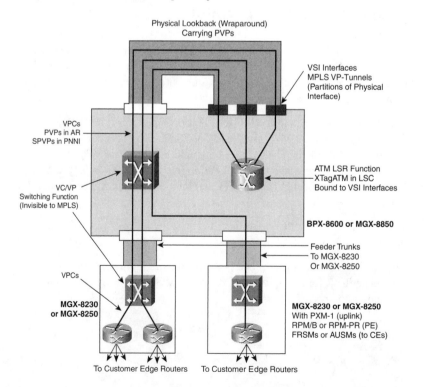

Even though only two MGX-8230 or MGX-8250 feeder shelves are shown, multiple MGX edge concentrators are supported with the same model.

MPLS Troubleshooting

The forthcoming sections cover various MPLS troubleshooting scenarios. In particular, LSR bringup problem cases and LVC starvation scenarios will be categorized and analyzed.

Troubleshooting LSR Bringup Problems

Essentially, two parameters need to match between the VSI master and the VSI slaves: **Controller ID** and **VSI base-vc**.

The debug command **debug vsi errors** in the LSC shown in Example 6-84 helps you troubleshoot LSC switch communication problems.

Example 6-84 *Debugging VSI Errors*

```
LSC#debug vsi errors
02:18:37: VSI Master: parse error (unexpected controller id) in IFC GET CNFG
TRAP rcvd on ATM1/0:0/42 (slave 2)
02:19:29: VSI Master: got NAK reason 105 (unknown ctrler id) in GEN ERROR RSP
```

This example shows two different manifestations of the same problem.

The first VSI error is a VSI master parser error. The VSI master receives a VSI message from the VSI slave that has an unexpected controller ID. That is, the VSI slave sends a VSI message to the master with a controller ID different from the one the master has configured. This error is generated in the master (LSC) when any VSI message (such as **IFC GET CNFG**) is received.

The second VSI error is generated at the VSI slave. The VSI master receives a special VSI message generated at the slave: **GEN ERROR RSP**. This message indicates that the VSI slave (BXM/UXM) had an error, which is shown in the NAK reason (error code). NAK reason 105 indicates an unknown controller ID.

Running Out of LVCs

Running out of LVCs is one of the most common problems in ATM MPLS networks. The immediate symptom is an error message on the LSC that says **TCATM-XCONNECT_FAILED**. This means that there are insufficient LCNs for LVCs on one of the LSR interface's partitions:

```
00:30:32: %TCATM-4-XCONNECT_FAILED: 10.10.10.3/32 Head end terminating on XTagATM3102
00:30:32: %TCATM-4-XCONNECT_FAILED: 10.10.10.5/32 Tail end terminating on XTagATM3102
```

The following is a quote from Josh Gahm, an early LSC developer, regarding LVC starvation:

A network that has run out of cross-connects is basically non-functioning. There really isn't any way to control which [LVCs] get created and which don't. This may result in permanent black holes and/or wildly overloaded control processors. The network needs to be designed from the start such that it won't run out of connections.

In the case of LVC starvation, ordinary IP traffic (meaning traffic that uses an MPLS label stack of one entry) is sent on the untagged VC (the control-vc, 0/32 by default). This traffic is processed by the LSCs and is not switched in the ATM switching fabrics, causing high CPU utilization. This traffic has poor performance and can affect LDP and IP routing traffic (because it shares the same VC). For services that use MPLS, such as MPLS VPNs, traffic is discarded (causing black holes) because there is no LVC to the destination eLSR.

We cannot stress enough the importance of a well-designed and dimensioned network. It's important to calculate the worst-case scenario for the number of LVCs in each link and configure the guaranteed LCN field (minLCN) to at least that value.

Identifying the Problem

To see whether the problem is LCN resource starvation in the VSI partition, we can use the command **dspvsipartinfo** on IGX-8400 or BPX-8600 platforms. See Example 6-85. Zero available LCNs indicates that the partition is starved for channel resources.

Example 6-85 *Checking Resource Usage from the Switch*

```
i8430-2a          TN    Cisco         IGX 8430  9.3.11    Apr. 15 2001 05:09 GMT


                 VSI Resources Status for trunk   31.2 Partition 3

Minimum Lcns       :      100    Minimum BW   (cps) :   100000
Maximum Lcns       :      300    Maximum BW   (cps) :   100000
Used Lcns          :      300    Used BW      (cps) :        0
Available Lcns     :        0    Available BW (cps) :   100000
Start VPI          :        2    End VPI            :        2






This Command: dspvsipartinfo 31.2 3


Hit DEL key to quit:
```

In an AXSM card on an MGX-8850 platform, we can use the command **dspload**. Refer to Example 6-86. In the **dsload** command output, the available channel lines are of prime importance.

Example 6-86 *Checking Resource Usage from the Switch*

```
P3-m8850.1.AXSM.a > dspload 1 2
       +---------------------------------------------+
       |    I N T E R F A C E    L O A D    I N F O  |
       +---------------------------------------------+
       | Maximum Channels          : 0001024         |
       | Guaranteed Channels       : 0000512         |
       | Igr Maximum Bandwidth     : 0096000         |
       | Igr Guaranteed Bandwidth  : 0096000         |
       | Egr Maximum Bandwidth     : 0096000         |
       | Egr Guaranteed Bandwidth  : 0096000         |
       | Available Igr Channels    : 0000000         |
       | Available Egr Channels    : 0000000         |
       | Available Igr Bandwidth   : 0096000         |
       | Available Egr Bandwidth   : 0096000         |
       +---------------------------------------------+
       |           E X C E P T -- V A L U E S        |
       +---------------------------------------------+
       | SERV-CATEG | VAR-TYPE | INGRESS  | EGRESS   |
       +---------------------------------------------+
       | TAG0       | Avl Bw   | 0027840  | 0027840  |      ! 96000 * 29 %
       | TAG1       | Avl Bw   | 0033600  | 0033600  |      ! 96000 * 35 %
       | TAG2       | Avl Bw   | 0033600  | 0033600  |      ! 96000 * 35 %
       | TAG3       | Avl Bw   | 0000960  | 0000960  |      ! 96000 * 1 %
       +---------------------------------------------+

P3-m8850.1.AXSM.a >
```

It is interesting to note that the command **dspload** in AXSM cards shows except values for bandwidth resources and not for LCN resources. This is because the LSC sends only bandwidth CoS values in the **interface policy parameter** VSI message. In contrast, the PNNI controller sends both bandwidth and LCN CoS values (as discussed in Chapter 11, "Advanced PNNI Configuration") .

From the LSC, we can use the command **show controllers vsi descriptor**. Refer to Example 6-87. The same information can be gathered using **VSI Master MIB**.

Example 6-87 *Checking Resource Usage from the LSC*

```
LSC#show controller vsi descriptor

Phys desc: 0.31.2.0
Log intf:  0x001F0200
```

continues

Example 6-87 *Checking Resource Usage from the LSC (Continued)*

```
Interface: XTagATM3102
IF status: up              IFC state: ACTIVE
Min VPI:   2              Maximum cell rate:  100000
Max VPI:   2              Available channels: 0
Min VCI:   32             Available cell rate (forward):  100000
Max VCI:   65535          Available cell rate (backward): 100000
```

To totally identify the problem, we can use the commands **show mpls atm-ldp summary**, discussed earlier, to check for bindings in the Bindwait state, and **show mpls atm-ldp bindwait** to identify the destination.

LCN Usage

The best way to solve problems is not to have them in the first place. To help design our network, we recommend reviewing the VSI Partitioning section in Chapter 2 titled "Hard, Soft, and Dynamic Resource Partitioning" and using two simple commands.

First, as shown in Example 6-88, we can use **dspcd** to identify the port groups (PGs) in each card. Even though the output of this command is slightly different on BPX-8600, IGX-8400, and MGX-8850 platforms, the port group information is present in all of them.

Example 6-88 *Identifying Port Groups*

```
P1-b8620        TN    Cisco        BPX 8620  9.3.30   Dec. 9 2001  21:22 GMT

Detailed Card Display for BXM-155 in slot 3
Status:          Active
Revision:        FBN              Backcard Installed
Serial Number:   688728            Type:        LM-BXM
Top Asm Number:  28215802          Revision:    BA
Queue Size:      228300            Serial Number: 759169
Supp:4 Pts, OC3, FST, VcShp        Top Asm Number:
Supp: VT,ChStLv 1,VSI(Lv 3,ITSM)   Supp: 4 Pts,OC3,MMF,RedSlot:NO
Supp: APS(FW), F4F5
Support: LMIv 1,ILMIv 1,NbrDisc,XL
Supp: OAMLp,TrfcGen,OAM-E
#Ch:16320,PG[1]:8160,PG[2]:8160    ! Port Group 1: Intfs 1 and 2.
PG[1]:1,2,PG[2]:3,4,               ! Port Group 2: Intfs 3 and 4.
#Sched_Ch:16384 #Total_Ch:16320


Last Command: dspcd 3


Next Command:
```

In this case, we have a BXM card with two port groups, each of which has a maximum of 8160 LCNs to be shared. PG 1 spans ports 1 and 2, and PG 2 includes ports 3 and 4. For every first-level label allocated to a next-hop destination, an LCN is used. As you will see, enabling multi-vc multiplies the number of LVCs by the number of classes used. LSC redundancy also multiplies the number of LVCs used.

Second, as shown in Example 6-89, we can use the command **dspchuse** on the BPX-8600 and IGX-8400 platforms to check the current LCN usage.

Example 6-89 *Identifying Port Groups and Checking LCN Reservations*

```
P1-b8620        TN    Cisco           BPX 8620  9.3.30    Dec. 9 2001  21:22 GMT

                    Channel Management Summary for Slot 3

                  max    used   avail   netw   pvc cnfg  f4-f5  vsi mgmt  vsi cnf

card 3:         16320   2582   13738    814     256        0       0       1512
port grp 1:      8160   1798    6362    542     256        0       0       1000
port grp 2:      8160    784    7376    272       0        0       0        512


              pvc cnfg  pvc used  nw used  f4-f5   vsi mgmt  vsvsi min   vsi max

phy if 1:         0         0       271       0        0       ---      --- ! -+
phy trk:          0         0                ---                         !  |
part 1:                                                        500      1000 !  | PG1
phy if 2:       256         0       271       0        0       ---      --- !  |
phy trk:        256         0                ---                         ! -+
phy if 3:         0         0       272       0        0       ---      --- ! -+
vtrk 1:           0         0                ---                         !  |
part 1:                                                        256       512 !  |
vtrk 2:           0         0                ---                         !  | PG2
part 1:                                                        256       512 !  |
phy if 4:         0         0         0       0        0       ---      --- ! -+




Last Command: dspchuse 3


Next Command:
```

From this output, we can calculate the sum of the **vsi min** and find the largest **vsi max**.

Solving the Problem

The solution to the running out of LVCs problem is straightforward: Increase the number of LCN resources for the VSI partition. We do this with **cnfrsrc** on the IGX-8400 and

BPX-8600 platforms and with **cnfpart** for AXSM cards on MGX-8850 platforms. The drawback is that this procedure affects service.

After that, we will monitor that there are no more bindings in the Bindwait state and that there are available LCN resources in the VSI slave.

Reducing the Number of LVCs

You can reduce the number of LVCs in an ATM MPLS network in several ways. For large networks with several eLSRs and destinations, the only option might be to enable VC-merge, but an expert network design plays a more important role.

By default, ATM LSRs also perform eLSR functions. They request LVCs for the IP destinations they learn (building headend LVCs), and they accept LVC requests from other eLSRs (creating tailend LVCs). Those headend and tailend LVCs terminating in the LSC normally are not needed. From the MPLS architectural model, the objective is to set up shortcut LVCs among all eLSRs, but not to the LSRs.

Disabling Headend LVCs

We can instruct the LSCs not to request bindings for the destinations in the routing table, effectively disabling headend LVCs from the LSC.

This does not mean that those destinations are unreachable from the LSC, but in the absence of LVC to a destination, the untagged VC (control-vc) forwards that traffic. It is important to note that in normal operation, LVCs originating on and terminating in the LSC are not used for customer traffic.

To achieve this, we can use the global configuration command **mpls atm disable-headend-vc** in all the LSCs.

If we look at a destination (in this case, a remote eLSR), we see that the LSC is setting up a headend LVC. See Example 6-90.

Example 6-90 *Identifying Headend VCs*

```
P1_LSC_c7204#show mpls atm-ldp bindings 172.27.1.133 32
 Destination: 172.27.1.133/32
    Headend Switch XTagATM1341 (2 hops) 27/37  Active, VCD=75
    Transit XTagATM332 10/76 Active -> XTagATM1341 27/43 Active
    Transit XTagATM331 9/76 Active -> XTagATM1341 27/49 Active
    Transit XTagATM133 2/128 Active -> XTagATM1341 27/55 Active
    Transit XTagATM133 2/134 Active -> XTagATM1341 27/61 Active
    Transit XTagATM133 2/140 Active -> XTagATM1341 27/67 Active

P1_LSC_c7204#conf t
```

We then disable headend eLSR functionality in the LSC. Refer to Example 6-91.

Example 6-91 *Disabling Headend Edge Functionality in the LSC*

```
P1_LSC_c7204#conf t
Enter configuration commands, one per line.  End with CNTL/Z.
P1_LSC_c7204(config)#mpls atm disable-headend-vc
P1_LSC_c7204(config)#^Z
```

We see in Example 6-92 that the headend LVC is removed.

Example 6-92 *Displaying Bindings with Headend Functionality Disabled*

```
P1_LSC_c7204#show mpls atm-ldp bindings 172.27.1.133 32
Destination: 172.27.1.133/32
    Transit XTagATM332 10/76 Active -> XTagATM1341 27/43 Active
    Transit XTagATM331 9/76 Active -> XTagATM1341 27/49 Active
    Transit XTagATM133 2/128 Active -> XTagATM1341 27/55 Active
    Transit XTagATM133 2/134 Active -> XTagATM1341 27/61 Active
    Transit XTagATM133 2/140 Active -> XTagATM1341 27/67 Active

P1_LSC_c7204#
```

We then perform this procedure on all the LSCs in the network.

Disabling Tailend LVCs and LVCs to Numbered Links

At this point, we have disabled half of the eLSR functionality in the LSCs. The other portion consists of the tailend VCs.

We can see in Example 6-93 that these LVCs exist by default, checking for bindings for the local loopback address. There are eight tailend LVCs.

Example 6-93 *Checking for Tailend Bindings*

```
P1_LSC_c7204#show mpls atm-ldp bindings 172.27.1.1 32
Destination: 172.27.1.1/32
    Tailend Switch XTagATM331 9/34 Active -> Terminating Active, VCD=19
    Tailend Switch XTagATM332 10/34 Active -> Terminating Active, VCD=15
    Tailend Switch XTagATM133 2/34 Active -> Terminating Active, VCD=23
    Tailend Switch XTagATM133 2/46 Active -> Terminating Active, VCD=24
    Tailend Switch XTagATM133 2/58 Active -> Terminating Active, VCD=25
    Tailend Switch XTagATM1341 27/34 Active -> Terminating Active, VCD=73
    Tailend Switch XTagATM1341 27/58 Active -> Terminating Active, VCD=77
    Tailend Switch XTagATM1341 27/64 Active -> Terminating Active, VCD=78

P1_LSC_c7204#
```

We can find the headend LVC in an eLSR corresponding to one of the tailend LVCs. Refer to Example 6-94.

Example 6-94 *Identifying a Headend LVC*

```
PE_m8250_RPMB_10#show mpls atm-ldp bindings 172.27.1.1 32
 Destination: 172.27.1.1/32
    Headend Router Switch1.1 (2 hops) 10/34  Active, VCD=155

PE_m8250_RPMB_10#
PE_m8250_RPMB_10#show mpls atm summary

Total number of destinations: 9

ATM label bindings summary
      interface   total   active   local   remote   Bwait   Rwait  IFwait
       Switch1.1     13      13       6       7        0       0       0
       Switch1.2      2       2       1       1        0       0       0
PE_m8250_RPMB_10#
```

Now we can go ahead and filter the LSC loopback addresses in this eLSR using the global configuration command **mpls ldp request-labels for *{ACL}*>**.

NOTE   The command **mpls ldp advertise-labels** has no effect on LC-ATM interfaces. This is because the label distribution method is Downstream on Demand (upstream-controlled).

We define an access list that denies the LSC loopback address and permits everything else (see Example 6-95). We call it BlockThis.

Example 6-95 *Creating an ACL Blocking LSC Loopback Addresses*

```
PE_m8250_RPMB_10#conf t
Enter configuration commands, one per line.  End with CNTL/Z.
PE_m8250_RPMB_10(config)#ip access-list standard BlockThis
PE_m8250_RPM(config-std-nacl)#deny 172.27.1.0 0.0.0.127
PE_m8250_RPM(config-std-nacl)#permit any
PE_m8250_RPM(config-std-nacl)#exit
PE_m8250_RPMB_10(config)#
```

NOTE   When planning our network, we assigned the LSC and eLSR loopback addresses such that there was a block assigned only to LSCs and a different block of addresses for the eLSRs. That allowed us to simplify this access list.

Finally, we apply the access list to the addresses we do not want to set up LVCs to:

```
PE_m8250_RPMB_10(config)#mpls ldp request-labels for BlockThis
```

We can check that this eLSR now has three fewer remote bindings (see Example 6-96). These are the three LSC loopback addresses for which we are not requesting LVCs.

Example 6-96 *Checking the MPLS ATM Binding Summary*

```
PE_m8250_RPMB_10#show mpls atm summ
Total number of destinations: 6

ATM label bindings summary
      interface   total  active   local  remote  Bwait   Rwait  IFwait
      Switch1.1      10      10       6       4       0       0       0
      Switch1.2       2       2       1       1       0       0       0
PE_m8250_RPMB_10#
PE_m8250_RPMB_10#show mpls atm-ldp bindings 172.27.1.1 32

PE_m8250_RPMB_10#show mpls atm-ldp bindings 172.27.1.2 32

PE_m8250_RPMB_10#show mpls atm-ldp bindings 172.27.1.3 32

PE_m8250_RPMB_10#
```

In the LSC, we can see that the tailend LVC on VPI 10 (from the eLSR we just configured) is no longer there. Now there are seven tailend LVCs, as shown in Example 6-97.

Example 6-97 *Showing Tailend LVCs*

```
P1_LSC_c7204#show mpls atm-ldp bindings 172.27.1.1 32
 Destination: 172.27.1.1/32
    Tailend Switch XTagATM331 9/34 Active -> Terminating Active, VCD=19
    Tailend Switch XTagATM133 2/34 Active -> Terminating Active, VCD=23
    Tailend Switch XTagATM133 2/46 Active -> Terminating Active, VCD=24
    Tailend Switch XTagATM133 2/58 Active -> Terminating Active, VCD=25
    Tailend Switch XTagATM1341 27/34 Active -> Terminating Active, VCD=73
    Tailend Switch XTagATM1341 27/58 Active -> Terminating Active, VCD=77
    Tailend Switch XTagATM1341 27/64 Active -> Terminating Active, VCD=78

P1_LSC_c7204#
```

We need to perform this procedure in all eLSRs and LSRs (in the LSCs) in the network.

NOTE   It is extremely important that we do not deny any eLSR address in the access list. If we do so, MPLS VPN connectivity will be broken because an LVC will not be bound to an eLSR destination.

This mechanism of applying an access list to the destinations that MPLS should request a label for can also be applied to any numbered link in the network.

Enabling VC-Merge

In very large networks, the best alternative to save LCN resources might be to enable VC-merge, also called multipoint-to-point. VC-merging is enabled by default in the LSC. It can be disabled with the global configuration command **no mpls ldp atm vc-merge**. However, for VC-merge to be supported, it must also be enabled in the VSI slave.

On BPX-8600 platforms, BXM VC-merge capability is displayed with the command **dspcd**. The command **cnfcdparm** enables VC-merge on a per-card basis. The VC-merge state can be displayed with the command **dspcdparm**.

From the LSC, as shown in Example 6-98, the VC-merge capability negotiated through LDP in the ATM Session Parameters TLV can be displayed using the command **show mpls atm-ldp capability**.

Example 6-98 *Identifying the VC-Merge Capability*

```
P3_LSC_m8850_RPM-PR_9#show mpls atm-ldp capability

                VPI           VCI           Alloc    Odd/Even  VC Merge
XTagATM1144     Range         Range         Scheme   Scheme    IN   OUT
  Negotiated    [27 - 27]     [33 - 65535]  BIDIR              -    -
  Local         [27 - 27]     [33 - 65535]  BIDIR    ODD       EN   EN
  Peer          [27 - 27]     [33 - 65535]  BIDIR    EVEN      -    -
P3_LSC_m8850_RPM-PR_9#
```

In a VC-merging environment, VC-merged bindings can be displayed with the command **show mpls atm-ldp bindings vc-merged**. See Example 6-99.

Example 6-99 *Showing VC-Merged Bindings*

```
P3_LSC_m8850_RPM-PR_9#show mpls atm-ldp bindings vc-merged
 Destination: 172.27.1.128/32
    Transit XTagATM101 0/121 Active -> XTagATM1114 0/41 Active
    Transit XTagATM1111 0/113 Active -> XTagATM1114 0/41 Active
```

In this command output, we can see how two different VPI/VCI pairs from two different interfaces are cross-connected to the same interface and VPI/VCI in a multipoint-to-point fashion.

We can use the same command (see Example 6-100) to check VC-merged bindings if multi-vc is enabled in the eLSRs.

Example 6-100 *Showing VC-Merged Bindings with Multi-vc Enabled*

```
P3_LSC_m8850_RPM-PR_9#show mpls atm-ldp bindings vc-merged 172.27.1.128 32
Destination: 172.27.1.128/32
    Transit XTagATM101 0/53 Active -> XTagATM1114 0/41 Active, CoS=available
    Transit XTagATM1111 0/53 Active -> XTagATM1114 0/41 Active, CoS=available
    Transit XTagATM101 0/55 Active -> XTagATM1114 0/45 Active, CoS=standard
    Transit XTagATM1111 0/55 Active -> XTagATM1114 0/45 Active, CoS=standard
    Transit XTagATM101 0/57 Active -> XTagATM1114 0/47 Active, CoS=premium
    Transit XTagATM1111 0/57 Active -> XTagATM1114 0/47 Active, CoS=premium
    Transit XTagATM101 0/59 Active -> XTagATM1114 0/49 Active, CoS=control
    Transit XTagATM1111 0/59 Active -> XTagATM1114 0/49 Active, CoS=control
```

When VC-merge is used in conjunction with multi-vc, only LVCs with the same service type are merged.

Tracing an LVC

Another common task in ATM MPLS networks is to trace the path an LVC is taking. The **traceroute** command does not work in cell-based MPLS networks because LSRs switch cells. To perform that function, three steps need to be carried out based on router functionality:

- **Headend router**—Displays the FEC binding
- **Transit router**—Displays the cross-connects
- **Tailend router**—Displays the ATM VC

The headend router and tailend router always exist in an LVC. The transit router does not exist if there is a direct connection between eLSRs. Therefore, the first and third steps always need to be performed.

On the headend router (see Example 6-101), we check the binding using the command **show mpls atm-ldp bindings** for the specific destination and mask.

Example 6-101 *Tracing an LVC – Headend Router*

```
PE_Headend#show mpls atm-ldp bindings 10.1.2.0 24
 Destination: 10.1.2.0/24
    Headend Router ATM1/0/0.1 (2 hops) 6/38 Active, VCD=8, CoS=available

PE_Headend#
```

From the command **show mpls atm-ldp bindings**, we gather the output interface and VPI/VCI pair that the LVC is taking. We do this for the FEC defined by the destination, subnet mask, and service class.

The next step (see Example 6-102) is to check the cross-connects in all LSRs that the LVC is traversing as a transit binding. Using the command **show mpls ldp neighbor**, we verify that the neighbor P_Transit is connected to ATM1/0/0.1 in PE_Headend. To check the cross-connect, we use the command **show xtagatm cross-connect** using the VPI and VCI from the headend router and the interface connecting to the headend router.

Example 6-102 *Tracing an LVC – Transit Router*

```
P_Transit#show xtagatm cross-connect descriptor 0.9.2.0 | incl 6/38
0.9.2.0      6/38        ->      0.1.3.0      6/35      UP
P_Transit#
```

NOTE The command **show xtagatm cross-connect** can be used with the parameter **traffic** to gather LVC traffic statistics from the LSC. These traffic counters reside in the VSI slave and are retrieved using VSI statistics commands. From AXSM VSI slaves in MGX platforms, the command **dspchancnt** can be used to get traffic counters.

We follow the LVC through the different LSRs it might traverse until we reach the tailend eLSR router. In this case, there is only one LSR between the eLSRs. At that instance (see Example 6-103), we check the ATM VC at the tailend router using the command **show atm vc**.

Example 6-103 *Tracing an LVC – Tailend Router*

```
PE_Tailend#show atm vc | include 35
1/0.1      5              6    35  TVC   MUX    UBR   155000                UP
PE_Tailend#
```

This is the last step, and with the information gathered, we know the path taken by this LVC. The path is shown in Figure 6-15.

Figure 6-15 *Tracing an LVC*

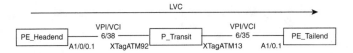

Summary

This chapter focused on basic MPLS provisioning and troubleshooting in a multiservice switching network. Beginning with the generic configuration of an LSR and eLSR, this chapter transitioned to the configuration specifics of the BPX-8600/MGX-8250, IGX-8400, and MGX-8850 implementations of eLSR and LSR bringups. With the configurations in place, the "MPLS Troubleshooting" section explained how to solve problems such as LVC depletion and LVC tracing.

CHAPTER 7

Practical Applications of MPLS

This is the last chapter of the MPLS section. Here we will cover two MPLS applications. We will implement basic MPLS Virtual Private Networks (VPNs) and DiffServ Quality of Service (QoS) on the MPLS network we built in the previous chapter.

Basic MPLS VPN Configuration

Now that we have our MPLS Network with P (Provider) and PE (Provider Edge) routers, we can configure a basic MPLS VPN between two sites. This section is not intended to be a complete MPLS VPN study, only an example of setting a basic VPN. For further details on MPLS VPNs, we recommend *MPLS VPN Architectures,* published by Cisco Press.

We start from our MPLS network, where we are running OSPF (Open Shortest Path First) as the Interior Gateway Protocol (IGP). We physically connect two CE (Customer Edge) routers to two PE routers. We chose the two RPM cards in slot 10 of the MGX-8250 and the MGX-8850 as PE routers. The topology is shown in Figure 7-1.

Figure 7-1 *Basic MPLS VPN*

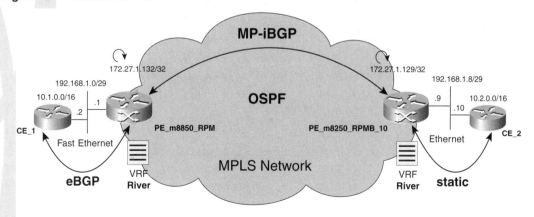

Configuring the VPN Routing Forwarding

A VPN routing and forwarding instance has its own routing context that includes the IP routing table, protocols and variables, a Cisco Express Forwarding (CEF) table used for forwarding, and interfaces, as well as rules to control the import and export of routes from and to the VPN routing table. We start by creating the VPN Routing Forwarding (VRF) and assigning a route distinguisher (RD) to it. The RD solves the problem of overlapping IP addresses, making the VPNv4 addresses globally unique by prepending the RD to the IPv4 address. An RD does not represent a VPN; it identifies a VRF. The key concept is not to think of a VRF as a VPN. One or more VRFs can be associated with a VPN. We also specify import and export route targets (MP-BGP extended community, which in essence identify networks belonging to a VPN or VPN membership), allowing the insertion of routes into a VRF. Finally, we assign that VRF to the interfaces facing the CEs. Please see Example 7-1.

Example 7-1 *Creating and Assigning VPN Routing Forwarding*

```
PE_m8850_RPM#conf t
Enter configuration commands, one per line.  End with CNTL/Z.
PE_m8850_RPM(config)#ip vrf River
PE_m8850_RPM(config-vrf)#rd 100:27
PE_m8850_RPM(config-vrf)#route-target both 100:27
PE_m8850_RPM(config-vrf)#exit
PE_m8850_RPM(config)#interface fastEthernet 1/1
PE_m8850_RPM(config-if)#ip vrf forwarding River
PE_m8850_RPM(config-if)#ip address 192.168.1.1 255.255.255.248
PE_m8850_RPM(config-if)#no shut
PE_m8850_RPM(config-if)#^Z
PE_m8850_RPM#
```

Note that we assign the Fast Ethernet (FE) IP address after assigning a VRF to an interface. This is done because an existing IP address is removed when the interface command **ip vrf forwarding** {*VRF Name*} is entered. That IP address is now a connected route in the VRF river, not in the global routing table.

We do the same with the other provider edge router. As shown in Example 7-2, we also configure the CE interfaces:

Example 7-2 *Configuring CE Interfaces*

```
CE_1#conf t
Enter configuration commands, one per line.  End with CNTL/Z.
CE_1(config)#interface FastEthernet 0/0
CE_1(config-if)#ip address 192.168.1.2 255.255.255.248
CE_1(config-if)#^Z
CE_1#
```

You can see that the CEs are not VPN-aware. They do not have any special configuration.

Configuring Multiprotocol iBGP

The next step in basic MPLS VPN configuration is to configure a multiprotocol iBGP session between the PEs. Multiprotocol BGP (MP-BGP) is BGP supporting address families other than IPv4 addresses. We will use AS 100 in this example. After configuring the iBGP neighbor, we also configure the VPNv4 address family so that the iBGP neighbors exchange 96-bit VPNv4 prefixes (64-bit RD + 32-bit IPv4 prefixes). See Example 7-3.

NOTE   VPNv4 prefixes are exchanged only between PE routers.

Example 7-3 *Configuring MP-iBGP*

```
PE_m8850_RPM#conf t
Enter configuration commands, one per line.  End with CNTL/Z.
PE_m8850_RPM(config)#router bgp 100
PE_m8850_RPM(config-router)#neighbor 172.27.1.129 remote-as 100
PE_m8850_RPM(config-router)#neighbor 172.27.1.129 update-source loopback 0
PE_m8850_RPM(config-router)#address-family vpnv4
PE_m8850_RP(config-router-af)#neighbor 172.27.1.129 activate
PE_m8850_RP(config-router-af)#neighbor 172.27.1.129 send-community extended
PE_m8850_RP(config-router-af)#neighbor 172.27.1.129 next-hop-self
PE_m8850_RP(config-router-af)#exit-address-family
PE_m8850_RPM(config-router)#end
PE_m8850_RPM#
```

Within the VPNv4 address family, we configure extended communities in order to send route targets. Other extended communities are Site of Origin (SoO) and OSPF route type.

We also configure next-hop-self to disable next-hop processing on the VPNv4 address family for MP-iBGP, because we will be running eBGP on a PE-CE link. If we do not disable next-hop processing of BGP updates, we could be causing a black hole when the IP address of a CE (in a VPN) running eBGP becomes a VPNv4 BGP next hop.

Again we perform the corresponding configuration in the other PE.

NOTE   In this case we are directly configuring a BGP neighbor. In large networks the iBGP sessions among all PE routers should be accomplished by using route reflectors for scalability.

We can check that the session is up as shown in Example 7-4.

Example 7-4 *MP-iBGP Summary*

```
PE_m8250_RPMB_10#show ip bgp vpnv4 all summary
BGP router identifier 172.27.1.129, local AS number 100
BGP table version is 1, main routing table version 1

Neighbor        V     AS MsgRcvd MsgSent   TblVer  InQ OutQ Up/Down   State/PfxRcd
172.27.1.132    4    100      20      20        1    0    0 00:09:44            0
PE_m8250_RPMB_10#
```

Many optimizations can be made to the BGP protocol to improve its convergence. These optimizations are beyond the scope of this section. However, one is worth mentioning given its simplicity and improvements.

By default, iBGP sessions use a maximum data segment of 536 bytes, which limits the amount of data it can transport in a single packet. This limit is defined as the TCP Maximum Segment Size (MSS), which is the largest packet that TCP can send. The command **show ip bgp neighbors** includes the status of the underlying TCP connection. This can also be seen with **show tcp**. Refer to Example 7-5:

Example 7-5 *Checking the BGP Maximum Data Segment*

```
PE_m8850_RPM#show ip bgp neighbors | i Datagrams
Datagrams (max data segment is 536 bytes):
PE_m8850_RPM#
```

The MSS default of 536 bytes means that TCP segments the data in the transmit queue into 536-byte chunks before passing packets to the IP layer.

Transmission Control Protocol (TCP) path Maximum Transmission Unit (MTU) discovery (defined in RFC 1191 and RFC 1435) can be configured in all PE routers using the global command **ip tcp path-mtu-discovery** to dynamically increase the maximum BGP data segment (MSS) and to include more advertisements in one packet. Please refer to Example 7-6. This reduces TCP/IP overhead and speeds BGP convergence. TCP path MTU discovery finds the smallest MTU size among all the links between the endpoints of the TCP session, which in our example is 4470 bytes of all ATM links:

Example 7-6 *Enabling TCP Path MTU Discovery*

```
PE_m8850_RPM#conf t
Enter configuration commands, one per line.  End with CNTL/Z.
PE_m8850_RPM(config)#ip tcp path-mtu-discovery
PE_m8850_RPM(config)#^Z

PE_m8850_RPM#show ip bgp neighbors | i Datagrams
Datagrams (max data segment is 4430 bytes):
PE_m8850_RPM#
```

The new maximum data segment value of 4430 bytes is the default ATM MTU of 4470 bytes minus 40 bytes of IP and TCP header.

Configuring the PE-CE Running Static Routing

You have several options for routing protocols between the PE and CE routers. RIPv2, OSPF, eBGP, EIGRP, and static are some of those. In this case we demonstrate static routing in one PE-CE link and eBGP in the other.

Static is the simplest case. The CE configuration consists only of configuring a static default route. We start by adding a loopback interface in the CE_2 router, as seen in Example 7-7:

Example 7-7 *CE Static Routing Configuration*

```
CE_2(config)#interface loopback 0
CE_2(config-if)#ip address 10.2.1.1 255.255.255.255
CE_2(config-if)#exit
CE_2(config)#ip route 0.0.0.0 0.0.0.0 Ethernet 0
```

In concept, the PE configuration is the same as in a non-MPLS VPN environment, except that the configuration applies to the VRF context. This is shown in Example 7-8. We will add a VRF static route. Then we need to redistribute this static route into the IPv4 address family so that it is propagated to the MP-iBGP peers. We also redistribute the connected VRF interfaces so that the remote CE can reach the IP address of the PE-CE link (192.1.1.8/29):

Example 7-8 *PE VRF Static Routing Configuration*

```
PE_m8250_RPMB_10#
PE_m8250_RPMB_10#conf t
Enter configuration commands, one per line.  End with CNTL/Z.
PE_m8250_RPMB_10(config)#ip route vrf River 10.2.0.0 255.255.0.0 192.168.1.10
PE_m8250_RPMB_10(config)#router bgp 100
PE_m8250_RPMB_(config-router)#address-family ipv4 vrf River
PE_m8250_RP(config-router-af)#redistribute connected
PE_m8250_RP(config-router-af)#redistribute static
PE_m8250_RP(config-router-af)#^Z
PE_m8250_RPMB_10#
```

NOTE   Per-VRF parameters are configured in the routing contexts (**address-family ipv4 vrf** {*VRF Name*} or **ip route vrf** {*VRF Name*} for static routing). All non-BGP per-VRF routes are redistributed into the per-VRF BGP context to be propagated by MP-BGP to other PEs.

We can check that both the connected and the static are in the PE VRF routing context table using **show ip route vrf** {*VRF Name*}, as shown in Example 7-9:

Example 7-9 *Checking the VRF Routing Table*

```
PE_m8250_RPMB_10#show ip route vrf River
Codes: C - connected, S - static, I - IGRP, R - RIP, M - mobile, B - BGP
       D - EIGRP, EX - EIGRP external, O - OSPF, IA - OSPF inter area
       N1 - OSPF NSSA external type 1, N2 - OSPF NSSA external type 2
```

continues

Example 7-9   *Checking the VRF Routing Table (Continued)*

```
           E1 - OSPF external type 1, E2 - OSPF external type 2, E - EGP
           i - IS-IS, L1 - IS-IS level-1, L2 - IS-IS level-2, ia - IS-IS inter area
           * - candidate default, U - per-user static route, o - ODR
           P - periodic downloaded static route

Gateway of last resort is not set

      10.0.0.0/16 is subnetted, 1 subnets
S        10.2.0.0 [1/0] via 192.168.1.10
      192.168.1.0/29 is subnetted, 1 subnets
C        192.168.1.8 is directly connected, Ethernet1/1
PE_m8250_RPMB_10#
```

The remote PE has learned these VRF routes through iBGP. We will use **show ip bgp vpnv4** as well as **show ip route vrf**. See Example 7-10:

Example 7-10   *Checking PE VPNv4 Prefixes*

```
PE_m8850_RPM#show ip bgp vpnv4 vrf River summary
BGP router identifier 172.27.1.132, local AS number 100
BGP table version is 5, main routing table version 5
2 network entries and 2 paths using 370 bytes of memory
1 BGP path attribute entries using 60 bytes of memory
1 BGP extended community entries using 24 bytes of memory
0 BGP route-map cache entries using 0 bytes of memory
0 BGP filter-list cache entries using 0 bytes of memory
BGP activity 2/0 prefixes, 2/0 paths, scan interval 15 secs

Neighbor        V    AS MsgRcvd MsgSent   TblVer  InQ OutQ Up/Down  State/PfxRcd
172.27.1.129    4   100      32      30        5    0    0 00:19:56            2
PE_m8850_RPM#
PE_m8850_RPM#show ip route vrf River
Codes: C - connected, S - static, I - IGRP, R - RIP, M - mobile, B - BGP
       D - EIGRP, EX - EIGRP external, O - OSPF, IA - OSPF inter area
       N1 - OSPF NSSA external type 1, N2 - OSPF NSSA external type 2
       E1 - OSPF external type 1, E2 - OSPF external type 2, E - EGP
       i - IS-IS, L1 - IS-IS level-1, L2 - IS-IS level-2, ia - IS-IS inter area
       * - candidate default, U - per-user static route, o - ODR
       P - periodic downloaded static route

Gateway of last resort is not set

      10.0.0.0/16 is subnetted, 1 subnets
B        10.2.0.0 [200/0] via 172.27.1.129, 00:00:37
      192.168.1.0/29 is subnetted, 2 subnets
B        192.168.1.8 [200/0] via 172.27.1.129, 00:01:37
C        192.168.1.0 is directly connected, FastEthernet1/1
PE_m8850_RPM#
```

We can also check the VRF CEF forwarding table with the command **show mpls forwarding-table vrf** {*VRF Name*}.

Configuring the PE-CE Running eBGP

Now it's time to configure the eBGP PE-CE link. In the CE, the BGP private ASN is 65027. We start by configuring a loopback interface in CE_1. We then configure BGP and add the 10.1.0.0/16 network to indicate to BGP that the network originates on this router. CE_1 generates a network entry for 10.1.0.0/16. See Example 7-11:

Example 7-11 *CE eBGP Configuration*

```
CE_1#conf t
Enter configuration commands, one per line.  End with CNTL/Z.
CE_1(config)#interface loopback 0
CE_1(config-if)#ip address 10.1.1.1 255.255.255.255
CE_1(config-if)#exit
CE_1(config)#router bgp 65027
CE_1(config-router)#no auto-summary
CE_1(config-router)#neighbor 192.168.1.1 remote-as 100
CE_1(config-router)#network 10.1.0.0 mask 255.255.0.0
CE_1(config-router)#redistribute static
CE_1(config-router)#exit
CE_1(config)#ip route 10.1.0.0 255.255.0.0 null 0
```

We need the static route to a null interface to get CE_1 to generate 10.1.0.0/16. This is because that static route generates a matching entry in the CE_1 routing table and therefore can be sent in BGP updates.

We now configure the PE portion. In this case also, the PE-CE IP routing configuration should go into the IPv4 address family (routing context) for the VRF we are configuring. VRF-specific eBGP neighbors are configured under the address-family, and MP-BGP neighbors are configured in the BGP routing process. The routes to the CEs should be installed in the VRF routing table and not anywhere else.

In general, CE neighbors need to be specified within the per-VRF context and not global BGP. All non-BGP per-VRF routes have to be redistributed into a per-VRF context to be propagated by MP-BGP to other PE routers. This is shown in Example 7-12:

Example 7-12 *PE VRF eBGP Configuration*

```
PE_m8850_RPM#conf t
Enter configuration commands, one per line.  End with CNTL/Z.
PE_m8850_RPM(config)#router bgp 100
PE_m8850_RPM(config-router)#address-family ipv4 vrf River
PE_m8850_RP(config-router-af)#redistribute connected
PE_m8850_RP(config-router-af)#neighbor 192.168.1.2 remote-as 65027
PE_m8850_RP(config-router-af)#neighbor 192.168.1.2 activate
PE_m8850_RP(config-router-af)#^Z
PE_m8850_RPM#
```

Again, it is important here to redistribute connected so that remote CEs can reach the PE-CE link's IP address. That address will be the next hop for the propagated CE routes.

At this point, as shown in Example 7-13, we can check the summary of BGP VPNv4 for our VRF to verify that both iBGP and eBGP neighbors are up.

Example 7-13 *Checking VRF VPNv4 Summary*

```
PE_m8850_RPM#show ip bgp vpnv4 vrf River summary
BGP router identifier 172.27.1.132, local AS number 100
BGP table version is 11, main routing table version 11
5 network entries and 5 paths using 925 bytes of memory
6 BGP path attribute entries using 360 bytes of memory
1 BGP AS-PATH entries using 24 bytes of memory
1 BGP extended community entries using 24 bytes of memory
0 BGP route-map cache entries using 0 bytes of memory
0 BGP filter-list cache entries using 0 bytes of memory
BGP activity 9/30 prefixes, 9/4 paths, scan interval 15 secs

Neighbor        V    AS MsgRcvd MsgSent   TblVer  InQ OutQ Up/Down   State/PfxRcd
172.27.1.129    4   100      48      54       11    0    0 00:02:58             2
192.168.1.2     4 65027      16      20       11    0    0 00:02:44             2
PE_m8850_RPM#
```

We can also check the details of the BGP VPNv4 table for the VRF river using the command **show ip bgp vpnv4 vrf** {*VRF Name*}. See Example 7-14:

Example 7-14 *Checking the VRF BGP VPNv4 Table*

```
PE_m8850_RPM#
PE_m8850_RPM#show ip bgp vpnv4 vrf River
BGP table version is 11, local router ID is 172.27.1.132
Status codes: s suppressed, d damped, h history, * valid, > best, i - internal
Origin codes: i - IGP, e - EGP, ? - incomplete

   Network          Next Hop            Metric LocPrf Weight Path
Route Distinguisher: 100:27 (default for vrf River)
*> 10.1.0.0/16      192.168.1.2              0             0 65027 i
*>i10.2.0.0/16      172.27.1.129             0    100      0 ?
*> 192.168.1.0/29   0.0.0.0                  0         32768 ?
*>i192.168.1.8/29   172.27.1.129             0    100      0 ?
PE_m8850_RPM#
```

Here we see how 10.1.0.0/16 is learned from the CE with eBGP and 10.2.0.0/16 is learned from the remote PE through iBGP. We also see why it is important to redistribute connected in the IPv4 address family for the VRF, because 192.168.1.2 is the next hop.

Finally, we can check the VRF routing table using **show ip route vrf**. See Example 7-15:

Example 7-15 *Checking the VRF Routing Information Base*

```
PE_m8850_RPM#
PE_m8850_RPM#show ip route vrf River
Codes: C - connected, S - static, I - IGRP, R - RIP, M - mobile, B - BGP
       D - EIGRP, EX - EIGRP external, O - OSPF, IA - OSPF inter area
       N1 - OSPF NSSA external type 1, N2 - OSPF NSSA external type 2
       E1 - OSPF external type 1, E2 - OSPF external type 2, E - EGP
       i - IS-IS, L1 - IS-IS level-1, L2 - IS-IS level-2, ia - IS-IS inter area
       * - candidate default, U - per-user static route, o - ODR
       P - periodic downloaded static route

Gateway of last resort is not set


     10.0.0.0/8 is variably subnetted, 2 subnets, 1 masks
B       10.2.0.0/16 [200/0] via 172.27.1.129, 00:02:50
B       10.1.0.0/16 [20/0] via 192.168.1.2, 00:01:21
     192.168.1.0/29 is subnetted, 2 subnets
B       192.168.1.8 [200/0] via 172.27.1.129, 00:02:50
C       192.168.1.0 is directly connected, FastEthernet1/1
PE_m8850_RPM#
```

From each PE we can ping both CEs and the CEs between each other. This connectivity test is shown in Example 7-16.

Example 7-16 *VPN Connectivity Test*

```
PE_m8850_RPM#ping vrf River 10.1.1.1

Type escape sequence to abort.
Sending 5, 100-byte ICMP Echos to 10.1.1.1, timeout is 2 seconds:
!!!!!
Success rate is 100 percent (5/5), round-trip min/avg/max = 1/1/1 ms
PE_m8850_RPM#ping vrf River 10.2.1.1

Type escape sequence to abort.
Sending 5, 100-byte ICMP Echos to 10.2.1.1, timeout is 2 seconds:
!!!!!
Success rate is 100 percent (5/5), round-trip min/avg/max = 1/1/1 ms
PE_m8850_RPM#
```

From the PEs we need to use the **vrf** form of the **ping** command, because those are VRF destinations (as opposed to global destinations).

As a final note, the global routing table in the PE routers has not changed since we began configuring MPLS VPNs. As shown in Example 7-17, the global routing table is only populated by the IGP in the MPLS network, which is OSPF in our case, plus the connected prefixes. In other words, VRF routes are not included in the global routing table.

Example 7-17 *Checking the PE Global Routing Table*

```
E_m8250_RPMB_10#show ip route
Codes: C - connected, S - static, I - IGRP, R - RIP, M - mobile, B - BGP
       D - EIGRP, EX - EIGRP external, O - OSPF, IA - OSPF inter area
       N1 - OSPF NSSA external type 1, N2 - OSPF NSSA external type 2
       E1 - OSPF external type 1, E2 - OSPF external type 2, E - EGP
       i - IS-IS, L1 - IS-IS level-1, L2 - IS-IS level-2, ia - IS-IS inter area
       * - candidate default, U - per-user static route, o - ODR
       P - periodic downloaded static route

Gateway of last resort is not set

     172.27.0.0/32 is subnetted, 9 subnets
O       172.27.1.132 [110/26] via 172.27.1.1, 1d20h, Switch1.1
O       172.27.1.133 [110/27] via 172.27.1.1, 1d20h, Switch1.1
O       172.27.1.128 [110/2] via 172.27.1.128, 1d20h, Switch1.2
C       172.27.1.129 is directly connected, Loopback0
O       172.27.1.130 [110/4] via 172.27.1.1, 1d20h, Switch1.1
O       172.27.1.131 [110/32] via 172.27.1.1, 1d20h, Switch1.1
O       172.27.1.1 [110/2] via 172.27.1.1, 1d20h, Switch1.1
O       172.27.1.2 [110/3] via 172.27.1.1, 1d20h, Switch1.1
O       172.27.1.3 [110/25] via 172.27.1.1, 1d20h, Switch1.1
PE_m8250_RPMB_10#
```

Quality of Service Configuration

As you know, MPLS QoS is different in frame-based MPLS networks and cell-based MPLS networks.
The difference comes from how DiffServ information is conveyed to LSRs. In frame-based MPLS,
where there is an MPLS shim header and the experimental bits are visible, the PHB Scheduling Class
(PSC) and drop priority are inferred from the EXP (experimental) field, giving birth to E-LSP (EXP
Inferred PSC Label Switch Paths). This contrasts with cell-based MPLS, where the class is inferred
exclusively from the label (and the drop priority is inferred from the CLP bit for cell-mode MPLS) and
thus is called L-LSP (Label-Only Inferred PSC Label Switch Paths).

Enabling and Configuring Multi-vc

The first part of the ATM MPLS QoS configuration consists of enabling Multi-vc. Multi-vc is an eLSR's
ability to set up multiple parallel L-LSPs (with different QoS treatment in the MPLS network) to a
destination. This capability is shown in Figure 7-2.

Figure 7-2   *QoS in ATM MPLS Networks: Multi-vc*

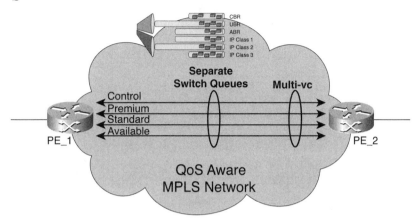

In ATM MPLS eLSRs (PE routers in MPLS VPN jargon), packets are mapped to different L-LSPs based on the precedence bits in the IP header. A maximum of four parallel LVCs to a specific destination are set up. Those LVCs have well-known names: available, standard, premium, and control. Four classes, one for each well-known LVC, are defined in ATM MPLS eLSRs, as shown in Table 7-1. These classes are fixed and cannot be changed.

Table 7-1   *IP Precedence-to-Class Mappings in eLSRs*

Class	IP Precedence
0	0/4
1	1/5
2	2/6
3	3/7

The default mapping from these classes to the LVCs (CoS-to-LSP mapping at the edge) is shown in Table 7-2. Different classes can be mapped to different LVCs by configuration.

Table 7-2   *Class-to-L-LSP Default Mapping*

Class	LVC
0	Available
1	Standard
2	Premium
3	Control

In our example, we will set up only three parallel LVCs: available, standard, and premium. The reason why we left control out will be more evident later in this chapter when we configure the ATM LSRs. The QoS design we will configure in our MPLS VPN network is shown in Figure 7-3.

Figure 7-3 *Sample DiffServ-Aware ATM MPLS Network*

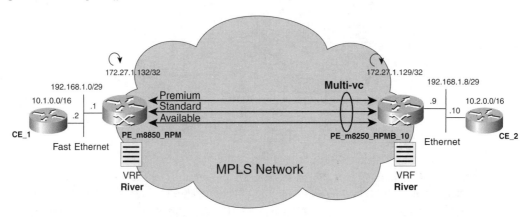

The first step in our configuration is to establish the class-to-LVC mapping. We will set up the mapping shown in Table 7-3.

Table 7-3 *Class-to-L-LSP Mapping in Our Example*

Class	LVC
0 1	Available
2	Standard
3	Premium

To achieve the class-to-LVC mappings, we define a match-all access list and a cos-map and combine the two in a global mpls prefix-map (see Example 7-18).

Example 7-18 *Using the* **mpls cos-map** *and* **mpls prefix-map** *Commands*

```
PE_m8850_RPM_10#conf t
Enter configuration commands, one per line.  End with CNTL/Z.
PE_m8850_RPM_10(config)#access-list 1 permit any
PE_m8850_RPM_10(config)#mpls cos-map 1
PE_m8850(config-mpls-cos-map)#class 0 ?
  available  Transmit on available Label-VC
  control    Transmit on control   Label-VC
  premium    Transmit on premium   Label-VC
  standard   Transmit on standard  Label-VC

PE_m8850(config-mpls-cos-map)#class 0 available
PE_m8850(config-mpls-cos-map)#class 1 available
```

Example 7-18 *Using the* **mpls cos-map** *and* **mpls prefix-map** *Commands (Continued)*

```
PE_m8850(config-mpls-cos-map)#class 2 standard
PE_m8850(config-mpls-cos-map)#class 3 premium
PE_m8850(config-mpls-cos-map)#exit
PE_m8850_RPM_10(config)#mpls prefix-map 1 access-list 1 cos-map 1
PE_m8850_RPM_10(config)#
```

In the general case, we can have several CoS maps configuring from one to four parallel LVCs and can apply them to different destinations based on the ACL. With this we can selectively enable or disable multi-vc for different destinations.

At this point, as shown in Example 7-19, we have only one LVC to the remote PE destination.

Example 7-19 *Checking ATM LDP Bindings Without Multi-vc*

```
PE_m8850_RPM_10#show mpls atm-ldp bindings 172.27.1.129 32
 Destination: 172.27.1.129/32
     Headend Router Switch1.27 (3 hops) 0/36  Active, VCD=209
PE_m8850_RPM_10#
```

To finish the multi-vc configuration, we need to enable multi-vc in the MPLS subinterfaces. See Example 7-20.

Example 7-20 *Enabling Multi-vc*

```
PE_m8850_RPM_10#conf t
Enter configuration commands, one per line.  End with CNTL/Z.
PE_m8850_RPM_10(config)#interface switch 1.27
PE_m8850_RPM_10(config-subif)#mpls atm multi-vc
PE_m8850_RPM_10(config-subif)#^Z
PE_m8850_RPM_10#
```

You can see in Example 7-21 that we have three LVCs to the remote PE destination.

Example 7-21 *Checking ATM LDP Bindings with Multi-vc*

```
PE_m8850_RPM_10#show mpls atm-ldp bindings 172.27.1.129 32
 Destination: 172.27.1.129/32
     Headend Router Switch1.27 (3 hops) 0/40  Active, VCD=238, CoS=available
     Headend Router Switch1.27 (3 hops) 0/42  Active, VCD=236, CoS=standard
     Headend Router Switch1.27 (3 hops) 0/44  Active, VCD=234, CoS=premium

PE_m8850_RPM_10#
```

Configuring the Core Behaviors

In the VSI slaves, four CoS buffers (CoSBs) (QBins in BXM and UXM slaves) are reserved for MPLS. As you know from Chapter 3, "Implementations and Platforms," DiffServ queuing is achieved by VSI and the Service Class Templates (SCTs), providing bandwidth subpartitioned for different classes of service. Each COS (control, premium, standard, and available) is mapped to one of those CoSBs in the general parameters of the SCT. So different traffic classes are queued separately, and each queue (for each class) receives a bandwidth allocation. This mapping is shown in Table 7-4.

Table 7-4 *Class-to-L-LSP Mapping in Our Example*

Class	IP Precedence	AXSM CoSB Number	BXM and UXM CoSB Number
0	0/4	7	10
1	1/5	8	11
2	2/6	9	12
3	3/7	10	13

This can be seen with the command **dspsct** on the BPX-8650, IGX-8400, and MGX-8850 platforms. More details can be found in the "Service Class Templates and QoS" section in Chapter 3.

This is the MPLS-only SCT in BPX-8600 and IGX-8400 slaves, which is the default SCT used. See Example 7-22:

Example 7-22 *Checking Service Class Templates in BPX and IGX Switches*

```
b8620-7b         TN    Cisco         BPX 8620  9.3.30    Dec. 9 2001   18:43 GMT

              Service Class Map for MPLS1 Template

Service Class    Qbin    Service Class    Qbin    Service Class    Qbin

     Default      13
     Signaling    10
     Tag0         10
     Tag1         11
     Tag2         12
     Tag3         13
     Tag4         10
     Tag5         11
     Tag6         12
     Tag7         13
     TagAbr       14

Last Command: dspsct 1


Next Command:
```

On MGX-8850 platforms, the default SCT which is SCT 0 does not have MPLS support. Example 7-23 shows SCT number 4 in an AXSM slave.

Example 7-23 *SCT Number 4*

```
m8850-7a.1.AXSM.a > dspsct gen 4 port


+----------------------------------+
¦ SCT - VERSION ¦ FIRMWARE - VERSION ¦
¦ 0000000000001 ¦ 000000000000000001 ¦
+----------------------------------+

+--------------------------------------------------------------------+
Service Class Template [4] : General Parameters                      (
+--------------------------------------------------------------------+
¦ SERV-TYPE ¦ COSB_NUM ¦ CAC_TYPE ¦ UPC_ENB  ¦ CLP-SELEC ¦   GCRA-1      )
+--------------------------------------------------------------------+
¦ VSI-SIG   ¦ 00000016 ¦  B-CAC  ¦GCRA 1 & 2¦ 000000002 ¦   DISCARD (
¦ CBR.1     ¦ 00000003 ¦  B-CAC  ¦GCRA1-ENB ¦ 000000003 ¦   DISCARD )
¦ VBR-RT.1  ¦ 00000004 ¦  B-CAC  ¦GCRA 1 & 2¦ 000000002 ¦   DISCARD (
¦ VBR-RT.2  ¦ 00000004 ¦  B-CAC  ¦GCRA 1 & 2¦ 000000001 ¦   DISCARD )
¦ VBR-RT.3  ¦ 00000004 ¦  B-CAC  ¦GCRA 1 & 2¦ 000000001 ¦   DISCARD (
¦ VBR-nRT.1 ¦ 00000005 ¦  B-CAC  ¦GCRA 1 & 2¦ 000000002 ¦   DISCARD )
¦ VBR-nRT.2 ¦ 00000005 ¦  B-CAC  ¦GCRA 1 & 2¦ 000000001 ¦   DISCARD (
¦ VBR-nRT.3 ¦ 00000005 ¦  B-CAC  ¦GCRA 1 & 2¦ 000000001 ¦   DISCARD )
¦ UBR.1     ¦ 00000006 ¦ LCN_CAC ¦GCRA1-ENB ¦ 000000003 ¦   DISCARD (
¦ UBR.2     ¦ 00000006 ¦ LCN_CAC ¦GCRA1-ENB ¦ 000000003 ¦ DSCD/SET-CLP )
¦ ABR       ¦ 00000001 ¦  B-CAC  ¦GCRA1-ENB ¦ 000000003 ¦   DISCARD (
¦ CBR.2     ¦ 00000003 ¦  B-CAC  ¦GCRA1-ENB ¦ 000000003 ¦   DISCARD )
¦ CBR.3     ¦ 00000003 ¦  B-CAC  ¦GCRA 1 & 2¦ 000000001 ¦   DISCARD (
¦ TagCOS-0c ¦ 00000007 ¦ LCN_CAC ¦DISABLED  ¦ 000000001 ¦   DISCARD )
¦ TagCOS-1c ¦ 00000008 ¦ LCN_CAC ¦DISABLED  ¦ 000000001 ¦   DISCARD (
¦ TagCOS-2c ¦ 00000009 ¦ LCN_CAC ¦DISABLED  ¦ 000000001 ¦   DISCARD )
¦ TagCOS-3c ¦ 00000010 ¦ LCN_CAC ¦DISABLED  ¦ 000000001 ¦   DISCARD (
¦ TagCOS-4c ¦ 00000007 ¦ LCN_CAC ¦DISABLED  ¦ 000000001 ¦   DISCARD )
¦ TagCOS-5c ¦ 00000008 ¦ LCN_CAC ¦DISABLED  ¦ 000000001 ¦   DISCARD (
¦ TagCOS-6c ¦ 00000009 ¦ LCN_CAC ¦DISABLED  ¦ 000000001 ¦   DISCARD )
¦ TagCOS-7c ¦ 00000010 ¦ LCN_CAC ¦DISABLED  ¦ 000000001 ¦   DISCARD (
+--------------------------------------------------------------------+

m8850-7a.1.AXSM.a >
```

To differentiate between the two IP Precedence values in the same CoSB or QBin, we use the Cell Loss Priority (CLP) bit in the ATM cell header as a drop priority.

From the LSC, we can see the service type of the different parallel LVCs (TagCOS0 to TagCOS7) and infer the CoSB used from the SCT. To do this we use the command **show xtagatm cross-connect** for a specific interface and VPI/VCI, as shown in Example 7-24.

Example 7-24 *Using* **show xtagatm cross-connect** *for a Specific Interface*

```
m8850_RPM_9#show xtagatm cross-connect interface xTagATM 101 0 40
Phys desc:   10.1
Interface:   XTagATM101
```

continues

Example 7-24 *Using* **show xtagatm cross-connect** *for a Specific Interface (Continued)*

```
Intf type:    extended tag ATM
VPI/VCI:      0/40
X-Phys desc: 1:1.4:40
X-Interface: XTagATM1144
X-Intf type: extended tag ATM
X-VPI/VCI:    27/742
Conn-state:  UP
Conn-type:   input
Cast-type:   point-to-point
Rx service type:   Tag COS 0
Rx cell rate:      n/a
Rx peak cell rate: 10000
Tx service type:   n/a
Tx cell rate:      n/a
Tx peak cell rate: n/a

m8850_RPM_9#show xtagatm cross-connect interface xTagATM 101 0 42
Phys desc:    10.1
Interface:    XTagATM101
Intf type:    extended tag ATM
VPI/VCI:      0/42
X-Phys desc: 1:1.4:40
X-Interface: XTagATM1144
X-Intf type: extended tag ATM
X-VPI/VCI:    27/740
Conn-state:  UP
Conn-type:   input
Cast-type:   point-to-point
Rx service type:   Tag COS 1
Rx cell rate:      n/a
Rx peak cell rate: 10000
Tx service type:   n/a
Tx cell rate:      n/a
Tx peak cell rate: n/a

m8850_RPM_9#show xtagatm cross-connect interface xTagATM 101 0 44
Phys desc:    10.1
Interface:    XTagATM101
Intf type:    extended tag ATM
VPI/VCI:      0/44
X-Phys desc: 1:1.4:40
X-Interface: XTagATM1144
X-Intf type: extended tag ATM
X-VPI/VCI:    27/738
Conn-state:  UP
Conn-type:   input
Cast-type:   point-to-point
Rx service type:   Tag COS 2
Rx cell rate:      n/a
Rx peak cell rate: 10000
Tx service type:   n/a
Tx cell rate:      n/a
```

Example 7-24 *Using* **show xtagatm cross-connect** *for a Specific Interface (Continued)*

```
Tx peak cell rate: n/a
m8850_RPM_9#


m8850_RPM_9#show xtagatm cross-connect interface xTagATM 101 0 32
Phys desc:    10.1
Interface:    XTagATM101
Intf type:    extended tag ATM
VPI/VCI:      0/32
X-Phys desc: 9.1
X-Interface: n/a
X-Intf type: switch control port
X-VPI/VCI:    0/134
Conn-state:   UP
Conn-type:    input/output
Cast-type:    point-to-point
Rx service type:   Tag COS 7
Rx cell rate:      n/a
Rx peak cell rate: 353208
Tx service type:   Tag COS 7
Tx cell rate:      n/a
Tx peak cell rate: 353208
m8850_RPM_9#
```

It is very important to note that the control-vc for untagged traffic (VPI/VCI 0/32 by default) uses the service type Tag CoS 7. So it shares the CoSB with the control class LVCs.

We will now configure the per-hop behaviors (PHBs) and, specifically, the PSC by setting the weights to implement class-based weighted fair queuing (CBWFQ) among the CoS buffers. We do that by configuring the weight as a percentage for each class in each LC-ATM interface, as shown in Example 7-25. That is, the total weight for the four classes in an XTagATM interface is 100 and corresponds to the full interface bandwidth:

Example 7-25 *Configuring the ATM CoS Weights*

```
m8850_RPM_9#conf t
Enter configuration commands, one per line.  End with CNTL/Z.
m8850_RPM_9(config)#interface xTagATM 101
m8850_RPM_9(config-if)#mpls atm cos control 1
m8850_RPM_9(config-if)#mpls atm cos available 29
m8850_RPM_9(config-if)#mpls atm cos standard 35
m8850_RPM_9(config-if)#mpls atm cos premium 35
m8850_RPM_9(config-if)#^Z
m8850_RPM_9#
```

So for this XTagATM interface, the premium class and the standard class have the same relative weight, the available class has a weight of 29/35 relative to each of the previous two (that is, a 29/100 weight relative to the 100 percent of interface bandwidth), and the control class has a weight of 1/100 relative to the whole interface.

NOTE	Even though no LVC is using the class control, we configure a relative weight of 1 percent for the CBWFQ PSC. This is because the control-vc carrying LDP and IGP updates (untagged traffic) uses the control CoS buffer.

We now configure all XTagATM interfaces in all LSCs to use these weights.

Configuring the Edge Behaviors

From the DiffServ model, we still need to configure the edge behaviors. We start by configuring the PE router interfaces toward the P routers (facing the MPLS network) by enabling CBWFQ, low-latency queue (LLQ), and weighted random early discard (WRED). We use the Cisco IOS Modular QoS CLI (MQC) provisioning mechanism to accomplish this task.

NOTE	These MQC features require CEF to be enabled. MQC features work only for packets traversing the router on the Cisco IOS CEF switching path. A common mistake is to test this with packets originating in the tested router. This does not work, because those packets are process-switched.

We first define our classes under **class-map** statements (see Example 7-26).

Example 7-26 *PE* **class-map** *Configuration*

```
PE_m8850_RPM_10#conf t
Enter configuration commands, one per line.  End with CNTL/Z.
PE_m8850_RPM_10(config)#class-map match-all voip
PE_m8850_RPM_10(config-cmap)#match mpls experimental 3
PE_m8850_RPM_10(config-cmap)#class-map match-all dlsw
PE_m8850_RPM_10(config-cmap)#match mpls experimental 2
PE_m8850_RPM_10(config-cmap)#class-map match-all avail1
PE_m8850_RPM_10(config-cmap)#match mpls experimental 4 5
PE_m8850_RPM_10(config-cmap)#class-map match-all avail2
PE_m8850_RPM_10(config-cmap)#match mpls experimental 0 1
PE_m8850_RPM_10(config-cmap)#exit
```

We define four classes (voip, dlsw, avail1, and avail2), and we match on the MPLS experimental bits copied from the IP Precedence bits.

It is important to note that we are matching on MPLS experimental bits. The following is an excerpt from RFC 3035:

...when a labeled packet is transmitted on an LC-ATM interface, where the VPI/VCI (or VCID) is interpreted as the top label in the label stack, the packet MUST also contain a **shim header**. [...] The label value of the top entry in the shim (which is just a "placeholder" entry) MUST be set to 0 upon transmission, and MUST be ignored upon reception. The packet's outgoing TTL, and its CoS, are carried in the TTL and CoS fields respectively of the top stack entry in the shim.

Then we define the policy to apply to those classes in a **policy-map**, building the Policy block in the edge behavior (see Example 7-27).

Example 7-27 *PE **policy-map** Configuration*

```
PE_m8850_RPM_10(config)#policy-map To-Network
PE_m8850_RPM_10(config-pmap)#class voip
PE_m8850_RPM_10(config-pmap-c)#priority 128
PE_m8850_RPM_10(config-pmap-c)#class dlsw
PE_m8850_RPM_10(config-pmap-c)#bandwidth 256
PE_m8850_RPM_10(config-pmap-c)#random-detect
PE_m8850_RPM_10(config-pmap-c)#class avail1
PE_m8850_RPM_10(config-pmap-c)#bandwidth 192
PE_m8850_RPM_10(config-pmap-c)#random-detect
PE_m8850_RPM_10(config-pmap-c)#class avail2
PE_m8850_RPM_10(config-pmap-c)#bandwidth 64
PE_m8850_RPM_10(config-pmap-c)#random-detect
PE_m8850_RPM_10(config-pmap-c)#set atm-clp
PE_m8850_RPM_10(config-pmap-c)#exit
PE_m8850_RPM_10(config-pmap)#exit
```

In these policies we are specifying CBWFQ, a LLQ for the voip class, and WRED. We are differentiating between the avail1 and avail2 classes by a different weight for CBWFQ, but the two classes use the same LVC in the core (Available). To differentiate them in the core, we set the ATM cell header CLP bit in the class avail2.

NOTE   To differentiate the drop priority (CLP bit) in the core, the CoS buffers descriptor template in the SCT needs to be configured for CLP Hysteresis as the Discard Selection, not Frame Discard. Frame Discard is the default. You need to change it so that cells with the CLP bit set or clear are treated differently.

On IGX-8400 and BPX-8600 platforms, the discard selection method can be changed between EPD (a frame-discard method) and CLP hysteresis (a selective cell-discard method) using the command **cnfqbin**. With these platforms, changing the discard selection mechanism in one QBin does not change the default values in the template.

A different way of setting the CLP bit is with **set-clp-transmit** as a **policing** action also using MQC.

We apply the policies for those classes as an output service policy in the PE interface toward the LSR, as shown in Example 7-28.

Example 7-28 *PE* **service-policy** *Configuration*

```
PE_m8850_RPM_10(config)#interface switch 1.27
PE_m8850_RPM_10(config-subif)#service-policy output To-Network
PE_m8850_RPM_10(config-subif)#^Z
PE_m8850_RPM_10#
```

Finally, we need to classify the traffic. We can do so either in the CE router as an output policy toward the PE or in the PE router as an input policy from the CE. In most cases, the CE router is a managed device. Therefore, in our example we will classify traffic in egress on the CE router toward the PE. If the CE router is not owned, the classification can be done at ingress in the PE router following the same procedure and applying the policy as a service-policy input.

For the classification, as shown in Example 7-29, we will use an access list (ACL) to identify VoIP traffic, a protocol match in MQC to categorize DLSW traffic, and other protocol matches to recognize some specific protocols (Telnet, HTTP, POP3, and SMTP are considered low-priority for this exercise). We will group those packets in the avail2 class and use class-default as the avail1 class. The class-default includes all packets not matched by any of the other classes.

Example 7-29 *CE* **class-map** *Configuration*

```
CE_1#conf t
Enter configuration commands, one per line.  End with CNTL/Z.
CE_1(config)#access-list 100 permit udp any any range 16384 32767
CE_1(config)#class-map match-any voip
CE_1(config-cmap)#match access-group 100
CE_1(config-cmap)#class-map match-any dlsw
CE_1(config-cmap)#match protocol dlsw
CE_1(config-cmap)#class-map match-any avail2
CE_1(config-cmap)#match protocol telnet
CE_1(config-cmap)#match protocol http
CE_1(config-cmap)#match protocol pop3
CE_1(config-cmap)#match protocol smtp
CE_1(config-cmap)#exit
```

After the classification is done, we mark those packets in those classes in a policy map, as shown in Example 7-30. We will mark IP Precedence that at the PE router will be used to identify these classes copied to the EXP bits.

Example 7-30 *CE* **policy-map** *Configuration*

```
CE_1(config)#policy-map To-PE
CE_1(config-pmap)#class voip
CE_1(config-pmap-c)#set ip precedence 3
CE_1(config-pmap-c)#class dlsw
CE_1(config-pmap-c)#set ip precedence 2
CE_1(config-pmap-c)#class avail2
CE_1(config-pmap-c)#set ip precedence 1
CE_1(config-pmap-c)#class class-default
```

Example 7-30 *CE* **policy-map** *Configuration (Continued)*

```
CE_1(config-pmap-c)#set ip precedence 4
CE_1(config-pmap-c)#^Z
CE_1#
```

Finally, we apply that policy to the FE interface in the output direction, as shown in Example 7-31.

Example 7-31 *CE* **service-policy** *Configuration*

```
CE_1(config)#interface fastEthernet 0/0
CE_1(config-if)#service-policy output To-PE
```

We perform a similar configuration in the other PE and CE routers, setting up the other direction.

MQC can also serve a different purpose than QoS support. To provide for counters and statistics on the different frames received by a PE with different EXP bits values, you can configure a **policy-map** that only sets the experimental bits to the same value as matched by the **class-map**, and then apply the **service-policy** in the input direction in the ATM interface facing the P router. This **service-policy** does not modify the packets, but enables per-COS counters to be displayed using the command **show policy-map interface**. So a **service-policy** in the output direction serves as congestion management towards the P router, and a **service-policy** in the input direction collects counters from the P router.

QoS Example Summary

As a summary of our QoS example, we can go backwards and draw a table with the network behavior. It would look similar to Table 7-5.

Table 7-5 *QoS Summary in Our Example*

Class	CE		PE		LVC
	Classif	**Marking**	**Classif**	**Marking**	
VoIP	VoIP ACL	IP Prec 3	MPLS Exp 3	LLQ WRED	Premium (Class 3; Precedence 3/7)
DLSW	DLSW protocol	IP Prec 2	MPLS Exp 2	CBWFQ WRED	Standard (Class 2; Precedence 2/6)
Avail2	Telnet, HTTP, POP3, and SMTP protocols	IP Prec 1	MPLS Exp 0 1	CBWFQ WRED CLP => 1	Available (Classes 0/1; Precedence 0/1/4/5)
Avail1	The rest	IP Prec 4	MPLS Exp 4 5	CBWFQ WRED	

This table helps you understand the complete picture. It also helps you map an IP packet entering the CE router to the LVC it will use while traversing the ATM LSRs.

Summary

This chapter concentrated on the configuration aspects of two applications of MPLS—VPNs and QoS. MPLS overcomes the problem of overlapping IP address ranges by introducing the route descriptor field that unambiguously identifies a service provider's end customer routes within the MPLS core. With the address problem resolved, this chapter moved on to describe the per-switch configuration of multi-protocol iBGP, required to transport customer VPN routes and associated traffic across the core. Besides iBGP, an explanation of the eBGP configuration is required to pass the VPN customers' private routes to the provider's edge routers.

After the VPN configuration, a QoS example was offered. Frame-based MPLS's method of conveying DiffServ information to LSRs using a shim header was contrasted with the cell-based MPLS method using the label and CLP bit. CoS enables QoS, so this chapter finished with a discussion of how to enable multi-vc functionality to support up to four classes of service along a given path. In addition, the CLI syntax required to categorize incoming traffic into the proper CoS was given.

CHAPTER 8

PNNI Explained

ATM is a fabulous implementation of a transport protocol. It provides differentiated Quality of Service (QoS), offers dynamic circuit provisioning, and initiates traffic rerouting around network failures. Many of these automatic features would not be possible without the inter-ATM switch routing and signaling protocol called Private Network-to-Network Interface (PNNI). PNNI is defined by the ATM Forum. The complete specification, af-pnni-0050.000, is available for download at the Forum's Web site. Many popular broadband consumer services such as high-speed data access via cable and DSL and next-generation wireless networks all rely on PNNI's services in the ATM core of the network to route massive amounts of end-user traffic. This reliance is because of the distributed control, scalable, resilient, easy to maintain, and standards-based nature of PNNI.

The discussion of PNNI must include a section on ATM end-system addressing for completeness because a PNNI-controlled ATM switch cannot deliver information to end systems without addresses.

ATM End-System Addressing

For two or more ATM devices to communicate, it is critical that they be able to locate and identify one another. An ATM address serves a dual role by functioning as an identifier and a locator. Like a license plate, the address signifies the name of the ATM device. Like a zip code, PNNI uses a portion of the ATM End System Address (AESA) to lock in the ATM device's position within the network topology.

Address Structures

Before we delve into the specifics of address structure, be aware that ATM addresses are different and independent from the VPI and VCI information included in the ATM cell header. VPI and VCI are not addresses but identifiers, implying the virtual identifier's local significance. An ATM address is significant network-wide, but VPIs and VCIs are significant only across a given point-to-point ATM link. In addition, any time ATM address information must be transported across an ATM network, as is the case in an ATM switched virtual circuit (SVC) call setup message, the ATM address must be segmented and transported inside the same discrete ATM cells that traffic all other data across the ATM infrastructure.

The ATM Forum has standardized the length of ATM addresses at 20 bytes. Currently, there are three different standardized formats. Each format defines all the fields in the entire 20-byte address structure. A fourth format known as the *local format* defines only the first byte of the address and leaves the definition of the remaining 19 bytes to the private organization that chooses to use this format.

Structured ATM address formats fall into one of two categories — the E.164 address category, used in telephone networks and administered by the ITU-T, or the AESA format. You might have heard the term Network Service Access Point (NSAP) address instead of AESA. NSAPs are address types defined by the International Organization for Standardization (ISO), and AESAs are types of NSAPs. At present, the two standardized AESA types are Data Country Code (DCC) and International Code Designator (ICD). A third AESA type called local allows any network administrator to define the AESA format for private use.

Regardless of the format, the first byte of an ATM address signifies the format in use. Appropriately, this first byte is called the Address Format Identifier (AFI). The AFI values for each of the four ATM address formats are shown in Table 8-1.

Table 8-1   *AFI Values*

AFI in Hexadecimal	ATM Address Format
39	DCC
47	ICD
45	E.164
49	Local

The three structured address formats — DCC, E.164, and ICD — all define the format of the remaining 19 ATM address bytes. All three include the following common fields in the address:

- **Initial Domain Identifier (IDI)** — Varies depending on the value of the AFI. It defines the country, if the AFI is DCC, or the organization, when the AFI is ICD, where the AESA resides. The IDI contains a valid E.164 telephone number when the AFI is set to hexadecimal 45. See the next section for more on the E.164 IDI format.

- **Domain-Specific Part (DSP)** — Made up of two subfields:
  - **High-Order Domain-Specific Part (HO-DSP)** — Agencies such as ANSI that administer NSAP allocation and format use this subfield to define the length and fields within the entire DSP. Typically, the defining organization

dictates the content of the first few fields in the HO-DSP and lets the network administrator define the rest. ANSI administers DCC values for the U.S., and the British Standards Institute administers ICD values and ICD-based AESA formats on behalf of ISO. The 4-byte E.164 HO-DSP might be set to all 0s.

— **End System Identifier (ESI)** — This 6-byte field typically contains the IEEE 802.2 media access control (MAC) address, but it is not required. For an end system such as a router or host computer, the ESI links the AESA to one of the end system's physical interfaces.

- **Selector Byte (SEL)** — End systems use this byte for purposes defined by the network administrator. This field is not used in PNNI routing decisions.

Specifics of the E.164 IDI Field

The IDI portion of the E.164 AESA format carries ISDN numbers in one of three predefined formats: geographic area, global services, and network. Figure 8-1 shows these E.164 IDI formats. All three formats begin with a one-, two-, or three-digit country code (CC) that identifies the telephone number's destination country. Other parts of the geographic format include the national destination code (NDC) and the subscriber number (SN). Appending the SN to the NDC produces a nationally significant telephone number. CCs are allocated to providers that offer a worldwide telephone number to each of their subscribers. These global service providers allocate a global service number (GSN) for each subscriber.

Figure 8-1 *IDI Format of E.164 AESA*

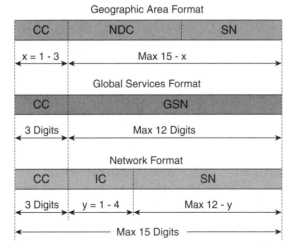

Service providers that are operating networks that span two or more countries and do not have a common telephone numbering plan receive their own CC as well as a two-digit identification code (IC).

The ISDN number inside each E.164 IDI is padded on the left with 0s until the number is equivalent to 15 binary coded decimal (BCD) numbers. Each BCD number is 4 bits long. 4 binary 1 bits (1111) are appended to the 0-padded ISDN number to lengthen it to 8 bytes—the size of the IDI field.

Figure 8-2 shows the ATM address structure of each of the four formats.

Figure 8-2 *ATM Address Formats*

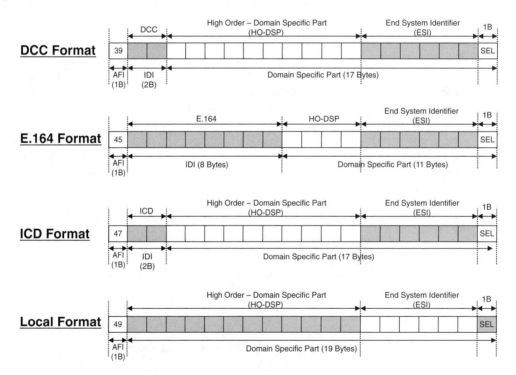

Using ILMI for Automatic Address Registration

To alleviate the cumbersome task of manually typing a 20-byte ATM address into an end station such as a router, the Integrated Local Management Interface (ILMI) protocol defines a method for automatic address registration and configuration between the ATM end system on one side of the User Network Interface (UNI) and an ATM switch on the opposite side of the UNI.

NOTE   ILMI stands for Interim Local Management Interface in the ATM Forum "ATM User-Network Interface 3.1 (UNI 3.1)" specification, implying its temporary scope. In version 4.0, ILMI was specified in a separate document called "Integrated Local Management Interface (ILMI) Specification Version 4.0," in which ILMI was renamed to stress the expectation of using ILMI procedures "indefinitely."

The ATM end system requests the network portion of its AESA from the switch. Assuming that the end system attaches to the switch via a dedicated physical link, the switch and end system use the standard full-duplex ILMI VCC of VPI=0, VCI=16 to communicate. In turn, the ATM switch relays the first 13 bytes of the address to the end device. These 13 bytes include the AFI, IDI, and HO-DSP. The end system prefixes these 13 bytes to its own ESI and SEL to form a complete AESA. Typically, the SEL is set to 0. The end station's newly formed AESA is relayed back to the switch via the ILMI protocol, and the switch uses this AESA to route SVC calls to the corresponding end station. Note that with E.164 addressing, the ATM switch supplies the full 20-byte ATM address to the end system.

The ATM address is deregistered from the end system and the switch whenever the ILMI link between them fails. Figure 8-3 depicts the address registration process. Deregistration implies that the ATM end system can be relocated and attached to another PNNI-controlled switch without tedious reconfiguration.

Figure 8-3   *ATM End System Automatic Address Registration Process Using ILMI*

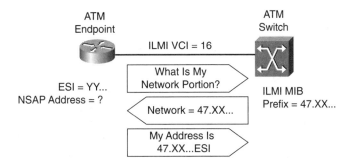

In an effort to leverage existing standards, the ILMI protocol employs Simple Network Management Protocol (SNMP) to categorize and relay pertinent information regarding these four areas:

- The physical and ATM layers
- Virtual connections
- Address registration
- Service registration across the physical or virtual ATM link

The ATM interface Management Information Base (MIB) is a collection of SNMP Object Identifiers (OIDs) that represent all the relevant attributes associated with these four areas. One device relays information across the ILMI link to the other by attaching that information to the proper OID, encapsulating that OID/information pair inside an SNMP protocol data unit (PDU) and moving that PDU down through the various ATM adaptation layers to the physical layer for transport. Here are some of the configuration attributes relayed across the ILMI link during interface auto-configuration:

- ATM address prefix and scope and address registration
- ATM interface configuration
- VPI/VCI ranges on UNI and NNI interfaces
- UNI/NNI user/network side and version discovery
- Keepalives

ATM Call Signaling

PNNI was devised to distribute, organize, and store routing information to all destinations within a given ATM network. PNNI-controlled ATM switches build SVCs and Soft Permanent Virtual Circuits (SPVCs) using this routing information. To initiate, destroy, and restore these virtual circuits (VCs), a set of signaling protocol messages was invented to control the state of a given circuit. State diagrams dictate the actions performed on a given connection based on the received inputs. Well-defined and standardized state diagrams allow signaling to occur between equipment built by different manufacturers.

SVC Signaling Specifications

ATM signaling occurs when one device wants to communicate its request or observation to another. Signaling occurs between switches and endpoints across a UNI interface or among switches inside a PNNI network. For signaling to be effective and efficient, the messages used in the signaling exchange must have an agreed-upon format, and they must be exchanged in a predefined order or protocol based on received inputs. The standards that define the message formats and protocol exchanges are shown in Figure 8-4.

The standards are stacked on top of one another to emphasize that signaling messages are built from the top down. Figure 8-4 shows the Signaling ATM Adaptation Layer (SAAL) enveloping the Service-Specific Convergence Sublayer (SSCS), Service-Specific Connection-Oriented Protocol (SSCOP), and Common Part Convergence Sublayer (CPCS) because these comprise the sublayers of the SAAL. Figure 8-5 describes the function of each SAAL sublayer.

Figure 8-4 *Parts of the ATM Signaling Stack*

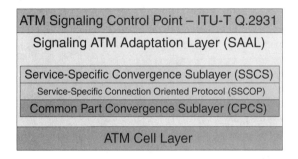

Figure 8-5 *Functions of the SAAL Sublayers*

SSCF (Q.2130)	**Service-Specific Coordination Function** – Session Management of Q.2930 to Q.2110
SSCOP (Q.2110)	**Service-Specific Connection-Oriented Protocol** – SSCOP Signaling Segments Exist Between Adjacent Switches – Ensures Reliable Q.2931 Signaling Message Transport – Has Error Recovery, Flow Control, Error and Status Reporting
CPCS (I.363)	**Common Part Convergence Sublayer** – Encapsulates Signaling Messages in AAL5 Frames – Protects Signaling Messages Using AAL5's CRC

Layer-to-Layer Communication

The Q.2931 signaling layer is the source (origination point) and sink (termination point) of all signaling messages. This layer must exchange signaling messages or datagrams with peer layers in other PNNI nodes as well as with ATM end systems connected via UNI links. Messages originated in the Q.2931 layer flow downward through the SAAL sublayers to the ATM cell layer, where the formatted and packaged signaling message is segmented into cells and is transmitted over the physical layer. During this downward flow, the higher layer of the signaling stack communicates with the next lower layer using primitives. The four communication primitives are as follows:

- **Request**—Sent from a higher layer to a lower layer, asking to perform a service.
- **Indication**—A notification sent to an upper layer by a lower layer regarding a request-related service.

- **Response**—Used by an upper layer to acknowledge the receipt of the indication.
- **Confirm**—Used to signify service completion.

Figure 8-6 shows the primitive exchange between the Q.2931 and SAAL layers in two ATM switches.

Figure 8-6   *Sample Primitive Exchange Between the Layers*

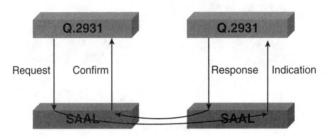

Reference Points

A few terms need defining before we can discuss end-to-end ATM virtual circuit creation and destruction. These terms clearly distinguish one end of the virtual circuit from the other. The *calling party* is the circuit endpoint that initiates the call; the *called party* is the endpoint that receives the call. Because circuits often route through multiple nodes on the way to the called party, the *preceding node* is any node that receives the virtual circuit setup/connect request and forwards it to another node. A *succeeding node* receives a setup/connect request from the preceding node. Finally, messages from the calling party to the called party flow in the "downstream" direction. Conversely, messages going the opposite way flow in the "upstream" direction. All these reference points are shown in Figure 8-7.

Figure 8-7   *Reference Points*

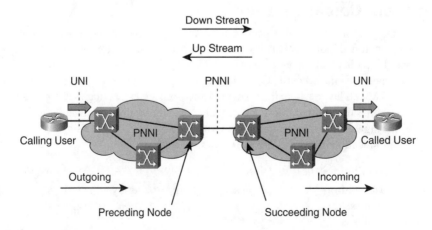

We also need to define direction (user to network or network to user) and significance (local or global), as shown in Figure 8-8.

Figure 8-8   *Definition of Direction and Significance*

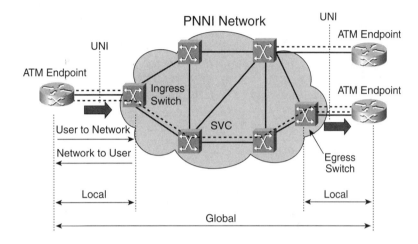

Common Signaling Messages

Virtual circuits connected using the PNNI signaling procedures outlined in Section 6 of the PNNI 1.0 specification use the same procedures as those outlined in the ITU-T's Q.2931 signaling specification for B-ISDN DSS2 networks. The ATM UNI interface specification references this same document for virtual circuit creation and deletion. The Q.2931 document outlines a set of signaling messages used to set up and tear down connection as well as to verify the status of those connections. The following list categorizes these messages:

- Call establishment messages
  - **Setup**—Initiates call establishment
  - **Call Proceeding**—The establishment phase has begun
  - **Connect**—The call is accepted by the receiver
- Call clearing messages
  - **Release**—Initiates call clearing
  - **Release Complete**—Acknowledges call clearing
- Miscellaneous messages
  - **Notify**—For the user to inform the network of the call
  - **Status**—Reports information regarding a particular call
  - **Status Enquiry**—Requests information regarding a particular call

Now that the signaling message types are defined, it is important to know whether the calling or called party can initiate the message. In addition, if the calling and called party are separated by multiple ATM switches, it is important to know if a message has end-to-end/global significance or if it is only locally significant between two directly connected switches. Table 8-2 lists the message types, the initiators, and the significance of each message.

Table 8-2 *Direction and Significance of Point-to-Point Signaling Messages*

Message Type	Message Name	Direction To	Significance
Establishment	Setup	Both	Global
Establishment	Call Proceeding (Proc.)	Both	Local
Establishment	Connect	Both	Global
Establishment	Connect Acknowledge (Ack.)	Both	Local
Clearing	Release	Both	Global
Clearing	Release Complete (Comp.)	Both	Local
Miscellaneous	Status	Both	Local
Miscellaneous	Status Enquiry	Both	Local
Miscellaneous	Restart	Both	Local
Miscellaneous	Restart Acknowledge (Ack.)	Both	Local

The additional message types for point-to-multipoint connections are shown in Table 8-3.

Table 8-3 *Messages for Point-to-Multipoint Connections*

Message Type	Message Name	Direction To	Significance
Multipoint	Add party	Both	Global
Multipoint	Add party acknowledge	Both	Global
Multipoint	Add party reject	Both	Global
Multipoint	Drop party	Both	Global
Multipoint	Drop party acknowledge	Both	Local

Signaling Channels

Before SVC signaling between two nodes commences, a dedicated AAL5 channel must be built between them. Signaling information regarding all dynamic ATM virtual circuit connections that traverse these two nodes passes over this channel.

Every node in a PNNI network has a dedicated AAL5 signaling channel with each of its directly attached neighbor nodes with which it intends to construct dynamic ATM virtual circuits. Typically, VPI 0 and VCI 5 pass SVC signaling information if the entire VPI range is available. Optionally, if two nodes are linked via a VPC, associated signaling is employed, and the signaling VPI is the same as the VPC VPI value. The signaling VCI value remains 5 on the link unless the network administrator reconfigures it on the two nodes connected by the VPC.

As soon as the signaling channels are established between neighbor nodes, dynamic VC setup can occur. Point-to-point call initiation begins when the preceding switch (in the case of an SPVC) or an originating end system (in the case of an SVC call) sends a setup message. The setup message and other Q.2931 messages contain information elements (IEs), which are formatted blocks within the message that enclose the parameters necessary to process the call. The setup message sent from a PNNI-controlled switch must contain at least the called party address, ATM traffic descriptor, Designated Transit List (DTL) (described in the section "PNNI Overview"), and broadband bearer capability information elements. The general signaling message format is shown in Figure 8-9.

Figure 8-9   *General Signaling Message Format*

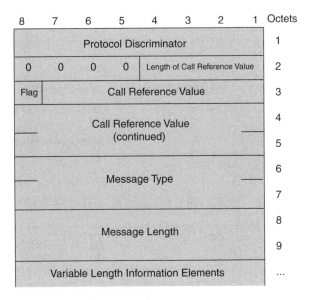

Each IE has the format shown in Figure 8-10.

Section 6 of the PNNI 1.0 specification, "PNNI Signaling Specification," contains detailed information on the mandatory IEs contained in and the optional IEs allowed in each PNNI signaling message. Section 6 also defines the purpose and coding of every field in each IE used in PNNI signaling.

Figure 8-10 *General IE Format*

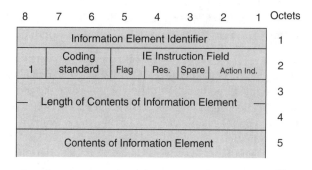

Call Control Timers

Timers start based on the sending or receiving of specific signaling messages. These timers ensure that the PNNI node's finite state machines for initiating and receiving dynamic virtual circuits are actively processing a call or are ready to. Table 8-4 lists the pertinent timers.

Table 8-4 *Q.2931 Call Timers*

Timer	Default	Start Reason	Stop Reason
T301	3 minutes	**ALERT** received	**CONNECT** received
T303	4 seconds	**SETUP** sent	**CONNECT, CALL PROC.**, or **RELEASE COMP.** received
T308	30 seconds	**RELEASE** sent	**RELEASE COMP.** or **RELEASE** received
T310	30 seconds	**CALL PROC.** received	**CONNECT, RELEASE**, or **ALERT** received
T313	4 seconds	**CONNECT** sent	**CONNECT ACK.** received
T316	2 minutes	**RESTART** sent	**RESTART ACK.** received
T317	—	**RESTART** received	Internal call references cleared
T322	4 seconds	**STATUS ENQUIRY** sent	**STATUS, RELEASE**, or **RELEASE COMP.** received

SVC Call Creation

Call or virtual circuit creation begins when the preceding side sends a setup message to the succeeding side. When the preceding side sends a setup message, it simultaneously starts timer T303. A call proceeding, connect, or release complete message must be received from the adjacent downstream node before this timer expires. Otherwise, this timer is reset and the setup message is retransmitted for a second time. A second T303 expiration forces the initiating ATM switch to send a release complete message to the succeeding node.

Typically, timer T303 is stopped by the receipt of a call proceeding message, and another timer, T310, starts at the preceding node. The receipt of a connect message stops T310 and moves the virtual circuit to the active state, but the receipt of an alerting message instead stops T310, commences timer T301, and transitions the virtual circuit state to alerting. Finally, the receipt of a connect message moves the call state from alerting to active and stops T301 if it was activated.

SVC Call Deletion

The preceding or succeeding side of an active SVC begins the deletion procedure by sending a release message toward the opposite side of the connection. Each transit or tandem PNNI-controlled switch receiving the release message forwards it to the subsequent downstream node and transmits a release complete message upstream to the immediately preceding node.

Whenever a node transmits a release message, it starts timer T308 and waits to receive either a release complete or release message. Unless both sides of a virtual circuit initiate a release at the same time, the node transmitting the release message receives a release complete message from the succeeding node.

Please note that release messages have end-to-end significance, whereas release complete messages are significant only across the local link between two adjacent nodes. If T308 expires without the receipt of either message, the timer is reset, and the release message is retransmitted. As soon as the timer expires for a second time, the PNNI-controlled node releases the circuit without regard for the virtual circuit's state at downstream nodes.

Figure 8-11 shows complete Q.2931 message flows for both SVC creation and deletion.

Figure 8-11 *SVC Creation and Deletion*

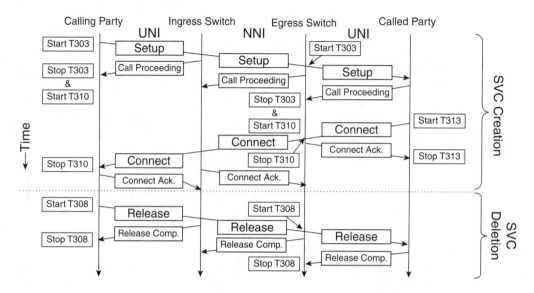

PNNI Overview

PNNI has two major components: signaling and routing. Figure 8-12 shows an overview of how signaling and routing combine to facilitate SVC or SPVC calls.

Figure 8-12 *PNNI: Combining Signaling and Routing*

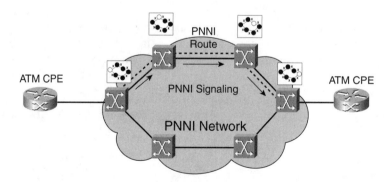

The major goals of PNNI are as follows:

- To calculate the first route that satisfies a connection's requirements
- To signal connection establishment along that route

The goal is never to calculate an optimal route. Doing so would slow connection establishment as multiple iterations of the source routing algorithm were invoked. Instead, the source node (typically, the one that directly connects to the ATM CPE) uses its routing algorithm to identify the first route available that satisfies all the requested connection parameters. Also, in hierarchical PNNI scenarios, the network topology is summarized using node and link aggregation, so it might be impossible to find the best route. Instead, the trade-off is great scalability achieved by hierarchy.

PNNI's routing component describes the ATM switch as a routing peer. It groups common peers into peer groups based on ID and ties common peers together using logical links. We will discuss these peers, groups, and links in detail shortly.

PNNI details a routing hierarchy of nodes. This hierarchy can have a single level or, at most, 104 levels. A simple single-level hierarchy or flat network requires all ATM switches to maintain complete routing information regarding the states of all the nodes and links in the network. All the ATM switches in the network are in a single peer group when the PNNI network comprises a single level.

As you probably suspect, the requisite amount of computational power and database memory per ATM switch grows as the number of nodes in the single-level hierarchy expands. Increasing the number of levels in the PNNI routing hierarchy helps stem this growth. A PNNI routing hierarchy allows an ATM peer to know the availability, reachability, and topology of its own PNNI peer group in detail and know that same information about other PNNI peer groups in general.

Dispersing routing information throughout a PNNI network would never occur without PNNI's signaling component. In addition, the signaling component initiates and builds ATM virtual circuits through the PNNI network. Making use of existing standards, PNNI's signaling component is derived from the ATM UNI 4.0 specification previously developed by members of the ATM Forum.

Taking its cue from a call setup request sent by a directly attached ATM UNI device, the signaling component within the ingress PNNI-controlled ATM switch builds a Designated Transit List (DTL). This DTL is a stack of instructions that tell all the downstream ATM switches how to route the virtual circuit through the PNNI network to the egress ATM switch. Because the DTL is implemented as a stack or "last in, first out" data structure, the top entry in the stack contains the route through the lowest-level peer group. Each subsequent stack entry contains the route through the next-level peer group. The bottom entry in the DTL stack contains the routing information needed to traverse the top level of the PNNI routing hierarchy. Note that in a multilevel DTL, only the lowest level or top stack entry contains detailed routing information such as node and link identifiers. The reason for this is that the ATM switch building the multilevel DTL has detailed topology information on only the node and link states in its peer group.

If ATM virtual circuit establishment fails at some node or link listed in the DTL, the node that constructed the DTL must modify it and replace the existing route with a new one that skirts the failure point. The process of returning the failed virtual circuit call attempt upstream to the originating ATM switch is called *crankback*. Regardless of whether the call attempt fails because of switch hardware problems, transmission link outage, or switch resource overloads, the call attempt is cranked back to the source of the DTL.

ATM virtual circuit connections are source-routed because the source switch that initiates the call establishment chooses the path through the network. The other ATM switches along the route just act on information listed in the received DTL.

Setting up Connections in the PNNI Routing Domain

PNNI's routing capabilities are widely known, but its call signaling capabilities are equally important. Whereas PNNI's routing capabilities spread an agreed-upon "road map" to destinations across a PNNI peer group, its call signaling capabilities allocate bandwidth on the paths connecting two or more destinations.

Although PNNI signaling is based on the UNI 4.0 specification, it adds additional features to facilitate expedient call routing through an ATM switch topology previously determined using PNNI's routing function. These additional features are as follows:

- DTLs that contain hop-by-hop routes through the hierarchical PNNI network to the ATM destination

- The ability to "crank back" or route around link failures in the PNNI-controlled ATM switch network

- Support for SPVCs

- Associated signaling to facilitate SPVC or SVC setup over any VPC logically connecting to PNNI-controlled ATM switches

The major PNNI IEs, including crankback and DTL, are described in Table 8-5.

Table 8-5   *Major PNNI Information Elements*

Information Element	Maximum Length	Maximum Number of Occurrences
Broadband bearer capability	7	1
QoS parameter	6	1
Calling party number	26	1
Called party number	25	1
ABR setup parameters	36	1
Called party soft PVC/PVCC	11	1
Calling party soft PVC/PVCC	10	1
Crankback	72	1
DTL	546	10

It is important to mention that one limiting factor in the width of a hierarchical PNNI network is the fact that the DTL IE (Designated Transit List Information Element) can have 10 maximum occurrences.

The following sections expound on each of these features.

DTL Construction

A complete source-originated route through a PNNI domain provides an ordered stack of DTLs. An individual DTL contains a sequential list of node ID and port ID pairs that unambiguously identify each succeeding next hop in the path from the source PNNI-controlled switch to the destination switch.

In flat and hierarchical PNNI networks, a PNNI-controlled ATM switch—called a *node* for short—has detailed network topology knowledge of nodes and connecting links within its PNNI peer group. However, a logical group node (LGN) representing a lower-level peer group at the next-higher level in a hierarchical network might or might not have only summary knowledge of other peer group topologies, depending on which node representation scheme is used. Simple node representation depicts a lower-level peer group as a single-cost hop at the next-higher-level peer group. This representation scheme is adequate when a lower-level peer group is small with few paths through the peer group. However, as the number of paths through the lower-level peer group starts to increase and the cost to traverse each path varies significantly, implementing complex node representation allows for more accurate routing.

At higher levels in the hierarchy, a complex node models a lower-level peer group as a nucleus with attached spokes or radii. Each spoke has an associated cost or metric and represents an exit point from the lower-level peer group. The complex logical node advertises the cost of traversing its radii to its fellow complex logical nodes within its higher-level peer group. Typically, a complex logical node advertises the same metric for all its radii unless some have significantly different costs. The complex node advertises an exception metric for spokes that have significantly different costs.

During source route calculation, at least two radii must be crossed to traverse a lower-level peer group represented as a complex node unless the complex node advertises exception bypass links. Exception bypass links directly connect two spokes in the complex logical node and, if available, enable routes that traverse a complex node in a single hop instead of passing through the nucleus.

Regardless of the node representation type implemented at upper layers of the routing hierarchy, if a node in one peer group must build a virtual circuit to a node in another peer group, the source routing node constructs a stack of DTLs with the most general DTL on the bottom of the stack and the most detailed on the top. The most general DTL needs nothing more than the node ID of the gateway switch used to reach the destination peer group, whereas the top DTL in the stack contains the detailed information necessary to traverse the source originating node's own peer group. Figure 8-13 shows an example of such a DTL using simple node representation.

Figure 8-13 *DTL Stack Manipulation Across Multiple Peer Groups*

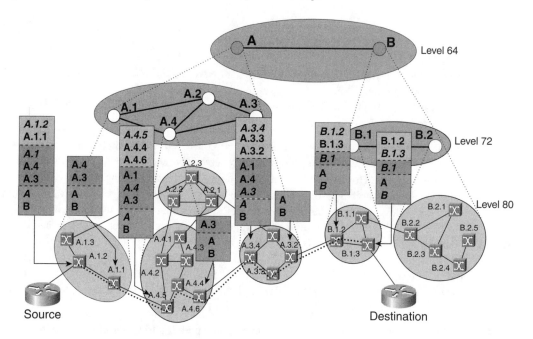

As soon as a virtual circuit setup message reaches the last node in the current DTL, that DTL is popped off the top of the DTL stack, and the message is forwarded according to the DTL now at the top of the stack. A stack of DTLs means that the PNNI network is hierarchical and that the required route to the destination traverses multiple PNNI peer groups. If the DTL at the top of the stack does not contain the detailed information to navigate through the next peer group, the entry node into the new peer group must generate a new DTL and place it on top of the stack. This new DTL must contain the requisite detailed information—node and port ID pairs—to reach the destination node if this node resides in the same peer group as the entry node. Otherwise, the new DTL must contain a list of node and port ID pairs that let the VC setup message cross the current peer group and exit at the proper node.

DTLs are always constructed in a dynamic routing and switching environment. Any of the ATM switches and their associated links might inadvertently fail after the DTL is built. Likewise, other PNNI border nodes might route virtual circuits through a given interior transit switch and occupy resources previously thought to be free.

Either case can result in virtual circuit setup failure while trying to transit the hop-by-hop route inside a given DTL. When a setup failure occurs, the setup and accompanying DTL are returned or cranked back to the originating node. The originating node is responsible for finding a new path through peer groups and revising the routing information in the DTL. If a new route cannot be found, the PNNI node notifies the attached ATM end device initiating the virtual connection that this particular VC setup failed.

Crankback: Getting Around the Problem in the Network

When calls fail to reach their destination because of unforeseen faults in the PNNI routing domain, the routing nodes that originated the DTL containing the faulty node or link must find an alternative path to the destination or called party. The crankback information element (CBIE) is contained in release, release complete, and add party reject (solely for multipoint connections) call-clearing messages. The CBIE contains blocked node or blocked link information. The DTL-originating node receiving one of these three call-clearing messages containing a CBIE must take appropriate action and reroute the call setup message around the failure and toward the destination. Note that calls rejected by the called party are not cranked back and rerouted toward the called party.

Reasons for crankback are grouped into two main areas: reachability failures and resource errors:

- Reachability failures occur when a transit path to the destination cannot be found. These failures include destination unreachable, transit network unreachable, and next node unreachable.

- Resource errors occur when some of the call setup requirements cannot be fulfilled along the path to the called party. Resource errors include service category not supported, traffic or quality of service parameters not supported, and requested VPI/VCI not available.

Soft Virtual Circuits

Establishing soft PVPs and soft PVCs using PNNI routing and signaling lets network administrators specify two static endpoints of a point-to-point ATM connection without having to manually set up each switch cross-connect that is part of the connection. These connections are called soft because PNNI signaling software in the master node containing the calling endpoint must set up, release, and reestablish the path to the called endpoint. By avoiding the laborious task of building each ATM cross-connect by hand, soft PVPs and PVCs can take advantage of PNNI's rerouting and crankback mechanism and dynamically reroute around network failures that might occur in the network infrastructure connecting the two static endpoints.

Building a Network Road Map

The PNNI routing discussion commences with a survey of routing components common to both flat and hierarchical PNNI networks. Then we'll discuss the specifics of each network type.

Common PNNI Routing Components

Before we delve into the differences between flat and hierarchical PNNI networks, the following sections discuss node IDs, routing control channels, and the hello protocol. These three routing components are common to both network types.

Node IDs: Combining Levels and Addresses

Concatenating the level indicator byte with a fixed-value byte of decimal 160 and appending the node's 20-byte ATM end system address form an individual node's 22-byte node ID. The format of a logical group node's (LGN's) ID is slightly different. The level indicator now denotes the LGN's peer group level. Following the level indicator is the 14-byte peer group ID of the LGN's child peer group. Next come the 6-byte ESI of the physical node acting as LGN and, finally, a single octet set to 0. Figure 8-14 shows both node ID formats.

Figure 8-14 *Two Node ID Formats*

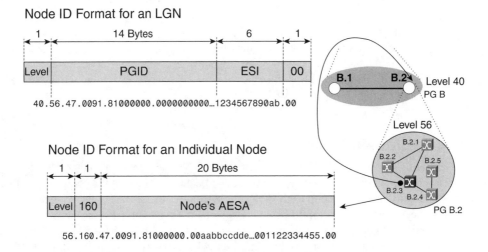

Routing Control Channels

Before two PNNI-controlled neighbor nodes can begin exchanging information, a communications channel must be opened between them. PVC or SVC connections called Routing Control Channels (RCCs) serve this purpose. When physical links connect two neighbor nodes, a PVC with VPI=0 and VCI=18 is used for the RCC. However, when a PVP connects two nodes, the RCC's VPI is the same as the PVP, and the VCI remains set at 18. LGNs connect with peer LGNs using SVC-based RCCs. The VPI and VCI values for these SVC-based RCCs are determined during the SVC setup process.

The Hello Protocol

Just as two strangers do not exchange personal information before they meet, PNNI nodes do not exchange topology information before using the hello protocol to negotiate PNNI version and peer group, node, and port IDs. Hello packets are exchanged over RCCs. RCCs are configured as PVCs if two nodes reside at the lowest layer of the routing hierarchy or are configured as SVCs if the nodes peer at any higher layer. If multiple links exist between two nodes, each link has its own RCC and hello exchange.

Nodes transmit hello packets periodically based on the expiration of an interval timer. Each link has its own timer, and that timer is reset each time a hello packet is sent. Besides timers, each node keeps a separate per-link database that contains link-state data as well as pertinent information about the neighbor node attached to the other end of the link. Independently, two node interfaces connected to the same link progress through a series of phases or states before topology exchange begins. The interfaces go from down to two-way inside (for nodes in common peer groups) or common outside (for nodes in different peer groups). The complete hello protocol Finite State Machine (FSM) is shown in Figure 8-15.

Figure 8-15 *Hello Protocol FSM*

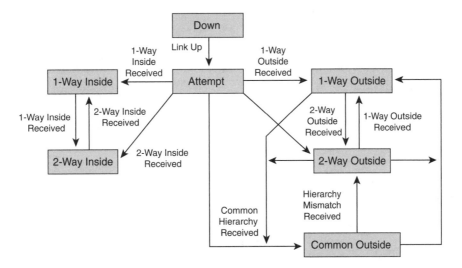

In the down phase, no PNNI packets are transmitted across the link. Typically, lower-level link failure forces interfaces into this state. As soon as physical connectivity is established, an interface moves to the one-way inside state when it receives a neighbor's hello packet that contains the same peer group ID but no values for remote port and node IDs. In this case, the received remote node and port IDs refer to the local node and port IDs of the node receiving the hello packet.

When a node receives a hello and sees its own node and port IDs listed in the values for the incoming packet's remote node and port ID, the receiving node is sure that one-way communication with the remote neighbor node is functioning. If this same hello packet contains a different peer group ID, the interface transitions to the one-way outside state instead. If the nodes reside in the same peer group, based on the data in the hello packets received, the transition to full two-way communication occurs when an interface receives a hello with the same peer group ID as well as values for remote node and port IDs. For links in a common peer group, this phase is called two-way inside. See Figure 8-16.

Figure 8-16 *Hello Protocol in a Single Peer Group*

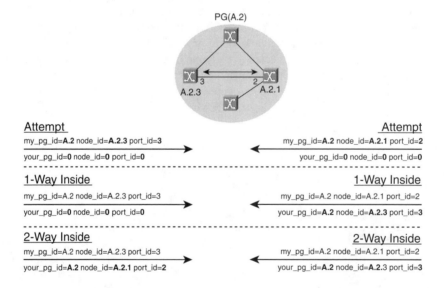

For links that connect nodes in different peer groups, as soon as a hello packet is received with remote port and node IDs, the two connected node interfaces move to the two-way outside phase, but full two-way communication is still not established. Full communication establishment occurs when the two nodes connected by this link agree to communicate on a common routing hierarchy level above the lowest level. At this point, these two nodes reach the common outside state. As soon as this phase is reached, full two-way communication begins and, in turn, topology exchange commences between these nodes in different peer groups. This is shown in Figure 8-17.

Figure 8-17 *Hello Protocol Between Peer Groups*

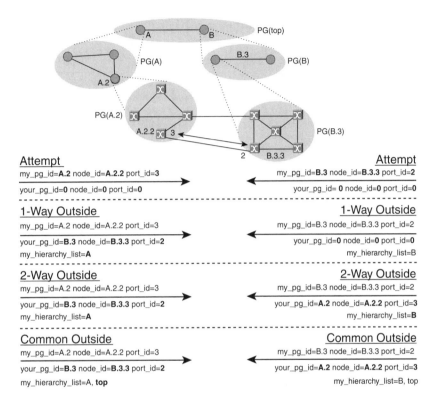

Attempt
my_pg_id=**A.2** node_id=**A.2.2** port_id=**3**

your_pg_id=**0** node_id=**0** port_id=**0**

1-Way Outside
my_pg_id=A.2 node_id=A.2.2 port_id=3

your_pg_id=**B.3** node_id=**B.3.3** port_id=**2**
my_hierarchy_list=**A**

2-Way Outside
my_pg_id=A.2 node_id=A.2.2 port_id=3

your_pg_id=**B.3** node_id=**B.3.3** port_id=**2**
my_hierarchy_list=**A**

Common Outside
my_pg_id=A.2 node_id=A.2.2 port_id=3

your_pg_id=**B.3** node_id=**B.3.3** port_id=**2**

my_hierarchy_list=A, **top**

Attempt
my_pg_id=**B.3** node_id=**B.3.3** port_id=**2**

your_pg_id= **0** node_id=**0** port_id=**0**

1-Way Outside
my_pg_id=B.3 node_id=B.3.3 port_id=2

your_pg_id=**0** node_id=**0** port_id=**0**
my_hierarchy_list=B

2-Way Outside
my_pg_id=B.3 node_id=B.3.3 port_id=2

your_pg_id=**A.2** node_id=**A.2.2** port_id=**3**
my_hierarchy_list=**B**

Common Outside
my_pg_id=B.3 node_id=B.3.3 port_id=2

your_pg_id=**A.2** node_id=**A.2.2** port_id=**3**

my_hierarchy_list=B, top

Routing Within a Flat PNNI Network

A PNNI network is considered flat when all the nodes in the PNNI domain have equal and complete knowledge of the topology. In short, every node knows the state of every connecting link and the health of every routing node in the network. As the network grows, so do the per-node memory requirements to hold all this link-state and node health information. A hierarchical PNNI network design slows the growth of memory requirements by segmenting the network and limiting a given node's view of the physical network. Hierarchical PNNI is discussed later in the "Routing in a Hierarchical PNNI Network" section.

How Nodes Share Topology Information

Before a flat or hierarchical PNNI network begins operating, each routing node must determine the operational state of each of its directly connected links and then discover all its neighbor nodes attached to its active links. As soon as nodes identify and establish communication with their directly connected neighbor nodes, the exchange of PNNI topology information begins. PNNI topology state packets (PTSPs) act as virtual shipping containers for packaging topology updates that contain link-state and node health information for transport throughout the PNNI routing network.

Besides carrying node and link-state information, PTSPs also carry information regarding the ATM address destinations that each routing node can reach. To identify the node originating the topology update within a PTSP and the scope of the update, the PTSP header contains an originating node ID and the originating node's peer group ID.

PTSPs transport one or more PNNI topology state elements (PTSEs) with each element packaging link, node, or reachable destination information. Each PTSE in a PTSP originates from the same routing node, where the 22-byte node ID inside the PTSP identifies the routing node. The 2-byte Type field in each PTSE header identifies the PTSE. Table 8-6 lists the types.

Table 8-6   *PTSE Types*

PTSE Name	PTSE Value
Nodal information group	96
Nodal information group	97
Outgoing resource availability	128
Incoming resource availability	129
Next-higher-level binding	192
Optional GCAC parameters	160
Internal reachable ATM addresses	224
Exterior reachable ATM addresses	256
Horizontal links	288
Uplinks	289
Transit network ID	304
System capabilities	640

Link, node, and reachability information can change at any moment during the continuous operation of a PNNI network; as such, PTSE contents are considered current only for a finite period. Each PTSE header has fields for sequence number and remaining life that determine which PTSE of a given type and from a given node is most recent and indicate how much longer a given PTSE's contents are valid. The remaining-life counter is uniformly decremented even after the PTSE is inserted into a node's topology database. Because a node cannot build a route using PTSEs with expired remaining-life counters, this constant decrementing and subsequent flushing ensures that routes are built using the most up-to-date information.

PTSE Flooding

All PTSEs must be disseminated throughout a PNNI peer group for every node in the group to have a clear and consistent view of the routing domain. *Flooding* is the process used to spread each node's PTSEs through the peer group. During flooding, every node acts as a PTSE relay agent for every other node. When a node receives a PTSE from one of its neighbor nodes, it bundles that PTSE—along with other PTSEs from the same neighbor— into a PTSP and forwards this PTSP to all its other neighbors. Clearly, the PTSEs originated by each node arrive at every other node in the peer group as long as each node follows this PTSE flooding algorithm. Figure 8-18 shows the flooding process.

Figure 8-18 *The Flooding Procedure*

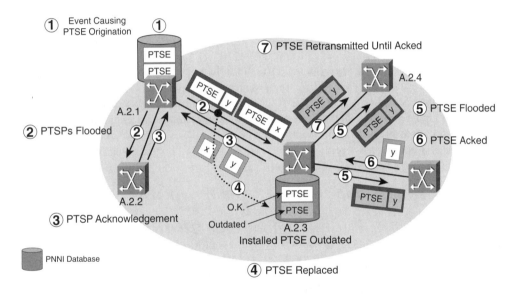

Acknowledgment, Aging, and Expiration of PTSEs

To ensure PTSE receipt, all receiving nodes send PTSE acknowledgments to the originating nodes. The identifier field in each PTSE is copied into an acknowledgment list destined for the node that originates all the PTSEs in the list. These acknowledgment lists can be sent immediately or after a delay. Typically, PTSEs received with no remaining life or dated sequence numbers are acknowledged immediately, but acknowledgments for valid PTSEs are often bundled and sent after a delay. This delay is bounded, though. An expiring acknowledgment timer forces a node to check its list of PTSEs needing acknowledgment and to send an update packet to the originating node. After expiration, the timer is reset. Just as any receiving node must keep track of all PTSEs it must acknowledge, transmitting nodes must keep track of all PTSEs sent but not acknowledged. Transmitting nodes periodically resend unacknowledged PTSEs based on internal timer expiration.

PTSE aging and expiration through either periodic or forced decrement of its remaining-life counter cause nodes to request PTSE refreshes or automatically transmit updated PTSE versions. Nodes might force the remaining-life counter to zero, thereby prematurely aging the PTSE in its topology database only if the node originates this PTSE. PTSEs are prematurely aged for a variety of reasons that all stem from changes in a node's hardware or link states that invalidate the contents of the PTSE. Premature aging of a PTSE is often followed by a triggered PTSE update.

Unscheduled Topology Updates

Asynchronously triggered updates are invoked when a node's information groups change. Information groups segment a node's characteristics into classes such as internally or externally reachable addresses, link parameters, resource availability, or the node's hardware state. Note that triggered updates can be encapsulated in PTSPs and flooded into the network at only a limited rate, thus preventing the CPU processing power of receiving nodes from being overwhelmed if a node has multiple neighbors simultaneously sending it triggered updates. Triggered PTSE updates are acknowledged in the same fashion as PTSEs received periodically. Figure 8-19 lists the pertinent parameters per information group that trigger PTSEs. Note that any change in some parameters, such as maximum cell rate in the resource availability information group, triggers a PTSE, whereas other parameters in the same group, such as average cell rate, require the absolute percentage change to reach a certain threshold before a node fires a PTSE update.

Figure 8-19 *Significant Events Per Information Group Triggering PTSE Updates*

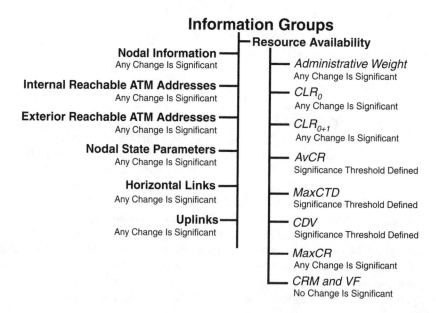

Database Synchronization Between Peer Nodes

Any two neighbor nodes are synchronized when they agree on the state of the network topology. Initial exchange of database summary packets initiates the synchronization process.

Database synchronization begins when the neighbor nodes reach the two-way inside state. The add port event is triggered by this state, and a node commences synchronization. Conversely, if neighbor nodes exit the two-way inside state for any reason, including physical link or RCC failure, the drop port event is launched, and a node removes all database information for the newly disconnected neighbor. Figure 8-20 shows all the states that any pair of neighbor nodes traverses as it moves toward full synchronization.

Figure 8-20 *Node States During the Synchronization Process*

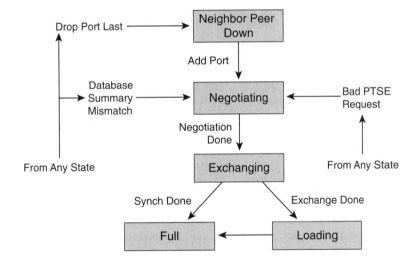

A database summary (DS) packet contains a complete list of the PTSEs that a node can originate. DS packets include the header portion of each PTSE. Initially, a neighbor node on one end of a two-way inside link ignites the database synchronization process by forwarding a DS packet toward its neighbor on the far end of the link. A single synchronization process exists between two nodes even if multiple two-way inside links exist between the pair. The node starting the process is considered the synchronization master for a given neighbor node pair. The slave acknowledges the master's DS packet by transmitting a DS packet of its own. Master nodes can have only one outstanding unacknowledged packet.

As soon as the DS packet exchange between two neighbor nodes is finished, the two nodes begin requesting all the PTSEs listed in the DS packets they just received. PTSEs are exchanged using PTSPs as previously described. Figure 8-21 shows the database synchronization process between the centrally located switch and its two adjacent neighbor nodes.

Figure 8-21 *Database Synchronization Process*

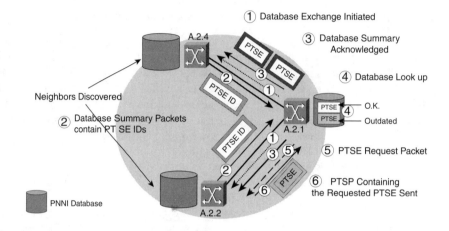

Routing in a Hierarchical PNNI Network

A rising number of nodes and increasing alternative paths in the routing domain force the network administrator to make choices to simplify the routing complexity in each peer group. To simplify per-peer group routing complexity, many network administrators choose to implement hierarchical PNNI. Hierarchical PNNI's multilevel routing scheme slows the growth of per-node memory requirements by limiting each node's view of the network topology. Levels of the hierarchy are configured, peer group leaders (PGLs) are elected, and peer groups are represented by LGNs at the next-higher level in the hierarchy. Figure 8-22 depicts a multilevel hierarchy, complete with PGLs and LGNs. As shown in the figure, an LGN can simultaneously act as an LGN for its lower-level peer group and serve as a PGL for its peer group.

Peer Group Leaders

Peer groups elect PGLs to perform the duties of LGN at the next-higher level in the routing hierarchy. Each level of the routing hierarchy above the lowest level has an LGN. An LGN is not an additional physical ATM switch in the network. Instead, it is a software data structure resident in the memory of one of the nodes in the lower-level peer group. The node-elected PGL at the lowest level serves as LGN and represents its lower-level peer group at the next-higher level, but it does not necessarily serve as PGL for this higher-level peer group. PGLs are chosen using a preconfigured leadership priority value in each node.

Figure 8-22 *LGNs and PGLs*

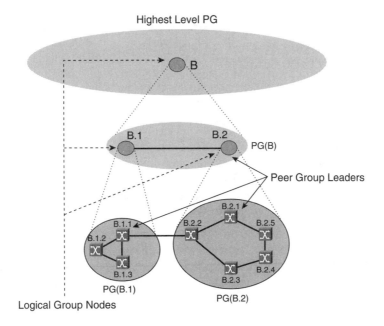

A node can have a different leadership priority value configured for each level in the hierarchy. The node with the highest leadership priority value becomes the PGL, but a node must exist in a peer group before it attempts to be the group leader. If all nodes in the lowest-level peer group have leadership priority values equal to zero, no PGL is elected. By default, Cisco multiservice switches have leadership priority values equal to zero.

Electing a PGL

Nodes in a peer group elect a leader using the leadership priority values and 22-byte node IDs found in the nodal PTSE they receive from every other peer in the group. As discussed, each node determines its preferred PGL by finding the node in its topology database that has the highest leadership priority. Cisco's multiservice switch family has a configurable leadership priority range of 0 to 200. If there is a tie for PGL based on leadership priorities, a node converts the 22-byte node IDs of the nodes that tied into unsigned integers and then selects the node with the highest integer value to become its preferred PGL. If a node calculates that it is PGL, it sets the "I am leader" bit in all transmitted nodal information group PTSEs. As soon as a node is elected PGL, it increments its leadership priority value by a network-wide constant. This action promotes stability within the peer group when multiple peers have similar leadership priority values.

Identifying the Hierarchy Levels

The most significant 13 bytes of the ATM address comprise all the possible levels in the PNNI routing hierarchy. From a routing perspective, these 13 bytes allow possibly 104 levels of network partitioning. The 104th level represents the lowest possible level, and the first level signifies the highest. The level indicator specifies, in bits, how many of the 104 most significant ATM end system address or node ID bits signify the routing hierarchy level. For example, if a node's level indicator equals 56, the most significant 56 bits in the ATM end system address identify this node's peer group ID.

The 1-byte level indicator combined with the first 13 bytes of the ATM address form the 14-byte peer group identifier (ID). Note that the first 13 bytes of the ATM address constitute all the possible 104 levels in the PNNI routing hierarchy. Each node generates a peer group ID separate from its 20-byte ATM address. If a node is located above the 104th level at level *x*, where *x* is less than 104, the rightmost 104–*x* bits within the peer group ID are set to 1. This peer group ID unambiguously denotes the node's hierarchy level and peer group.

Higher-level or "parent" peer groups must have level indicator values less than the child's level indicator. Conversely, child peer groups always have level indicator values greater than their parents do. For example, a node at level 56 could have a parent peer group above it at level 36 and a child peer group below it at level 80, as shown in Figure 8-23.

Figure 8-23 *Sample Hierarchy Levels*

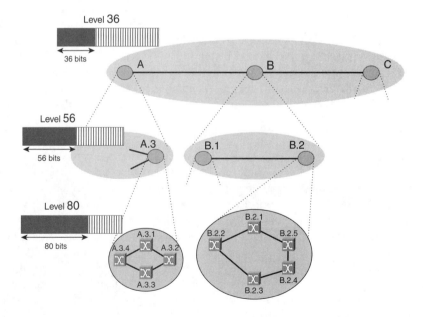

Note that peer group levels do not have a predefined starting point within the 104 possible choices. Likewise, peer group levels do not need to be consecutive in a given PNNI network. In short, no requirement states that the middle level in a routing hierarchy must start at x, where x is a number between 1 and 104, and lower levels must start at $x+1$, whereas higher levels must begin at $x-1$.

In fact, because PNNI node IDs, such as ATM end system addresses, are commonly displayed in hexadecimal format, it is easier to discern the peer group ID if the level indicators are always multiples of 4. This numbering convention ensures that the peer group ID, which is the number of significant bits of the node ID dictated by the level indicator, is always an integral number of hexadecimal characters.

LGN Hellos

Just as lowest-level PNNI nodes have RCCs connecting them to their peers, each LGN has SVCC-based RCCs that connect it to the other LGNs in its peer group. To recap, Figure 8-24 shows RCCs between physical nodes as well as LGNs.

Figure 8-24 *RCC Types*

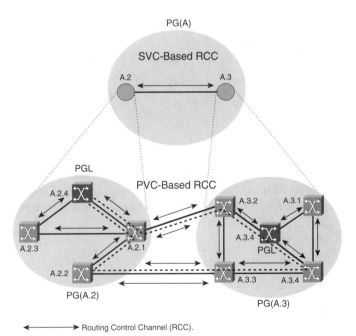

These RCCs serve the same purpose regardless of level: They provide passage for neighbor-to-neighbor hellos and PNNI routing exchanges via PTSPs. If a link fails in one of the paths connecting any two LGNs, the SVCC-based RCC might fail as well. However, the LGNs do not acknowledge this loss of connectivity until their own intranode timers expire without hello packet or PTSP receipt. As such, the SVCC might fail and be rebuilt with no significant loss of connectivity between the two LGNs.

Unlike hello packet exchanges between any two nodes at the lowest level of the routing hierarchy, hello packets between LGNs do not contain port IDs, and the remote port ID field is set to all 1s in these hello packets. An LGN does not send hello packets to another LGN until the hello packet contains a PTSE with the node ID of the remote LGN. An LGN receiving hello packets must discard them all until it has a PTSE from the LGN of its neighbor border node containing a remote node ID. At this point, the receiving node accepts the hello packet if the remote node ID field corresponds to the one listed in its uplink PTSE. (Uplink is defined in the next section.)

Information Exchange Across the Different Link Types

PNNI introduces three link types:

- **Horizontal link**—A link between any pair of logical or physical nodes in the same peer group

- **Outside link**—A link between any pair of physical nodes in different peer groups

- **Uplink**—A link between a border node and the LGN of its neighbor border node

Figure 8-25 depicts the three link types.

Horizontal links, often called *inside links,* exist in either flat or hierarchical topologies, whereas outside links are unique to hierarchical PNNI. These outside links exist only between pairs of physical border nodes. If the two lowest-level nodes determine that they reside in different peer groups during the hello exchange, they are both border nodes. Figure 8-26 shows border nodes. The entry and exit terms that preface two of the border nodes in Figure 8-26 are in relation to the call setup message traversing the peer groups. When traversing a peer group, a call setup message must enter at one border node and exit the peer group at another border node.

Figure 8-25 *Link Types*

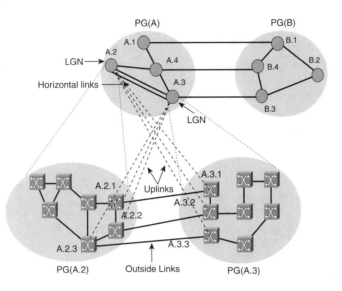

Figure 8-26 *Border Nodes*

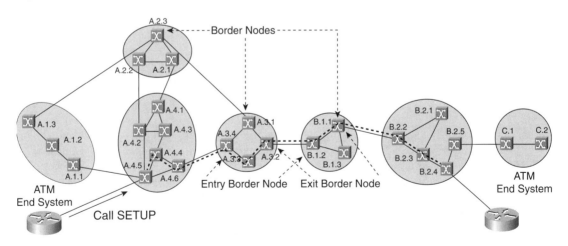

Unlike horizontal links, packets full of PTSEs are not exchanged. Instead, as soon as neighbors determine they reside in different peer groups, both neighbors transfer hello packets containing uplink information attributes (ULIAs). ULIAs contain the same topology state parameters as PTSPs, but these parameters contain the link state and condition of the border node at the far end of the outside link as well as the complete list of hierarchy levels above and below the remote border node's peer group. In addition to the list of hierarchy levels, a border node also advertises the connectivity and reachability information for all the hierarchy levels above it.

Using each other's list of hierarchy levels, both border nodes determine the nearest higher-level peer group they have in common. Next, each border node advertises a logical link to its neighbor border node's LGN that resides in this common higher-level peer group. This logical link is called an *uplink*. A border node only advertises uplinks in its own peer group. Lower-level nodes use the connectivity and reachability information advertised in uplinks to route connections to end systems in remote peer groups.

Link Aggregation

When several outside links connect two border nodes, the PNNI routing algorithm summarizes the links into a single aggregate link that is placed in the PGL's topology database if the component links have the same aggregation token values. These 32-bit tokens, exchanged via the hello protocol on outside links, correlate multiple parallel links connecting two border nodes. Because the default aggregation token for all links is 0, the PNNI routing algorithm automatically begins summarizing parallel links unless the administrator intervenes by configuring different token values on opposite ends of each parallel link.

Aggregate link information is fed upward to each border node's LGN. The LGN advertises only the aggregate logical link within its peer group. Additionally, if multiple logical links with the same aggregate token exist between two border LGNs, those links are summarized by LGNs in the next-higher layer of the routing hierarchy using the method just described. Link aggregation lets PNNI support expansive networks because upper-level or parent nodes do not know the detailed topology of their lower-level child peer groups.

Address Summarization

Besides links, LGNs representing the lowest-level peer groups must aggregate and summarize the ATM addresses of end systems directly attached to the PNNI-controlled switches inside those peer groups. By default, an LGN advertises a summary address equal to the peer group ID of the peer group it represents. This summary address is advertised in the "internally reachable ATM address" information group PTSEs distributed by the LGN. "Internally reachable" implies that the addresses originate from end systems or nodes within the same PNNI domain as the LGN. Addresses learned through connections to other PNNI networks are summarized just like internal addresses but use the "exterior reachable ATM address" information group PTSEs for distribution.

In most instances, all end systems and nodes in a peer group have the same peer group ID, so the summary address accurately accounts for all devices in the peer group. As such, all SVCs or SPVCs initiated to addresses inside the summary are sent to the LGN that advertises that summary address. If an end system has a unique AESA that does not align with its peer group's address format, that AESA cannot be summarized.

The address summarization process just described is repeated at every level in the routing hierarchy above the lowest level. This continual process of constructing summaries of summaries ensures that higher-level LGNs do not have to store and distribute unabridged lists of every AESA in the domain.

From the address summarization process, you can infer the importance of a consistent and structured addressing plan for address summarization and nodal aggregation, as well as for avoiding service affecting address changes down the road.

Another process called *scoping* limits a summary address's range of advertisement. Scope is a 1-byte field inside the internally reachable ATM address information group (IG) PTSE that dictates the highest PNNI routing level at which the summary addresses in that PTSE can be advertised. For example, if a PNNI domain has four hierarchy levels—80, 56, 40, and 36—and the PTSE generated by the LGN at level 56 has a scope of 40, the summary addresses in the same PTSE cannot be advertised by the parent LGN residing at level 36. Each summary address has a scope, and LGNs typically group summary addresses with equal scopes into the same PTSEs. A scope value of zero means that the address can be advertised at the highest level of the routing hierarchy.

Passing Down Higher-Level Topology Information

As each LGN receives PTSEs from its peer LGNs, it sends this topology information downward to the PGL of the lower-level peer group it represents. Remember, in most instances, an LGN is also the PGL of that lower-level peer group. The PGL immediately starts flooding this upper-level topology information to its peers. When the PGL is not in the lowest level of the routing hierarchy, the PGL is an LGN for some lower-layer peer group. In this case, the PGL floods the summarized higher-level topology state information it received from its LGN along with the topology information from its peer group down through its child peer groups.

The Use of IISP and AINI in a PNNI Network

Joining two PNNI networks always entails some inherent risks, such as keeping routing information and the network topology protected. To minimize this particular risk, two interconnect specifications address this problem. The initial interconnect specification, Interim Inter-Switch Protocol (IISP), applied to UNI 3.0 and UNI 3.1 protocols. Originally, IISP was conceived to define how two ATM routing nodes were joined prior to the release of the PNNI 1.0 specification, but IISP is now commonly used to join PNNI 1.0 networks controlled by different entities. IISP is also called PNNI 0. The follow-up specification, ATM Inter-Network Interface (AINI), provides support for the UNI 4.0 protocol. In both

IISP and AINI, no routing messages are exchanged between the two networks (or nodes), and static routes direct call signaling messages across interconnected networks.

On point-to-point ATM links implementing IISP, SVC, and SPVC, circuits are still built across these links using standard ITU-T Q.2931 signaling messages. IISP does not allow dynamic routing updates across this link; rather, all routing is done using static routes encoded into the switches on opposite sides of the IISP link. IISP performs this hop-by-hop routing using a per-switch routing database. Database routing entries contain three fields each—ATM address, ATM address length, and interface index. The ATM address is the same 20-byte address discussed earlier. ATM address length specifies how many bits of the associated ATM address are significant during computation of the longer match routing algorithm. Functionally, the ATM address length field is equivalent to the network mask used in IP routing algorithms. Finally, the interface index denotes the physical or logical port that the routed ATM cell should exit.

AINI only serves as a signaling standard for ATM connections built across network boundaries. It does not cover the distribution of connection routing information across these boundaries. AINI does specify interworking standards used to connect a network using the B-ISUP signaling protocol, with another network implementing the PNNI signaling protocol. AINI also defines the connection signaling used between two PNNI networks. Although AINI allows ATM service providers and enterprise network operators to use AESA or E.164 addressing techniques, it does not require providers and operators to support all four ATM addressing formats.

IISP is an asymmetric protocol. IISP has a network side and a used side. The network side is responsible for assigning VPI and VCI identifiers. Conversely, AINI is a symmetric protocol. An AINI interface joins two networks. Therefore, in an AINI interface, administrators must define by mutual agreement which side is the assigning network (responsible for allocating the connection identifier VPI and VCI values) and which side is the receiving network. The preceding and succeeding side can vary call by call.

Summary

This chapter began with a description of ATM end system addressing and then moved on to an explanation of ITU-T-compliant ATM signaling for SVC creation and deletion. Leveraging this signaling explanation, call signaling through a PNNI network was dissected. Next, PNNI routing was discussed. That section first defined the routing commonalities between flat and hierarchical PNNI networks and then went on to explain the distribution of routing information in both network types. A discussion of interconnecting two PNNI networks using IISP or AINI concluded this chapter.

CHAPTER 9

PNNI Network Design Goals

Now that the theoretical Private Network-to-Network Interface (PNNI) discussion is complete, the next step is to discuss PNNI network design goals and implementation specifics of Cisco's multiservice switches.

Peer Group Design Considerations

Providing adequate user services requires the PNNI network designer to estimate address range, UNI bandwidth, and NNI bandwidth requirements. If the network designer expects the PNNI network to expand into a hierarchical structure, he must consider the computing power and memory capacity of the switches designated to assume peer group leader (PGL) status.

Constructing a peer group requires considering the following:

- **Volume of topology updates**—As the number of nodes in the peer group increases, the volume of PNNI topology state packets (PTSPs) broadcast through the group begins consuming significant bandwidth and per-node processing power.

- **Convergence**—The process of generating, evaluating, and synchronizing all the PNNI topology state elements (PTSEs) sent and received in PTSPs lengthens the time it takes for all nodes to agree or converge on a common view of the network topology.

- **Span**—The PNNI standard dictates that the maximum number of entries in a Designated Transit List (DTL) be no more than 20, which implies that nodes within a peer may be no more than 20 hops apart. That limits the peer group diameter. It also mandates that the maximum number of DTL IEs in a setup message should not exceed ten, limiting the number of hierarchical levels.

- **Growth**—If a network is constructed as a single peer group and the number of contained nodes is near the per-peer group maximum, how will the addition of extra nodes be handled? On the other hand, if a hierarchical network begins at the lowest level, 104, what happens if lower levels need to be added in the future?

- **Resiliency**—What happens if a PGL fails? The network operator should ensure the outcome of the peer group leader election is deterministic by configuring proper leadership priorities. Network stability should also be sought by preventing PGL re-elections.

- **Administrative**—A peer group might need to exist within an administrative boundary in order to obey geographical, political, business, or security considerations.

- **Redundancy**—Redundancy considerations include minimizing single points of failure, provisioning parallel links, and creating multiple paths.

- **Clocking**—Synchronization and timing distribution must be designed carefully. You should choose automatic clock distribution protocols such as Network Clock Distribution Protocol (NCDP) in MGX-8850 and MGX-8950 nodes and AutoRoute (AR) in BPX-SES nodes whenever possible.

During the hierarchical design process, administrators must choose between the higher routing granularity offered by a peer group's complex node representation and the simplified routing and path selection inherent in simple node representation. Complex node representation allows the logical group node (LGN) to inject node and link-state information from the lower-level peer group it summarizes into the LGN's peer group. This additional information carried in PTSEs from LGNs configured for complex node representation lets nodes in other lower-level peer groups compute DTLs that better adhere to the traffic parameters requested by the end system. Simple node representation applies a single, constant cost to traverse any foreign peer group on the way to the destination.

You should do the following when selecting a PGL to carry out the duties of an LGN at all higher levels of the hierarchy:

- Select a switch that has more processing power and memory.

- Select a switch that is not destined to be a border node. Performing both roles puts undue strain on a single switch's processing power and memory limits. Border nodes perform DTL and uplink origination duties already.

- Select a switch that is a transit node and not an edge node that builds connections and associated DTLs.

- Select a switch with a lower number of signaling links (links running Service-Specific Connection-Oriented Protocol (SSCOP), such as PNNI and UNI).

- Select a switch with a lower number of connections based on traffic engineering (TE) or empirical results.

- Select a switch that has links to multiple nodes in the peer group. This minimizes the risk of the PGL's separating from the peer group if one of the PGL's links fails.

- Select a switch that is not an NMS gateway.

NOTE   It is worth noting that running ILMI does not impose any CPU load on the controller card. ILMI is a distributed application that runs in the line cards (AXSM or BXM cards), not on the controller card.

These same considerations apply when you choose peers to act as border nodes.

A single PGL is also a single point of failure. Depending on resiliency requirements, it may be desirable to configure primary and secondary peer group leaders (each chosen based on the above mentioned selection criteria).

Optimizing PNNI Network Parameters

Resource restrictions, routing summarization, and Quality of Service (QoS) parameters are all adjustable. Here are the some of the commonly tuned parameters:

- **Administrative weight (AW)**—This is a per-link routing metric used by routing nodes that construct DTLs to determine the cost to traverse a given path through the PNNI network. Lower total cost paths are preferred. By manipulating the AW of individual links, the administrator can influence switched virtual circuits (SVC) and Soft Permanent Virtual Path (SPVC) path selection. AW can be configured per ATM class of service. Other metrics and attributes can also be tuned, such as cell transfer delay (CTD) and cell loss ratio.

- **Aggregation token**—A default link aggregation token of 0 results in all parallel horizontal links appearing as one at higher levels of the hierarchy. The administrator may force the PNNI routing algorithm to differentiate between multiple parallel links by applying different aggregation tokens to each link or combinations of links.

- **Bandwidth overbooking factor**—This factor directly controls the available cell rate (AvCR). The booking factor determines the AvCR that a node advertises for a particular PNNI port. A booking factor of 1 percent reserves only 1 percent of the bandwidth requested in the SVC setup. This is overbooking because two connections could request bandwidth of 100 percent, but a configured booking factor percentage less than 100 percent, such as 50 percent, would allow both these connection to built along the same route even though they both need 100 percent of the bandwidth. In turn, connection setups prefer this link. Cisco multiservice switches support overbooking per interface or per Class of Service (CoS), such as constant bit rate (CBR) or Variable Bit Rate-Nonreal Time (VBR-NRT). Although this raises link utilization, over time more connections might crank back because of a lack of resources.

- **Cell delay variation (CDV)**—This represents a measurement of the cell transfer delay (CTD) variation over links and through nodes. The PNNI route selection algorithm does not choose routes with lower CDV. The CDV is used to determine which routes are eligible for selection. It is important not to mistake CDV with the GCRA term cell delay variation tolerance (CDVT).

- **Node transit restrictions**—To force a particular distribution of SVCs or SPVCs through the PNNI domain, the administrator can enable the node transit restriction on any physical or virtual node, denying it participation in virtual circuit (VC) routing. The node with transit restriction enabled may still terminate and originate VCs. This restriction is encapsulated in nodal PTSEs and is flooded to other peers.

- **Link selection**—Different algorithms for parallel link selection as well as policies can be configured to perform TE tasks.

MGX and service expansion shelf (SES) PNNI switches select routes through the PNNI network using the following parameters:

- Destination address
- Administrative weight (AW)
- Maximum cell rate (MaxCR)
- AvCR
- CTD
- CDV
- CLR0
- CLR0+1

Table 9-1 shows the connection parameters used by the MGX and SES to fulfill end-user service requests.

Table 9-1 *Connection Parameters Per Service Class*

Service Class	Destination Address	AW	MaxCR	AvCR	CTD	CDV	CLR0	CLR0+1
CBR	Required	Required	Optional	Required	Required	Required	Required	Required
VBR-RT	Required	Required	Optional	Required	Required	Required	Required	Required
VBR-NRT	Required	Required	Optional	Required	Required		Required	Required
ABR	Required	Required	Required	Required				
UBR	Required	Required	Required					

Address Planning

If the planned PNNI network is destined to interconnect with other PNNI routing domains that are owned by other corporations or governments, it behooves the designer to secure globally unique ATM End System Address (AESA) prefixes from a recognized standards body. Globally unique prefixes prevent address duplication in the interconnected networks.

Table 9-2 lists the three standardized AESAs and the authorities that control the distribution of addresses within each.

Table 9-2   *AESA Distribution Authorities*

AESA Format	Registration Authority
DCC	ISO National Administrative Authority
E.164	International Telecommunications Union
ICD	American National Standards Institute
	British Standards Institution

Typically, telephony service providers that provide intercountry services use E.164 addressing. The ICD format's 2-byte code designator field is used to assign an organization or company such as Cisco Systems, Inc. a globally unique identifier. Many enterprises choose ICD AESA formatting for this reason.

Within an AESA, the network designer is free to define the format of the remaining bytes of the HO-DSP, ESI, and selector portions. Typically, the 6-byte ESI is filled with the IEEE formatted MAC address of the ATM switch's Ethernet management interface. The user-defined hierarchy levels are carved out of the unrestricted HO-DSP bits.

For example, the DCC AESA format allows the private entity purchasing the DCC-formatted address to specify the least-significant 6 bytes of the 13-byte DCC prefix. Following the guideline to allot 4 AESA bits per hierarchy level, potentially 12 additional user-defined hierarchy levels can be constructed. Because AESAs are mostly displayed in hexadecimal format, the 4-bit addressing principal ensures that the AESA portion associated with each hierarchy level is easily identifiable, because each hexadecimal number represents 4 bits. (This principle is discussed in the "Identifying the Hierarchy Levels" section in Chapter 8, "PNNI Explained.")

Continuing with this DCC example, a network designer devising the address plan might set the lowest hierarchy level at 88 bits. Using the 4-bit level selection process, the 88-bit level allows up to four levels of lower-layer expansion. Lower-layer expansion is useful when a rapidly expanding business unit within a company subdivides. On the opposite end of 88, eight higher-level layers are available.

A well designed address structure is of paramount importance to optimize routing, simplify planning, and allow multi peer group migration. The structure can be geographically based (for example, 8 bits for the continent, next 8 bits for the country, and then the region, city, and so on) or administratively assigned. PNNI addressing allows a structure that scales globally.

Address Summarization and Suppression

Address summarization prevents every routing node in the entire PNNI domain from maintaining a dynamic list of all the AESAs of every other node and attached ATM end system. By default, every PNNI routing node, logical or physical, advertises a summary address equal to its level in the PNNI hierarchy. For example, a node residing in the lowest layer of the previous example would advertise the most significant 88 bits of its AESA as a summary address. All end systems attached to this node will have the identical 88-bit pattern in their AESAs.

If a routing node only performs local switching for a group of attached end systems or external PNNI networks, the network administrator can configure a suppression summary address in the node performing the local switching. This suppression summary address prevents the routing node from advertising those locally switched AESAs to other routing nodes, thus lowering routing traffic in the domain and lessening memory demands on other nodes.

PNNI routing follows a prefix longest-match algorithm similar to IP routing. A default prefix is defined as a prefix with a length of zero, and it matches all addresses. A default route can be configured on domain border nodes and redistributed into PNNI. This is particularly useful in stub areas.

Scope

PNNI routing nodes do not need clean, minty breath when they exchange hellos with their neighbors. The term *scope* refers to AESA advertisements. By limiting the areas where ATM end system addresses are advertised, scope provides the same benefits as summary addresses.

Why advertise source AESAs to areas that have no demand for connections to these end systems? Scope is typically discussed at the UNI interface. The UNI 4.0 specification outlines 15 scope levels that correspond to certain hierarchy levels in PNNI. This scope-to-hierarchy mapping ensures that AESAs are advertised to only a predefined hierarchy level. Table 9-3 shows this mapping.

Table 9-3   *Scope Mapping*

UNI Scope	PNNI Hierarchy Level
1 to 3	96
4 to 5	80
6 to 7	72
8 to 10	64
11 to 12	48
13 to 14	32
0	Global (0)

Multi Peer Group Considerations

The multi peer group (MPG) PNNI migration process can be divided into three steps:

- **Preparation**—A well-designed addressing scheme minimizes the preparation tasks.

- **Migration**—While PNNI is down during the migration, SSCOP remains up. This means that the connection states are maintained. Routing however is blocked since the PNNI node is unreachable for PNNI routing.

- Monitoring—After the migration, several validation tests need to be performed. Among the things you should monitor are the CPU real time and memory of the switching system, SVC and SPVC status, nodes and links aggregated as expected, and addresses summarized as planned. It is strongly discouraged to reduce the PTSE timers to age out outdated PTSEs. This can be detrimental to the node health and will induce instability.

MSS Node Types and Limitations in a PNNI Network

Using the enhanced PXM-1E or one of the PXM45 variants, the 8850 and 8950 switches all support PNNI. MGX 8850 and 8950 switches support the following PNNI features in the latest maintenance release of software version 2.1:

- UNI 3.0/3.1/4.0

- PNNI 1.0 with hierarchy and multiple peer group

- ILMI 4.0

- ATM Inter-Network Interface (AINI)

- Standard IISP with PNNI interworking

- Enhanced IISP for SPVC support

- Point-to-point ATM SVCC and SVPC

- Support for ABR, CBR, VBR, VBR-RT, VBR-NRT, and UBR

- Alternate call routing

- On-demand call routing

- Native E.164 and AESA (E.164, ICD, DCC) addressing

- Per-class of service, per-connection overbooking

- Address filtering

The BPX supports PNNI with the addition of the SES. PNNI feature support in the latest maintenance release of SES software version 1.1 is identical to the MGX 2.1 features just listed.

Unlike the MGX, the SES does not support NCDP for clock synchronization. AR serves this purpose.

Additional MGX and SES PNNI features supported in software release 3.0 include the following:

- **Priority connection routing**—High-priority connections establish and release before low-priority connections, improving high-priority reroute times.

- **Preferred connection routing**—This lets the administrator override PNNI path selection for a connection and manually configure the connection's route through the network.

- **Per-connection overbooking**—This lets the administrator configure the percent utilization per connection, providing more granularity than configuration percent utilization per interface.

Current MGX release 2.1 MSS PNNI limitations include the following:

- Ten levels of hierarchy

- 192 interfaces per node. Up to 100 of those are configured for IISP, PNNI, or UNI signaling because of SSCOP support on 100 interfaces.

- A 255-node limit per peer group. Any single node may have visibility to 254 other peer nodes. For LGNs, 255 includes the peer nodes in its group plus the nodes in its child peer group.

- 3000 AESAs per node

- 3200 PNNI links visible to any single node

- 100,000 total SVCs and/or SPVCs per node

PNNI POP Design and Scaling

Because of the exorbitant expense of dedicated optical transmission lines such as OC-3 and OC-12, most PNNI network designers opt for peripheral nodes to feed high-capacity core-switching nodes. Peripheral nodes offer end customers an array of ATM uplink speeds, whereas high-capacity switches such as the MGX 8950 do not. These peripheral nodes attach to the core using lower-speed electrical transmission lines such as E3 and T3 or some fractional optical bandwidth via VPC.

Although the peripheral nodes would connect to only a single core switch, or two core switches for redundancy, all nodes in the core switch complex typically require a redundant full mesh of links providing direct any-to-any connectivity. This mesh minimizes switching delay and maximizes failure recovery. A full mesh of links between the core switches is not required if an acceptable percentage of simultaneous peripheral traffic entering the core switch complex can successfully exit the complex and reach its destinations without being blocked because of lack of core bandwidth. A higher percentage of simultaneous peripheral traffic successfully passing through the core switch complex requires higher-bandwidth links between core switches. Higher-bandwidth links lessen the number of blocked ATM PVCs because of bandwidth starvation. However, these links might be an unjustifiable cost if the instantaneous amount of end-customer traffic does not frequently tax the bandwidth limits of the core switch complex.

Inserting high-capacity switches between the existing core switch complex and the ring of peripheral switches allows the MSS PNNI network to expand. These newly inserted switches act as aggregation switches for the peripheral nodes. If the peripheral nodes are redundantly linked to the nodes in the core, the new aggregation nodes can be inserted with minimal end-user downtime because of PNNI's automatic rerouting capabilities.

Redundancy

Redundancy considerations begin at the link level before progressing to the node level. Any SONET/SDH link connecting a pair of PNNI nodes may utilize Automatic Protection Switching (APS) for SONET or Multiplexed Switching Protection (MSP) for SDH within the same line card or across a pair of line cards in each connected node to provide redundancy at the physical layer. When APS/MSP is not an option, or when you're using an electric interface such as a T3, PNNI's rerouting capabilities make full use of parallel trunks connecting two nodes.

All Cisco MSS products offer full processor card redundancy. Optional Y-RED redundancy provides card and trunk redundancy for optical or electrical interfaces when optical protection switching using APS or MSP is not a viable option. Y-RED redundancy does incur the extra expense of additional line cards and loss of chassis slots for other purposes.

Core Bandwidth Concerns

When redundant trunks are implemented between the core switches in a PNNI network, the administrator's awareness of the average percentage load on each trunk is critical to ensuring connectivity when individual trunks fail. A target percentage load per trunk of 40 percent to 50 percent ensures that if one of two parallel trunks fails, the other one can withstand the full connection load of both trunks while repairs are completed. It is better for connections to reroute rather than fail because of lack of available bandwidth.

PNNI Migration

The two migration paths from legacy AR-based BPX networks into PNNI are

- **Deployment of SES controllers to enable PNNI on the BPXs**—At first, logically independent AR and PNNI networks exist on the same physical infrastructure. The AR connections can then be automatically migrated to standards-based PNNI SPVCs.

- **Deployment of an MGX-8850 or MGX-8950 PNNI core**—At first, hybrid connections (X-PVCs) can be created using AR-PNNI interworking. The PNNI core can then grow to eventually include the complete network. In this scenerio, the core either can be summarized as a logical group node (LGN) or can be exposed as individual core nodes. Exposing the core gives full visibility of the core and enables optimal routing accross peer groups.

A third migration path also exists if you combine the previous two paths into a mixed migration path.

Summary

This chapter discussed design considerations, including convergence, span, and growth. You must consider these points before constructing a PNNI peer group. With the initial peer group topology in place, this chapter discussed PNNI parameter optimization with regard to connection routing. The address planning section dove into the topics of hierarchical addressing using public address prefixes, followed by the addressing issues: summarization, suppression, and scope. Because a PNNI network designer is bound by the switch's capabilities and restrictions, this chapter included a section on MSS node types—BPX/SES and MGX—and their limitations. The final sections of this chapter covered the redundancy and bandwidth concerns associated with designing a PNNI network to provide end-user services. In case you are migrating from an AR system to a standardized PNNI routing, the final section discussed prudent migration strategies.

CHAPTER 10

PNNI Implementation and Provision

This and the following chapters cover configuring ATM Forum services in multiservice switching platforms. All the concepts from Chapters 8 and 9 are projected into actual equipment configurations.

In this chapter, you set up a complete PNNI network composed of four PNNI nodes. The PNNI nodes are implemented with four different platforms: a Service Expansion Shelf (SES) controller and BPX-8600 node, a PXM-45-based MGX-8850, an MGX-8950, and a Cisco IOS Software-based PNNI node. Network Clock Distribution Protocol (NCDP) provides timing distribution.

This PNNI network provides SVC- and SPVC-based ATM services among three routers.

PNNI Configuration in MSS

The objective of this chapter is to bring up the network shown in Figure 10-1. First, the four PNNI nodes are configured to set up the PNNI network.

Figure 10-1 *PNNI Network*

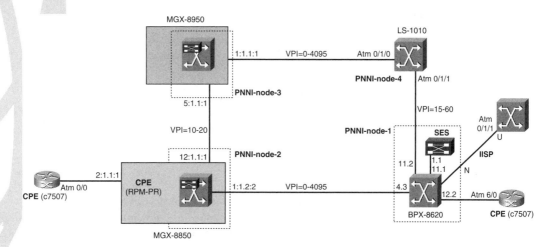

Two external Cisco 7500 routers will be connected through the network using switched virtual circuits (SVCs) and Soft Permanent Virtual Circuits (SPVCs). Connectivity to an RPM-PR-based CPE in the MGX-8850 will also be configured using SPVCs.

Your PNNI network is a single peer group (SPG) network at level 64. This means that the first 64 bits or 8 bytes form the peer group ID (PGID). The PNNI node's ATM End System Addresses (AESAs) are chosen so that there's a unique and common peer group ID at level 64. See Figure 10-2.

Figure 10-2 *PNNI Node's AESAs*

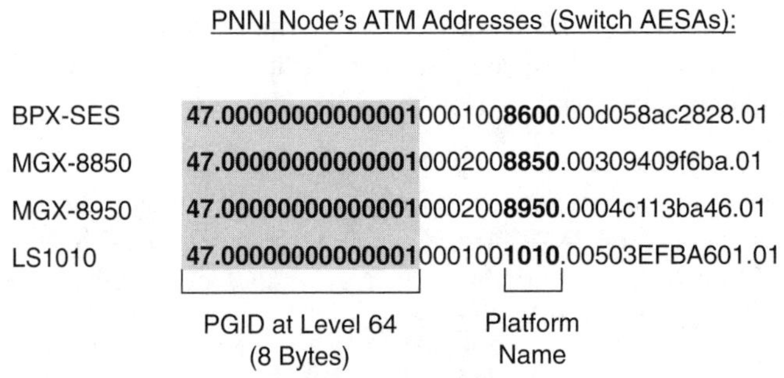

PNNI Node's ATM Addresses (Switch AESAs):

BPX-SES	**47.0000000000001**0001008**600**.00d058ac2828.01
MGX-8850	**47.0000000000001**0002008**850**.00309409f6ba.01
MGX-8950	**47.0000000000001**0002008**950**.0004c113ba46.01
LS1010	**47.0000000000001**0001001**010**.00503EFBA601.01

PGID at Level 64   Platform
(8 Bytes)   Name

For the purposes of this and the following chapters, AESAs with International Code Designator (ICD) format have been chosen, with a bogus ICD of 0x0000.

As shown in Figure 10-2, the number in the switch platform name is included in the switch ATM address for ease of identification.

Generic Configuration Model

Before we dive into the configurations, this section provides a generic approach to PNNI configuration in multiservice switching networks. Different flow charts show the generic configuration steps.

Near the end of this chapter, after each network element has been configured, the section "Summary of PNNI Configuration Commands" summarizes all the configuration commands in the different platforms.

Generic PNNI Node Bringup

The generic PNNI node bringup can be split into three main steps. Two of the three are performed in the controlled switch, and the remaining step is applied to the PNNI controller. A flow chart specifying these three steps is shown in Figure 10-3.

Figure 10-3 *Generic PNNI Node Bringup*

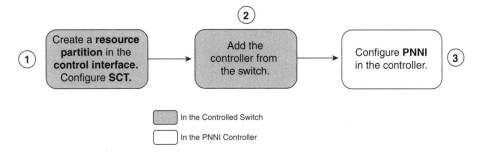

In a SES-controlled BPX-8600 PNNI node, the first step includes upping and configuring a trunk, adding a resource partition, and configuring the port Service Class Template (SCT). The second step includes adding the SES as a feeder and a VSI controller. In MGX-8850 and MGX-8950 PNNI implementations, the first step is not performed because the PNNI controller is included in the PXM-45 software image, so there is no external interface to bring up.

The PNNI configuration in Step 3, which is executed in the PNNI controller, is exactly the same in all platforms because PNNI software is the same among all multiservice switches. During this configuration step, the ATM address, the hierarchical level, the peer group ID, and the PGL designation are configured. They are particularly important in a PNNI multilevel scenario.

Generic Link Configuration

Figure 10-4 is a diagram for generic link bringup. Figure 10-4 is used for links connecting the PNNI node to both CPEs and PNNI neighbors.

In Figure 10-4, Steps 1 through 3 are used for both UNI and PNNI link configuration. The configuration of Steps 1 and 2 is the same in both cases because you are configuring the controlled switch that is unaware of network protocols. Step 3, which is executed on the PNNI controller, selects the signaling to be used. Links facing CPEs use UNI signaling, and links toward PNNI neighbors use PNNI signaling and routing.

PNNI configuration finishes on the third step, and CPE configuration continues in Steps 4 and 5.

It is worth noting that MGX PNNI platforms need an extra initial step—configuring the card SCT. Card SCTs are not present in BXM VSI slaves.

After Step 5 are two configuration paths, depending on the services provided. SPVCs are set up following Steps 6 and 7. SVCs are configured following Steps 8 and 9.

Figure 10-4 *Generic Link Bringup*

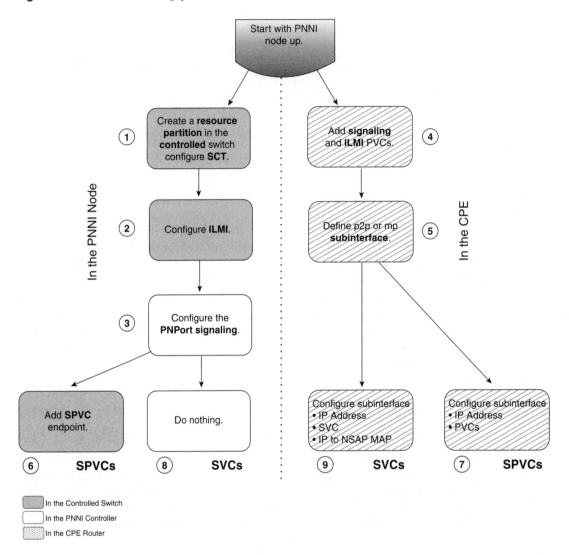

BPX-8600 and SES-Based PNNI Node

The BPX-8600-based PNNI node is the first one that is brought up. The PNNI node is formed by a BPX-8600 switch controlled by a SES PNNI controller. Figure 10-5 shows the details of this PNNI node.

Figure 10-5 *BPX-8600 SES-Based PNNI Node*

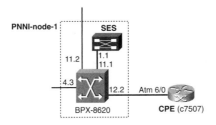

We will analyze this scenario in two steps. First you will bring up the PNNI node. Then in the next section, you will configure its interfaces that interconnect this node to other PNNI nodes.

PNNI Node Bringup: BPX + SES

To bring up the multiservice switching PNNI node, you need to start VSI communication between the controlled switch and the PNNI controller. You up the physical and ATM layers in the BPX-8600 control interface using the command **uptrk** to up a trunk between the BPX-8600 and the SES PNNI controller. You use 11.1 in the BPX-8600 as the control interface. It is important that the protocol for the control interface be run on the BCC controller card and not in the BXM card. This is achieved with the following trunk setting using the command **cnftrk**:

```
Protocol By The Card:  No
```

The statistical reserve protects VSI protocol traffic. It can be changed with the command **cnftrk**. The default value is appropriate.

In the SES, the control interface is always up, so you do not need to modify it. You add the SES as shelf type AAL/5. See Example 10-1.

Example 10-1 *Adding a Service Expansion Controller Shelf*

```
b8620-5a        TN    Cisco          BPX 8620   9.3.4L    Feb. 19 2002 20:12 GMT

                      BPX 8620 Interface Shelf Information

 Trunk    Name     Type     Part Id    Ctrl Id      Control_VC      Alarm
                                                  VPI  VCIRange
 11.1    SES-3a    AAL/5       -          -          -      -        OK
```

continues

Example 10-1 *Adding a Service Expansion Controller Shelf (Continued)*

```
Last Command: addshelf 11.1 x

Shelf has been added
Next Command:
```

This effectively establishes LMI communication between the BPX and the SES using VPI/VCI 3/31. It also creates the PVC 3/8 for IP Relay traffic.

You define a partition in the control interface using the command **cnfrsrc**. VSI will use this partition to create connections terminating in the SES controller. In this case, partition 1 is configured. It is important not to include VPI 3 in the partition's VPI range because this VPI is being used by LMI and IP Relay. The total bandwidth managed by VSI in the control interface is the total physical bandwidth minus the statistical reserve.

You now assign SCT 3 to the partition. SCT 3 includes MPLS and PNNI service types, as shown in Example 10-2. PNNI service types (also called ATM Forum service types) have policing off in SCT 3.

Example 10-2 *Assigning a Service Class Template to a Port*

```
b8620-5a          TN    Cisco          BPX 8620  9.3.4L    Feb. 19 2002 20:15 GMT

    Trunk: 11.1

    Service Class Template ID: 3

    VSI Partitions :
                    channels        bw              vpi
    Part  E/D   min    max    min      max      start   end      ilmi
    1     E     1000   4000   348207   348207   4       10       D
    2     D     0      0      0        0        0       0        D
    3     D     0      0      0        0        0       0        D




Last Command: cnfvsiif 11.1 3

Interface has active VSI partition(s): changing SCT will be service affecting
Next Command:
```

Finally, you add the SES as a controller to the BPX-8600. Refer to Example 10-3.

Example 10-3 *Adding the SES as a PNNI Controller*

```
b8620-5a        TN    Cisco          BPX 8620   9.3.4L    Feb. 19 2002 20:16 GMT

                         VSI Controller Information

Ctrl  Part    Control_VC       Trunk   Intfc Type  Name        IP Address
 Id    Id    VPI  VCIRange
  2     1     0    40-54        11.1    AAL/5       SES-3a      172.18.111.72




Last Command: addctrlr 11.1 2 1 0 40

Ctrlr with id 2 has been added to 11.1
Next Command:
```

You need to add the controller with controller-id 2 and control-vc equal to 0/40 because these values cannot be changed in the SES.

You now need to change the default PNNI node parameters in the PNNI software (in the SES). A **dsppnni-node** at this point shows the default values. See Example 10-4.

Example 10-4 *Default SES PNNI Values Using* **dsppnni-node**

```
SES-3a.1.PXM.a > dsppnni-node

node index: 1                        node name: SES-3a
    Level.............     56        Lowest.............     true
    Restricted transit..  off        Complex node........    off
    Branching restricted  on
    Admin status.......   up         Operational status..    up
    Non-transit for PGL election..     off
    Node id.............56:160:47.00918100000000d058ac2828.00d058ac2828.01
    ATM address..........47.00918100000000d058ac2828.00d058ac2828.01
    Peer group id........56:47.00.9181.0000.0000.0000.0000.00

SES-3a.1.PXM.a >
```

To configure the PNNI parameters, you need to disable PNNI in the node, configuring the operational status to administratively down. You now configure the ATM address and node and peer group IDs, as shown in Example 10-5.

Example 10-5 *Configuring the ATM Address, Node, and Peer Group IDs from the SES*

```
SES-3a.1.PXM.a > cnfpnni-node 1 -enable false

SES-3a.1.PXM.a > cnfpnni-node 1 -atmAddr 47.000000000000010001008600.
  00d058ac2828.01 -level 64

SES-3a.1.PXM.a > cnfpnni-node 1 -nodeId
64:160:47.000000000000010001008600.00d058ac2828.01

SES-3a.1.PXM.a > cnfpnni-node 1 -pgId 64:47.000000000000010000000000

SES-3a.1.PXM.a > cnfpnni-node 1 -enable true

SES-3a.1.PXM.a > dsppnni-node

node index: 1                         node name: SES-3a
    Level..............        64      Lowest.............       true
    Restricted transit..      off      Complex node.......       off
    Branching restricted       on
    Admin status........       up      Operational status..      up
    Non-transit for PGL election..     off
    Node id..............64:160:47.000000000000010001008600.00d058ac2828.01
    ATM address..........47.000000000000010001008600.00d058ac2828.01
    Peer group id........64:47.00.0000.0000.0001.0000.0000.00

SES-3a.1.PXM.a >
```

The peer group ID is derived from the node ID as follows

Step 1   Begin with the PNNI node level.

Step 2   Include a number of bits equal to the PNNI node level from the ATM address.

Step 3   Append trailing 0s to complete 13 bytes.

NOTE   Because of how the command line works, the parameters in the command **cnfpnni-node** could be input on a single line. It is easier and clearer to enter them in different commands.

You also need to configure the prefix used for SPVCs, as shown in Example 10-6. This is desirable so that all the PNNI node addresses can be summarized.

Example 10-6 *Configuring the SPVC Prefix*

```
SES-3a.1.PXM.a > cnfspvcprfx -prfx 47.000000000000010001008600

SES-3a.1.PXM.a >
```

At this stage, your BPX-SES-based PNNI node is ready to perform PNNI routing and signaling.

Configuring PNNI Interfaces

To configure interfaces in the PNNI node, you first need to configure the controlled switch using the commands **upln**, **addport**, **upport**, and **cnfport**. You configure interface 4.3 toward the MGX-8850-based PNNI node. It is important to enable Integrated Local Management Interface (ILMI) in the controlled switch (in the BXM card in this case) because ILMI is a distributed application in multiservice switching nodes, usually called dILMI (distributed ILMI). ILMI's distributed personality is allowed by the protocol passthrough VSI capabilities discussed in Chapter 2, "SCI: Virtual Switch Interface." ILMI switch configuration is shown in Example 10-7.

Example 10-7 *Enabling ILMI in a BPX or IGX VSI Slave*

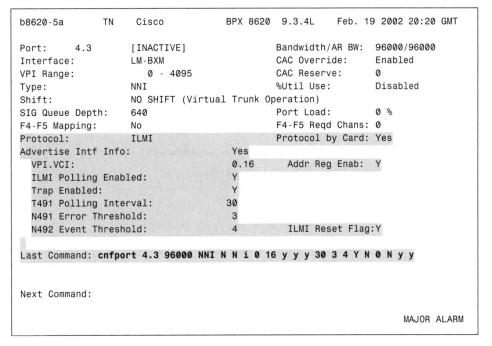

```
b8620-5a        TN    Cisco        BPX 8620   9.3.4L    Feb. 19 2002 20:20 GMT

Port:      4.3      [INACTIVE]            Bandwidth/AR BW:  96000/96000
Interface:         LM-BXM                CAC Override:     Enabled
VPI Range:          0 - 4095             CAC Reserve:      0
Type:              NNI                   %Util Use:        Disabled
Shift:             NO SHIFT (Virtual Trunk Operation)
SIG Queue Depth:   640                        Port Load:     0 %
F4-F5 Mapping:     No                    F4-F5 Reqd Chans: 0
Protocol:          ILMI                  Protocol by Card: Yes
Advertise Intf Info:                Yes
  VPI.VCI:                          0.16     Addr Reg Enab:  Y
  ILMI Polling Enabled:             Y
  Trap Enabled:                     Y
  T491 Polling Interval:            30
  N491 Error Threshold:             3
  N492 Event Threshold:             4        ILMI Reset Flag:Y

Last Command: cnfport 4.3 96000 NNI N N i 0 16 y y y 30 3 4 Y N 0 N y y


Next Command:

                                                        MAJOR ALARM
```

NOTE   For trunk interfaces in a BPX controlled switch, ILMI is enabled using the command **cnfvsipart**. ILMI needs to be configured to run in the BXM card using the command **cnftrk**.

After upping the port using **upport**, you add a resource partition using **cnfrsrc** and assign SCT 3 (without policing) to the VSI partition with the command **cnfvsiif**. At this point, a VSI interface trap is sent to the SES controller, and the pnport is visible from the SES to control. See Example 10-8.

Example 10-8   *Displaying the PnPorts*

```
SES-3a.1.PXM.a > dsppnports

...

DSPPNPORTSPer-port status summary

PortId     LogicalId    IF status     Admin status  ILMI state      #Conns

4.3          262912      up           up            NotApplicable    0

SES-3a.1.PXM.a >
```

It is significant to note that even though ILMI is running in the VSI slave, the state from the VSI master is **NotApplicable**. This is because by default, the pnport signaling is configured as NONE, and in this case, ILMI is not polled from the VSI master to the VSI slave. ILMI keepalives and neighbor discovery are running, but address registration isn't. To change this, you need to change the pnport signaling to something other than NONE, such as pnni10 or uni40. You will do this in the next section.

The SES has the PNNI system addresses shown in Example 10-9.

Example 10-9   *Displaying the PNNI System Addresses*

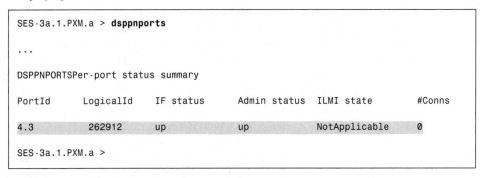

```
SES-3a.1.PXM.a > dsppnsysaddr

47.0000.0000.0000.0100.0100.8600.00d0.58ac.2828.01/160        <= Node address
Type:     host     Port id:     721152   Physical Desc: NA

47.0000.0000.0000.0100.0100.8600.00d0.58ac.2828.99/160        <= AESA PING address
Type:     host     Port id:     721152   Physical Desc: NA

47.0000.0000.0000.0100.0100.8600.0000.0004.0300.00/160        <= 4.3 SPVC address
Type:     host     Port id:     721152   Physical Desc: NA

SES-3a.1.PXM.a >
```

MGX-8850-Based PNNI Node

The PXM-45-based MGX-8850 is the first multiservice switch that houses the controller in the switch's controller card. This section investigates its configuration.

PNNI Node Bringup: PXM-45-Based MGX-8850

Because the PNNI controller software is bundled in the PXM-45 controller card image, you do not need to configure external interfaces to add the PNNI controller. The controller-switch communication uses internal software interfaces. To add the controller, you only need to use the command **addcontroller**. See Example 10-10.

Example 10-10 *Adding a PNNI Controller in an MGX-8850*

```
m8850-7a.7.PXM.a > addcontroller 2 i 2 7 PNNI_CTRLR

m8850-7a.7.PXM.a > dspcontroller
m8850-7a                      System Rev: 03.00   Feb. 20, 2002 01:20:46 GMT
MGX8850                                           Node Alarm: NONE
Number of Controllers:     1
Controller Name:           PNNI_CTRLR
Controller Id:             2
Controller Location:       Internal
Controller Type:           PNNI
Controller Logical Slot:   7
Controller Bay Number:     0
Controller Line Number:    0
Controller VPI:            0
Controller VCI:            0
Controller In Alarm:       NO
Controller Error:


m8850-7a.7.PXM.a >
```

You now configure PNNI as you did with the BPX-SES-based PNNI node. Refer to Example 10-11.

Example 10-11 *Configuring the PNNI Parameters*

```
m8850-7a.7.PXM.a > cnfpnni-node 1 -enable false

m8850-7a.7.PXM.a > cnfpnni-node 1 -atmAddr
  47.00000000000010002008850.00309409f6ba.01 -level 64

m8850-7a.7.PXM.a > cnfpnni-node 1 -nodeId
  64:160:47.00000000000010002008850.00309409f6ba.01
```

continues

Example 10-11 *Configuring the PNNI Parameters (Continued)*

```
m8850-7a.7.PXM.a > cnfpnni-node 1 -pgId 64:47.00000000000010000000000

m8850-7a.7.PXM.a > cnfpnni-node 1 -enable true

m8850-7a.7.PXM.a > dsppnni-node

node index: 1                     node name: m8850-7a
   Level..............   64      Lowest.............    true
   Restricted transit..  off     Complex node.......    off
   Branching restricted  on
   Admin status.......   up      Operational status..   up
   Non-transit for PGL election..   off
   Node id.............64:160:47.00000000000010002008850.00309409f6ba.01
   ATM address.........47.00000000000010002008850.00309409f6ba.01
   Peer group id.......64:47.00.0000.0000.0001.0000.0000.00


m8850-7a.7.PXM.a >
```

You also configure the SPVC prefix. See Example 10-12.

Example 10-12 *Changing the SPVC Prefix*

```
m8850-7a.7.PXM.a > cnfspvcprfx -prfx 47.00000000000010002008850

m8850-7a.7.PXM.a >
```

At this point, your MGX-8850-based PNNI node is configured and operational.

Configuring PNNI Interfaces

You now put together the MGX-8850 end of the PNNI link configuration to connect your two PNNI nodes.

As with the BPX-SES PNNI node, the line, port, and resource partition need to be configured in the controlled switch—in this case, AXSM cards. However, there is an extra first step shown in Example 10-13—configuring the card SCT in AXSM cards. This step needs to be performed with all the ports administratively down, so it is recommended that you do it before any configuration.

Example 10-13 *Changing the AXSM Card SCT*

```
m8850-7a.1.AXSM.a > cnfcdsct 3

m8850-7a.1.AXSM.a >
```

NOTE	The card SCT is configured in the AXSM slaves but does not exist in BXM slaves. In AXSM slaves, the card SCT contains the templates controlling ATM parameters that apply to traffic between the card you are configuring and other AXSM cards in the switch (going into and out of the midplane). SCT 0 comes bundled with the software image. However, card SCT 3 is highly recommended for the majority of applications because you do not want to perform policing going to the cross-point switch.

You start by upping line 1.2 toward the BPX-SES, adding an NNI port and a resource partition. Refer to Example 10-14. The controller-id needs to be 2, as you configured in the **addcontroller** command.

Example 10-14 *Adding a Port and VSI Partition in an AXSM Card*

```
m8850-7a.1.AXSM.a > upln 1.2

m8850-7a.1.AXSM.a > addport 2 1.2 96000 96000 3 2

m8850-7a.1.AXSM.a > addpart 2 1 2 1000000 1000000 1000000 1000000 0 4095 1 65535
    1000 4000

m8850-7a.1.AXSM.a > dspparts
if   part Ctlr egr       egr      ingr     ingr     min  max   min   max   min   max
Num  ID   ID   GuarBw    MaxBw    GuarBw   MaxBw    vpi  vpi   vci   vci   conn  conn
              (.0001%) (.0001%) (.0001%) (.0001%)
----------------------------------------------------------------------------------
 2    1    2 1000000 1000000 1000000 1000000      0 4095     1 65535    1000   4000

m8850-7a.1.AXSM.a >
```

The pnport can now be seen from PNNI software (see Example 10-15), as was learned with VSI interface commands.

Example 10-15 *Displaying the PnPorts*

```
m8850-7a.7.PXM.a > dsppnports
...
DSPPNPORTSPer-port status summary

PortId     LogicalId   IF status    Admin status  ILMI state     #Conns

7.35       17251107    up           up            NotApplicable  0
7.36       17251108    up           up            NotApplicable  0
7.37       17251109    up           up            NotApplicable  0
7.38       17251110    up           up            NotApplicable  0
1:1.2:2    16848898    up           up            NotApplicable  0

m8850-7a.7.PXM.a >
```

NOTE	The value 1:1.2:2 is the VSI interface physical descriptor ASCII value, and the value 16848898 is the decimal representation of the 32-bit Logical Interface Number (LIN). The command **dsppnportidmaps** shows the mapping between the two.
	In AXSM cards, the physical descriptor is formed as follows: *card:bay.line:port,* where *bay.line* refers to the physical line upped with the command **upln**, and *port* is the logical port created with the command **addport**.

You need to enable ILMI in the VSI slave (the AXSM card) using the command **upilmi**. Refer to Example 10-16.

Example 10-16 *Enabling ILMI in the AXSM VSI Slave*

```
m8850-7a.1.AXSM.a > upilmi 2 1
Warning: connections (if any) on port could get rerouted.
Do you want to proceed (Yes/No) ? Yes

m8850-7a.1.AXSM.a >
```

You can see that ILMI is running between the MGX-8850 and the BPX-SES PNNI nodes from the BPX-8600 controlled switch displaying neighbor discovery information, as shown in Example 10-17.

Example 10-17 *Displaying ILMI Neighbor Discovery from the Switch*

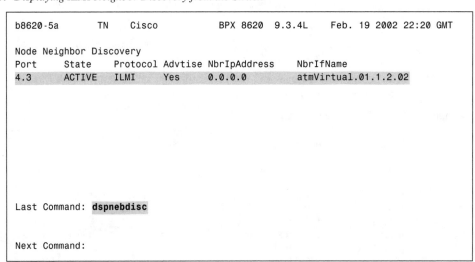

```
b8620-5a        TN    Cisco          BPX 8620  9.3.4L    Feb. 19 2002 22:20 GMT

Node Neighbor Discovery
Port    State    Protocol Advtise NbrIpAddress    NbrIfName
4.3     ACTIVE   ILMI     Yes     0.0.0.0         atmVirtual.01.1.2.02







Last Command: dspnebdisc


Next Command:
```

However, as mentioned in the preceding section, the PNNI controllers do not poll ILMI from the VSI slaves. To initiate dILMI VSI protocol passthrough, you need to configure the pnports to run pnni10 signaling. See Example 10-18.

Example 10-18 *Enabling Distributed ILMI*

```
SES-3a.1.PXM.a > dnpnport 4.3

SES-3a.1.PXM.a > cnfpnportsig 4.3 -nniver pnni10

SES-3a.1.PXM.a > uppnport 4.3
```

ILMI information is passed through to the controller. See Example 10-19.

Example 10-19 *Displaying ILMI Information from the Controller*

```
SES-3a.1.PXM.a > ddsppnilmi 4.3

Port:  4.3                  Port Type:  PNNI            Side:  symmetric
Autoconfig:  enable         UCSM: disable
Secure Link Protocol:  enable
Change of Attachment Point Procedures:  enable
Modification of Local Attributes Standard Procedure:  enable
Addressreg:  Permit All
VPI:       0                VCI:       16
Max Prefix:     16     Total Prefix:      0
Max Address:    64     Total Address:     0
Resync State:    0     Node Prefix: yes
Peer Port Id:    16848898   System_Id : 0.4.222.35.136.192
Peer Addressreg:  enable
Peer Ip Address : 0.0.0.0
Peer Interface Name : atmVirtual.01.1.2.02
ILMI Link State : UpAndNormal
ILMI Version : ilmi40

INFO:  No Prefix registered

INFO:  No ilmi address registered

SES-3a.1.PXM.a >
```

Finally, you also configure line 1.1 in AXSM slot 12 (pnport 12:1.1:1) as PNNI toward the MGX-8950-based PNNI node.

You follow the procedure to configure a PNNI link. You configure the card SCT, line, port, and VSI partition. This PNNI port is configured as Enhanced Virtual NNI (EVNNI) with a VPI range of 10 to 20, different than previous NNI ports. These steps are shown in Example 10-20.

Example 10-20 *Configuring SCT, Line, Port and VSI Partition*

```
m8850-7a.12.AXSM.a > dspports
ifNum Line Admin Oper. Guaranteed Maximum    SCT Id ifType   VPI    minVPI maxVPI
           State State Rate       Rate       (D:dflt         (VNNI, (EVNNI,EVUNI)
                                             used)           VUNI)
----- ---- ----- ----- ---------- ---------- ------ ------ ------ ------ ------
    1  1.1  Up    Up    600000     600000     3      EVNNI    0     10     20

m8850-7a.12.AXSM.a > dspparts
if   part Ctlr egr      egr      ingr     ingr     min max  min   max   min  max
Num  ID   ID   GuarBw   MaxBw    GuarBw   MaxBw    vpi vpi  vci   vci   conn conn
          (.0001%) (.0001%) (.0001%) (.0001%)
-----------------------------------------------------------------------------
1    1    2  1000000 1000000 1000000 1000000   10  20    1 65535  1000 4000

m8850-7a.12.AXSM.a >
```

The different types of ports that can be configured and their characteristics are summarized in Table 10-1.

Table 10-1 *AXSM Port Types*

Port Type	VPIs Used	Specified VPI Values	Allowed VPI Values
UNI	Fixed VPI range	None	Fixed 0 to 255
NNI	Fixed VPI range	None	Fixed 0 to 4095
VNNI	Fixed VPI value	Single VPI	VPI between 1 and 4095
VUNI	Fixed VPI value	Single VPI	VPI between 1 and 255
EVUNI	Configurable VPI range	minVPI and maxVPI	minVPI and maxVPI between 0 and 255
EVNNI	Configurable VPI range	minVPI and maxVPI	minVPI and maxVPI between 0 and 4095

In BXM slaves on a BPX-SES PNNI node, the ports are either UNI or NNI, as configured with the command **cnfport**. You specify the VSI VPI usage when you configure the resource partition. You can set up a single VPI, a VPI range, or a full VPI range. It is recommended that you configure single VPI ports as virtual trunks using the commands **uptrk** and **cnfrsrc**.

PNNI Neighbor Verification

You can verify PNNI connectivity by checking the PNNI neighbors. The neighbor state is **FULL**. In addition, you can display the PNNI node list with the command **dsppnni-node-list**, as shown in Example 10-21.

Example 10-21 *Displaying PNNI Neighbors and PNNI Node List*

```
m8850-7a.7.PXM.a > dsppnni-neighbor

node index    : 1

node name     : SES-3a
Remote node id: 64:160:47.0000000000000010001008600.00d058ac2828.01
Neighbor state: FULL
    Port count.........    1     SVC RCC index.......     0
    RX DS pkts.........    3     TX DS pkts..........     2
    RX PTSP pkts.......    3     TX PTSP pkts........     2
    RX PTSE req pkts....   1     TX PTSE req pkts....     1
    RX PTSE ack pkts....   1     TX PTSE ack pkts....     1


m8850-7a.7.PXM.a > dsppnni-node-list

node #  node id                                              node name   level
------  ---------------------------------------------------- ---------- -------
    1   64:160:47.0000000000000010002008850.00309409f6ba.01 m8850-7a      64

node #  node id                                              node name   level
------  ---------------------------------------------------- ---------- -------
    2   64:160:47.0000000000000010001008600.00d058ac2828.01 SES-3a        64

m8850-7a.7.PXM.a >
```

The PNNI link, a lowest-level horizontal link, is in state **twoWayInside** (see Example 10-22).
The complete state machine, with the name and meaning of all states, was covered in
Chapter 8, "PNNI Explained."

Example 10-22 *Displaying PNNI Links*

```
m8850-7a.7.PXM.a > dsppnni-link

node index    : 1
Local port id:   16848898        Remote port id:      262912
Local Phy Port Id: 1:1.2:2
    Type. lowestLevelHorizontalLink    Hello state....... twoWayInside
    Derive agg..........    0     Intf index..........  16848898
    SVC RCC index.......    0     Hello pkt RX.........        7
                                  Hello pkt TX.........        9
    Remote node name.......SES-3a
    Remote node id........64:160:47.0000000000000010001008600.00d058ac2828.01
    Upnode id.............0:0:00.00000000000000000000000.000000000000.00
    Upnode ATM addr........00.00000000000000000000000.000000000000.00
    Common peer group id...00:00.00.0000.0000.0000.0000.0000.00


m8850-7a.7.PXM.a >
```

You can also check the PNNI internal database using the command **dsppnni-idb** and the PTSE database using the command **dsppnni-ptse**.

From the MGX-8850 PNNI node, you can reach one summarized network prefix belonging to the BPX-SES-based PNNI node. See Example 10-23.

Example 10-23 *Displaying the PNNI Reachable Addresses*

```
m8850-7a.7.PXM.a > dsppnni-reachable-addr network

scope..............     0       Advertising node number     2
Exterior............    false
ATM addr prefix.....47.0000.0000.0000.0100.0100.8600/104
Transit network id..
Advertising nodeid..64:160:47.00000000000010001008600.00d058ac2828.01
Node name..........SES-3a

m8850-7a.7.PXM.a >
```

MGX-8950-Based PNNI Node

MGX-8950 also contains the PNNI controller in the PXM-45B controller card. The configuration steps for an MGX-8950-based PNNI node bringup and interface configuration are the same as in the MGX-8850 PNNI node we explored in the preceding section. For that reason, not every configuration command is repeated in the following sections.

PNNI Node Bringup: PXM-45-Based MGX-8950

The internal PNNI controller in the MGX-8950 switch looks like Example 10-24.

Example 10-24 *Displaying the PNNI Controller Details*

```
m8950-7b.7.PXM.a > dspcontroller
m8950-7b                        System Rev: 03.00   Feb. 20, 2002 03:16:52 GMT
MGX8950 (JBP-2)                                     Node Alarm: MAJOR
Number of Controllers:    1
Controller Name:          PNNI_8950
Controller Id:            2
Controller Location:      Internal
Controller Type:          PNNI
Controller Logical Slot:  7
Controller Bay Number:    0
Controller Line Number:   0
Controller VPI:           0
Controller VCI:           0
Controller In Alarm:      NO
Controller Error:

m8950-7b.7.PXM.a >
```

The final PNNI configuration for the network is as displayed in Example 10-25.

Example 10-25 *Displaying the PNNI Configuration Details*

```
m8950-7b.7.PXM.a > dsppnni-node

node index: 1                      node name: m8950-7b
   Level..............    64      Lowest.............      true
   Restricted transit..   off     Complex node.......      off
   Branching restricted   on
   Admin status........   up      Operational status..     up
   Non-transit for PGL election..    off
   Node id.............64:160:47.000000000000010002008950.0004c113ba46.01
   ATM address.........47.000000000000010002008950.0004c113ba46.01
   Peer group id.......64:47.00.0000.0000.0001.0000.0000.00


m8950-7b.7.PXM.a >
```

The steps to achieve this PNNI configuration are like those for the other two PNNI nodes.

Configuring PNNI Interfaces

After bringing up pnport 5:1.1:1, you have a lowest-level horizontal link in **twoWayInside** state. That link faces the MGX-8850 PNNI neighbor, as shown in Example 10-26.

Example 10-26 *Using the Command **dsppnni-neighbor***

```
m8950-7b.7.PXM.a > dsppnni-neighbor

node index    : 1

node name     : m8850-7a
Remote node id: 64:160:47.000000000000010002008850.00309409f6ba.01
Neighbor state: FULL
   Port count.........      1      SVC RCC index.......     0
   RX DS pkts..........     3      TX DS pkts..........     2
   RX PTSP pkts........     3      TX PTSP pkts........     3
   RX PTSE req pkts....     1      TX PTSE req pkts....     1
   RX PTSE ack pkts....     1      TX PTSE ack pkts....     1


m8950-7b.7.PXM.a >
```

Both MGX-8850 and BPX-SES PNNI nodes' summary addresses are network-reachable. See Example 10-27.

Example 10-27 *Using the Command* **dsppnni-reachable-addr**

```
m8950-7b.7.PXM.a > dsppnni-reachable-addr network

scope..............        0     Advertising node number       2
Exterior...........     false
ATM addr prefix.....47.0000.0000.0000.0100.0100.8600/104
Transit network id..
Advertising nodeid..64:160:47.00000000000010001008600.00d058ac2828.01
Node name..........SES-3a

scope..............        0     Advertising node number       3
Exterior...........     false
ATM addr prefix.....47.0000.0000.0000.0100.0200.8850/104
Transit network id..
Advertising nodeid..64:160:47.00000000000010002008850.00309409f6ba.01
Node name..........m8850-7a

m8950-7b.7.PXM.a >
```

PNNI Network Verification

You can check prefix reachability using the command **aesa_ping**, as shown in Example 10-28. You ping the BPX-SES PNNI node address from the MGX-8950 PNNI node.

Example 10-28 *Using the Command* **aesa_ping** *to Verify Reachability*

```
m8950-7b.7.PXM.a > aesa_ping 47.00000000000010001008600.00d058ac2828.01

Ping Got CLI message, index=0

PING: from PNNI - SOURCE ROUTE
DTL    1 : Number of (Node/port)elements    3

DTL 1:NODE 1: :64:160:71:0:0::2:0:137:80:

 Port 1:17111041

DTL 1:NODE 2: :64:160:71:0:0::2:0:136:80:

 Port 2:16848898

DTL 1:NODE 3: :64:160:71:0:0::1:0:134:0:

 Port 3:0

Port List : no of ports =    1
```

Example 10-28 *Using the Command* **aesa_ping** *to Verify Reachability (Continued)*

```
Port ID    1:17111041

m8950-7b.7.PXM.a >
```

Analyzing the command output, you see only one Designated Transit List (DTL) with three elements. This is because an SPG PNNI network is at level 64 (meaning that the 64 leftmost bits from the node ID form the peer group ID).

The only DTL has three nodes, which are at level 64. The node IDs convert the decimal numbers to hexadecimal, as shown in Table 10-2.

Table 10-2 *PNNI Node IDs in the* **aesa_ping** *Path*

Node	PNNI Node ID	Node
Node 1	64:160:47.0000...02008950...	MGX-8950
Node 2	64:160:47.0000...02008850...	MGX-8850
Node 3	64:160:47.0000...01008600...	BPX-8600

The ports are displayed in LIN decimal format. Port 1 is 17111041 (0x1051801 for slot 5 and port 1), and port 2 is 16848898 (0x1011802 for slot 1 and port 2).

The path taken and specified in the DTL is shown in Figure 10-6.

Figure 10-6 **aesa_ping** *Path and DTL Through the PNNI Network*

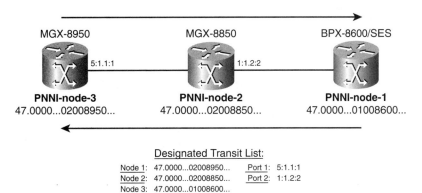

This is a good time to present the command **dsppnportidmaps**. Some commands such as **aesa_ping** display the pnports in LIN format. The command **dsppnportidmaps** can be used to display the mapping between the ASCII PnPort ID representation and the LIN representation, as shown in Example 10-29.

Example 10-29 *Using the Command* **dsppnportidmaps**

```
SES-3a.1.PXM.a > dsppnportidmaps

Port Id          Logical ID (Dec)  Logical ID (Hex)  OperStatus

4.3              262912            40300             up
11.2             721408            b0200             up
12.1             786688            c0100             up
12.2             786944            c0200             up
SES-3a.1.PXM.a >

m8850-7a.7.PXM.a > dsppnportidmaps

Port Id          Logical ID (Dec)  Logical ID (Hex)  OperStatus

7.35             17251107          1073b23           up
7.36             17251108          1073b24           up
7.37             17251109          1073b25           up
7.38             17251110          1073b26           up
10.1             17257217          1075301           up
1:1.2:2          16848898          1011802           up
2:1.1:1          16914433          1021801           up
12:1.1:1         17569793          10c1801           up
m8850-7a.7.PXM.a >
```

Cisco IOS Software-Based PNNI Node

For completeness, we include in the picture a Cisco IOS Software-based PNNI node. The
PNNI configuration is shown in Example 10-30, following our address scheme.

Example 10-30 *PNNI Configuration in Cisco IOS*

```
LS-1010#conf t
Enter configuration commands, one per line.  End with CNTL/Z.
LS-1010(config)#atm address 47.00000000000010001001010.00503EFBA601.01
LS-1010(config)#no atm address 47.0091.8100.0000.0050.3efb.a601.0050.3efb.a601.00
LS-1010(config)#atm router pnni
LS-1010(config-atm-router)#node 1 disable
LS-1010(config-pnni-node)#node 1 level 64 lowest
LS-1010(config-pnni-node)#node 1 enable
LS-1010(config-pnni-node)#^Z
LS-1010#show atm pnni local-node

PNNI node 1 is enabled and running
  Node name: LS-1010
  System address       47.00000000000010001001010.00503EFBA601.01
  Node ID        64:160:47.00000000000010001001010.00503EFBA601.00
  Peer group ID      64:47.0000.0000.0000.0100.0000.0000
  Level 64, Priority 0 0, No. of interfaces 1, No. of neighbors 0
...
```

Cisco IOS Software-based PNNI node interfaces have autoconfiguration set by default. No more configurations are needed on this side as long as ILMI is configured on the links. ILMI is used for link signaling and parameter negotiation.

To complete your PNNI network setup and form a square, you configure two pnports: port 11.2 using VPI range 15 to 60 in the BPX-SES, and port 1:1.1:1 using VPI range 0 to 4095 in the MGX-8950. Both pnports go toward the Cisco IOS Software PNNI node, and the configuration is exactly as you did before. ILMI and PNNI should be up. VSI ILMI configuration in the BXM and AXSM slaves should be set.

The current PNNI network topology is shown in Figure 10-7.

Figure 10-7 *PNNI Network*

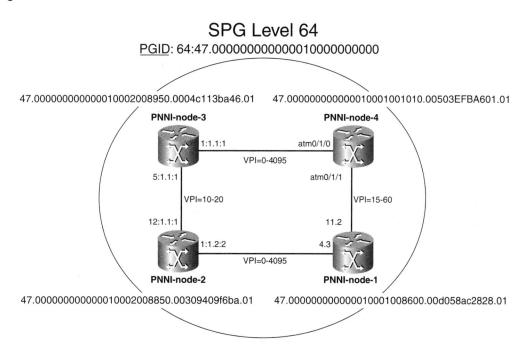

SPG Level 64
<u>PGID</u>: 64:47.0000000000000010000000000

47.000000000000010002008950.0004c113ba46.01   47.000000000000010001001010.00503EFBA601.01

PNNI-node-3   **PNNI-node-4**

1:1.1:1   atm0/1/0

VPI=0-4095

5:1.1:1   atm0/1/1

VPI=10-20   VPI=15-60

12:1.1:1   11.2

1:1.2:2   4.3

VPI=0-4095

PNNI-node-2   **PNNI-node-1**

47.000000000000010002008850.00309409f6ba.01   47.000000000000010001008600.00d058ac2828.01

To display the details of your PNNI network, display the PNNI database from the LS1010 using the command **show atm pnni database** (equivalent to the command **dsppnni-idb** in BPX-SES and MGX PNNI nodes). See Example 10-31.

Example 10-31 *Using* **show atm pnni database**

```
LS-1010#show atm pnni database

Node 1 ID 64:160:47.00000000000010001001010.00503EFBA601.00 (name: LS-1010)

  PTSE ID  Length  Type  Seq no.  Checksum  Lifetime  Description
  1        144     97    5        62436     2503      Nodal info
  2        52      224   6        56669     2490      Int. Reachable Address
  3        124     288   2        22664     2620      Horizontal Link

Node 9 ID 64:160:47.00000000000010001008600.00D058AC2828.01 (name: SES-3a)

  PTSE ID  Length  Type  Seq no.  Checksum  Lifetime  Description
  1        148     97    22       24725     2725      Nodal info
  12       52      224   39       7983      3462      Int. Reachable Address
  13       220     288   3        13020     2725      Horizontal Link
  16       220     288   1        33604     3480      Horizontal Link

Node 10 ID 64:160:47.00000000000010002008850.00309409F6BA.01 (name: m8850-7a)

  PTSE ID  Length  Type  Seq no.  Checksum  Lifetime  Description
  1        152     97    108      43988     2644      Nodal info
  8        52      224   38       31703     3001      Int. Reachable Address
  11       220     288   3        13681     2759      Horizontal Link
  13       220     288   2        18578     3070      Horizontal Link

Node 11 ID 64:160:47.00000000000010002008950.0004C113BA46.01 (name: m8950-7b)

  PTSE ID  Length  Type  Seq no.  Checksum  Lifetime  Description
  1        152     97    107      49822     3105      Nodal info
  8        52      224   4        4360      3104      Int. Reachable Address
  9        220     288   2        18653     3105      Horizontal Link
LS-1010#
```

You can see the PNNI neighbors and the links between them, as shown in Example 10-32.

Example 10-32 *Displaying PNNI Neighbors*

```
LS-1010#show atm pnni neighbor

Neighbors For Node (Index 1, Level 64)

  Neighbor Name: SES-3a, Node number: 9
  Neighbor Node Id: 64:160:47.00000000000010001008600.00D058AC2828.01
  Neighboring Peer State: Full
  Link Selection For CBR    : minimize blocking of future calls
  Link Selection For VBR-RT : minimize blocking of future calls
  Link Selection For VBR-NRT: minimize blocking of future calls
  Link Selection For ABR    : balance load
```

Example 10-32 *Displaying PNNI Neighbors (Continued)*

```
Link Selection For UBR   : balance load
 Port                      Remote Port Id        Hello state
 ATM0/1/1                      11.2              2way_in  (Flood Port)


Neighbor Name: m8950-7b, Node number: 11
Neighbor Node Id: 64:160:47.00000000000010002008950.0004C113BA46.01
Neighboring Peer State: Full
Link Selection For CBR   : minimize blocking of future calls
Link Selection For VBR-RT : minimize blocking of future calls
Link Selection For VBR-NRT: minimize blocking of future calls
Link Selection For ABR   : balance load
Link Selection For UBR   : balance load
 Port                      Remote Port Id        Hello state
 ATM0/1/0                      1.1               2way_in  (Flood Port)
LS-1010#
```

The links are in **twoWayInside** state.

You can see topology details for a specific PNNI node in the database using the command **show atm pnni topology** as shown in Example 10-33.

Example 10-33 *Displaying PNNI Topology Details*

```
LS-1010#show atm pnni topology node 11

Node 11 (name: m8950-7b, type: bpx/mgx, ios-version: 0.0)
Node ID..: 64:160:47.00000000000010002008950.0004C113BA46.01
Node AESA:      47.00000000000010002008950.0004C113BA46.01
Leadership Priority: 0, Claims PGL: No, Transit Calls: Allowed
Ancestor: No, Nodal Representation: Simple, Connected: Yes
More P2MP Branch Points: No, Non-Transit For PGL Election: No
Advertises ONHLs: No


   status  link type  local port    remote port    neighbor
   ~~~~~~  ~~~~~~~~~  ~~~~~~~~~~~~~  ~~~~~~~~~~~~~  ~~~~~~~~
    up      hrz       5.1            12.1           m8850-7a
    up      hrz       1.1            ATM0/1/0       LS-1010
LS-1010#
```

A Note on Feeder Shelves

It is very common among service providers to attach feeder nodes to PNNI switching nodes as aggregators or concentrators. The link in the PNNI routing node going to the feeder node terminates SPVC calls. The feeder concentrates IP, Frame Relay, and ATM users and terminates PVCs on its uplink. This scenario is shown in Figure 10-8.

Figure 10-8 *Feeder Nodes Attached to a PNNI Network*

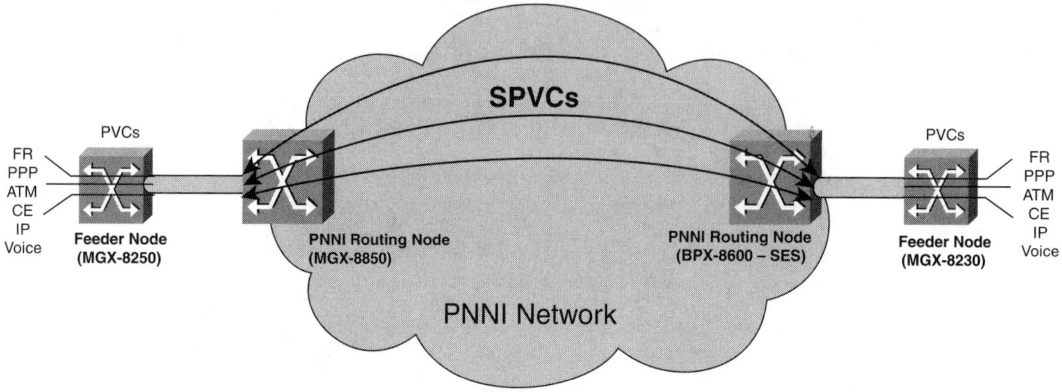

In order to configure this scenario, you must configure as **none** the PnPort signaling in the link from the PNNI routing node to the feeder shelf. This means that no signaling is running in that link and that only PVCs and SPVCs are supported.

You add the feeder from the controlled switch using the command **addshelf** from a BPX-8600 or the command **addfdr** from an AXSM card in an MGX-8850. LMI runs between the controlled switch and the feeder node. In addition, you need to set the parameter **cntlvc** in the **cnfpnportsig** command to **ip** so that VPI/VCI = 3/8 is created and the feeder can be managed.

Another special case as far as PnPort signaling configuration appears when VISM cards exist in a PNNI node. In that case, the PnPort signaling is configured to **self**. This allows the VISM card to set up voice SVCs.

Configuring NCDP

MGX-8850 and MGX-8950-based PNNI nodes as well as Cisco IOS Software-based PNNI platforms support NCDP as a network application for clock distribution.

In MGX-8850 and MGX-8950 PNNI switches, physical links are enabled as NCDP ports, and virtual links are disabled by default. See Example 10-34.

NOTE   You may need to enable NCDP using the command **cnfncdp -distributionMode 1**.

Example 10-34 *Displaying NCDP Ports*

```
m8950-7b.7.PXM.a > dspncdpports

PortId    Clock mode  Clock Vpi   Clock Vci   Admin Cost    Ncdp Vc
1:1.1:1   enable      0           34          10            up
5:1.1:1   disable     10          34          10            down

m8950-7b.7.PXM.a >
```

In your network scenario, you can enable NCDP on port 5:1.1:1 in the MGX-8950 and corresponding port 12:1.1:1 in the MGX-8850 (see Example 10-35) because they are directly connected. You need to understand that pnport 5:1.1:1 is a logical port using a range of VPIs only.

Example 10-35 *Enabling NCDP on Two Ports*

```
m8950-7b.7.PXM.a > cnfncdpport 5:1.1:1 -ncdp enable

m8950-7b.7.PXM.a > dspncdpports

PortId    Clock mode  Clock Vpi   Clock Vci   Admin Cost    Ncdp Vc
1:1.1:1   enable      0           34          10            up
5:1.1:1   enable      10          34          10            up

m8850-7a.7.PXM.a > cnfncdpport 12:1.1:1 -ncdp enable
```

Next you configure the external clock input number 1 in the MGX-8850 as an NCDP clock source with priority 10 (the default priority is 128). See Example 10-36.

Example 10-36 *Creating NCDP Clock Sources*

```
m8850-7a.7.PXM.a > cnfncdpclksrc 7.35 0 t1 -priority 10 -stratumLevel 2

m8850-7a.7.PXM.a > dspncdpclksrcs

PortId     Best clk src   Priority   Stratum level   Prs id          Health
7.35 (t1)  Yes            10         2               0(external)     Good
255.255    No             128        3               255(internal)   unknown

m8850-7a.7.PXM.a >
```

The NCDP clock source 255.255 refers to the node's internal oscillator.

You also configure the MGX-8950 internal oscillator as a second NCDP clock source with a lowest priority (highest priority number). Refer to Example 10-37.

Example 10-37 *Configuring a Secondary NCDP Clock Source*

```
m8950-7b.7.PXM.a > cnfncdpclksrc 255.255 255 -priority 100
```

The MGX-8950 gets clock from interface 5:1.1:1 facing the MGX-8850. You can see this using the command **dspncdp**, as shown in Example 10-38.

Example 10-38 *Displaying NCDP Status*

```
m8950-7b.7.PXM.a > dspncdp
Distribution Mode              : ncdp
Node stratum level             : 3
Max network diameter           : 20
Hello time interval            : 500
Holddown time interval         : 500
Topology change time interval  : 500
Root Clock Source              : 5:1.1:1
Root Stratum Level             : 2
Root Priority                  : 10
Last clk src change time       : Feb 20 2002 14:36:31
Last clk src change reason     : Topology Changed

m8950-7b.7.PXM.a >
```

You can enable NCDP in the Cisco IOS Software-based platform using the global configuration command **ncdp,** as shown in Example 10-39.

Example 10-39 *Enabling NCDP in Cisco IOS*

```
LS-1010#conf t
Enter configuration commands, one per line.  End with CNTL/Z.
LS-1010(config)#ncdp
LS-1010(config)#^Z
LS-1010#
2d11h: %CLOCKSW: Switching from System to ATM0/1/0
```

You can see the LS1010 clock status using the command **show ncdp status** (see Example 10-40). Note that the clocking root AESA corresponds to the MGX-8850 node.

Example 10-40 *Showing NCDP Status in Cisco IOS*

```
LS-1010#show ncdp status
  = ncdp switch information ==== enabled ==============
  revertive
  root clock source priority:      10
```

Example 10-40 *Showing NCDP Status in Cisco IOS (Continued)*

```
root clock source stratum level: 2
root clock source prs id:        0
stratum level of root switch:    3
clocking root address:           47000000000000010002008850003094009F6BA01
hop count:                       2
root path cost:                  20
root port:                       2 <ATM0/1/0>
max age:                         20
hello time:                      500
...
LS-1010#
```

To force the NCDP network application to change the network clock source and distribution paths, you unplug the external clock input cable. At this time, the external clock source in the MGX-8850 is declared bad. See Example 10-41.

Example 10-41 *Displaying NCDP Clock Sources*

```
m8850-7a.7.PXM.a > dspncdpclksrcs

PortId    Best clk src    Priority    Stratum level    Prs id          Health
7.35 (t1) No              10          2                0(external)     Bad
255.255   Yes             128         3                255(internal)   unknown

m8850-7a.7.PXM.a >
```

The next priority network clock source should become clock root, as shown in Example 10-42.

Example 10-42 *Displaying NCDP Status and Root Clock Source*

```
m8950-7b.7.PXM.a > dspncdp
Distribution Mode              : ncdp
Node stratum level             : 3
Max network diameter           : 20
Hello time interval            : 500
Holddown time interval         : 500
Topology change time interval  : 500
Root Clock Source              : 255.255
Root Stratum Level             : 3
Root Priority                  : 100
Last clk src change time       : Feb 20 2002 14:38:59
Last clk src change reason     : Root Aged Out

m8950-7b.7.PXM.a >
```

Because of the fact that the BPX controller card where the BPX clock inputs reside does not currently house a VSI slave, the SES-BPX-based node does not currently support NCDP. However, AutoRoute provides automatic clock distribution.

Attaching Customer Premises Equipment

In this section, you attach two external customer premises equipment (CPE) routers to your network, as shown in Figure 10-9. A Cisco 7505 router is connected to port 2:1.1:1 in the MGX-8850 PNNI switch, and a Cisco 7507 router is connected to port 12.2 in the BPX-SES PNNI node. Afterwards, you will configure SVCs and SPVCs.

Figure 10-9 *Attaching CPEs*

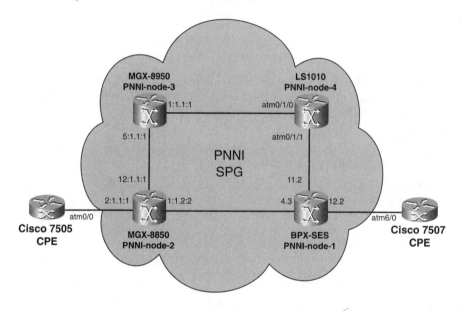

The CPEs are configured with an End System Identifier (ESI). They get their ATM addresses using the ILMI address registration process. Each ESI includes the platform number. The router AESAs are shown in Figure 10-10.

Figure 10-10 *CPEs' AESAs*

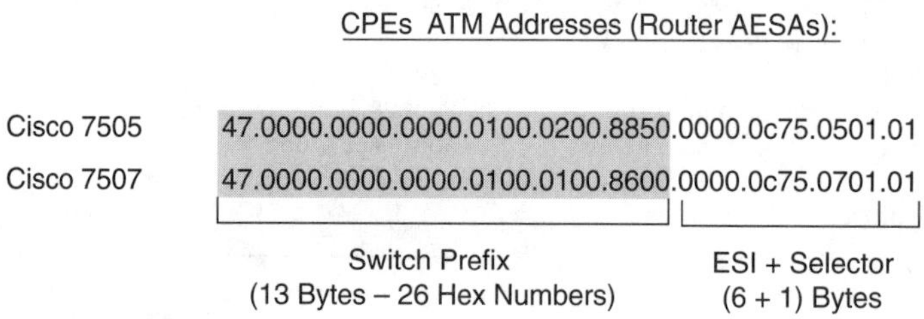

CPEs ATM Addresses (Router AESAs):

Cisco 7505   47.0000.0000.0000.0100.0200.8850.0000.0c75.0501.01
Cisco 7507   47.0000.0000.0000.0100.0100.8600.0000.0c75.0701.01

Switch Prefix
(13 Bytes – 26 Hex Numbers)

ESI + Selector
(6 + 1) Bytes

If you analyze the router AESAs shown in Figure 10-10, you can see that the C7505 router is connected to the MGX-8850 and the C7507 router is connected to the BPX-8600-SES.

You start the configuration by connecting and setting up an SVC-capable router to the BPX-SES PNNI node's port 12.2. On the BPX side, you use the commands **upln**, **addport**, and **upport** to up Layers 1 and 2 in the physical link. You use the command **cnfport** to enable ILMI and configure it to run on the BXM card. Subsequently, you configure the VSI slave by configuring a VSI resource partition with the command **cnfrsrc** and apply SCT 2 to the resource partition with the command **cnfvsiif**. SCT 2 includes policing for ATM Forum service types.

That is all the provisioning needed on the slave side. As you can infer, this configuration is independent of the controller application (MPLS, PNNI, or something else).

In the SES PNNI controller, you need to configure UNI signaling in the now-visible pnport 12.2 (see Example 10-43). As I mentioned earlier, this also starts dILMI passthrough between the VSI master and VSI slave.

Example 10-43 *Configuring UNI Signaling in a PnPort*

```
SES-3a.1.PXM.a > dnpnport 12.2

SES-3a.1.PXM.a > cnfpnportsig 12.2 -univer uni40

SES-3a.1.PXM.a > uppnport 12.2
```

On the router side, you start the provisioning by adding UNI signaling and ILMI PVCs (see Example 10-44). Using VCIs 5 and 16, respectively, you configure both UNI and ILMI and up the interface.

Example 10-44 *CPE Signaling and ILMI Configuration*

```
C7507-1a(config)#interface ATM 6/0
C7507-1a(config-if)#pvc Sig 0/5 qsaal
C7507-1a(config-if-atm-vc)#vbr-nrt 149760 2000 16
C7507-1a(config-if-atm-vc)#pvc ILMI 0/16 ilmi
C7507-1a(config-if-atm-vc)#vbr-nrt 1498 1498 16
C7507-1a(config-if-atm-vc)#exit
C7507-1a(config-if)#atm ilmi-enable
% ATM6/0 :ILMI enabling will take effect in the next interface restart
C7507-1a(config-if)#atm ilmi-keepalive 5
C7507-1a(config-if)#atm uni-version 4.0
C7507-1a(config-if)#no shut
C7507-1a(config-if)#
```

The signaling and ILMI virtual circuits have been configured with traffic parameters conforming to ATM Forum specifications. For the signaling VC traffic parameter, refer to ATMF specification af-sig-0061.000, Section 4.2. For ILMI PVC characteristics, refer to ATMF specification af-ilmi-0065.000, Section 5.1.

NOTE   On the PNNI switch, signaling and PNNI VCs use the signaling VSI service type, which has the highest priority. The signaling service type has a service type number of 0x0002. You can display the traffic parameter configuration with the command **dsppnctlvc** and modify it with the command **cnfpnctlvc**. PNNI RCC virtual circuits default to PCR = 906 CPS, SCR = 453 CPS, and MBS = 171 cells.

As discussed in Chapter 8, ATM signaling uses Service-Specific Connection-Oriented Protocol (SSCOP) as a reliable transport. SSCOP provides error recovery, flow control, and error and status reporting. You can check the SSCOP status, timers, and statistics in the router using the command **show sscop**, as shown in Example 10-45.

Example 10-45 *Using the Command* **show sscop**

```
C7507-1a#show sscop ATM 6/0
SSCOP details for interface ATM6/0
   Current State = Active,   Uni version = 4.0
   Send Sequence Number: Current = 95,   Maximum = 125
   Send Sequence Number Acked = 95
   Rcv Sequence Number: Lower Edge = 95, Upper Edge = 95, Max = 125
   Poll Sequence Number = 1157, Poll Ack Sequence Number = 1157
   Vt(Pd) = 0   Vt(Sq) = 0
   Timer_IDLE = 10 - Active

   Timer_CC = 1 - Inactive
   Timer_POLL = 1000 - Inactive
   Timer_KEEPALIVE = 5 - Inactive
   Timer_NO-RESPONSE = 45 - Inactive
   Current Retry Count = 0, Maximum Retry Count = 10
   AckQ count = 0, RcvQ count = 0, TxQ count = 0
   AckQ HWM = 1,  RcvQ HWM = 0, TxQ HWM = 1
   Local connections currently pending = 0
   Max local connections allowed pending = 0
   Statistics -
      Pdu's Sent = 1513, Pdu's Received = 1513, Pdu's Ignored = 0
      Begin = 3/0, Begin Ack = 0/3, Begin Reject = 0/0
      End = 2/0, End Ack = 0/2
      Resync = 0/0, Resync Ack = 0/0
      Sequenced Data = 112/112, Sequenced Poll Data = 0/0
      Poll = 1313/1396, Stat = 1396/1313, Unsolicited Stat = 0/0
      Unassured Data = 0/0, Mgmt Data = 0/0, Unknown Pdu's = 0
      Error Recovery/Ack = 0/0, lack of credit 0
C7507-1a#
```

Finally, you create a multipoint subinterface where you configure the IP address and subnet mask and the ESI for ILMI address registration. See Example 10-46.

Example 10-46 *Configuring the CPE Subinterface*

```
C7507-1a(config-if)#interface ATM 6/0.10 multipoint
C7507-1a(config-subif)#description Subinterface for SVCs towards BPX-SES switch.
C7507-1a(config-subif)#ip address 172.18.1.1 255.255.255.0
C7507-1a(config-subif)#atm esi-address 00000C750701.01
C7507-1a(config-subif)#
Feb 20 03:47:23.728: %LANE-6-INFO: ATM6/0: ILMI prefix add event received
C7507-1a(config-subif)#^Z
```

You can see in Example 10-47 that ILMI is up from the router. You see the following fields:

- The state is **UpAndNormal**.

- IP addresses and interfaces have been exchanged.

- The maximum VPI and VCI bits have been negotiated.

- The ATM address has been registered using the PNNI switch prefix.

Example 10-47 *Showing the ILMI Status*

```
C7507-1a#show atm ilmi-status

Interface : ATM6/0 Interface Type : Private UNI (User-side)
ILMI VCC : (0, 16) ILMI Keepalive : Enabled/Up (5 Sec 4 Retries)
ILMI State:        UpAndNormal
Peer IP Addr:      30.1.1.50        Peer IF Name:     atmVirtual.12.1.2.2
Peer MaxVPIbits:   0               Peer MaxVCIbits:  16
Active Prefix(s) :
47.0000.0000.0000.0100.0100.8600
End-System Registered Address(s) :
47.0000.0000.0000.0100.0100.8600.0000.0c75.0701.01(Confirmed)
C7507-1a#
```

You can also observe the router's ILMI exchanged parameters from the PNNI node. In particular, because you have a distributed ILMI application, you can check the parameters from the VSI slave using the command **dspnebdisc**, as shown in Example 10-48.

Example 10-48 *Displaying ILMI Neighbor Discovery from the Controlled Switch*

```
b8620-5a        TRM    Cisco          BPX 8620   9.3.4L    Feb. 20 2002 11:19 GMT

Node Neighbor Discovery
Port    State     Protocol Advtise NbrIpAddress    NbrIfName
4.3     ACTIVE    ILMI     Yes     0.0.0.0         atmVirtual.01.1.2.02
11.2    ACTIVE    ILMI     Yes     172.18.111.253  ATM0/1/1
```

continues

Example 10-48 *Displaying ILMI Neighbor Discovery from the Controlled Switch (Continued)*

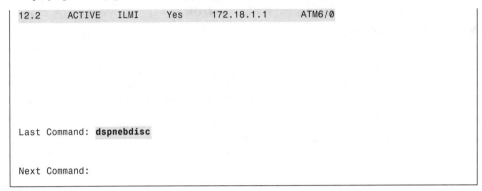

```
12.2      ACTIVE    ILMI     Yes      172.18.1.1      ATM6/0




Last Command: dspnebdisc


Next Command:
```

You can see the ILMI visibility from the SES controller shelf using the commands **dsppnport** and **dsppnilmi**. You can display the ILMI registered address using the command **dspilmiaddr** (see Example 10-49). That address is a local reachable address that is included using the command **dsppnni-reachable-addr local**.

Example 10-49 *Displaying ILMI Registered Address from the PNNI Controller*

```
SES-3a.1.PXM.a > dspilmiaddr 12.2
47.0000.0000.0000.0100.0100.8600.0000.0c75.0701.01        scope: LocalNetwork

SES-3a.1.PXM.a >
```

The other router is a Cisco 7505 connected to port 2:1.1:1 in the MGX-8850-based PNNI node.

On the AXSM card, you start by configuring the card SCT to 3. Exactly the same as with the PNNI links, you up a line and add a port and partition using the commands **upln**, **addport**, and **addpart**, respectively. In the **addport** command, you configure the SCT port to 2 because the interface is connected to an externally controlled router, and you want to turn on policing. You start ILMI in the slave using the command **upilmi**.

Finally, on the master side, you configure the pnport signaling to UNI. See Example 10-50.

Example 10-50 *Configuring UNI Signaling in the PnPort*

```
m8850-7a.7.PXM.a > dnpnport 2:1.1:1

m8850-7a.7.PXM.a > cnfpnportsig 2:1.1:1 -univer uni40

m8850-7a.7.PXM.a > uppnport 2:1.1:1
```

The router-side configuration is equivalent to the other router's configuration. On the 0/0.10 subinterface, you configure the IP address and ESI shown in Example 10-51.

Example 10-51 *CPE Subinterface Configuration for SVCs*

```
C7505-7a(config-if)#interface ATM 0/0.10 multipoint
C7505-7a(config-subif)#description Subinterface for SVCs towards MGX-8850 switch.
C7505-7a(config-subif)#ip address 172.18.1.2 255.255.255.0
10:56:22: %LANE-6-INFO: ATM0/0: ILMI prefix add event received
C7505-7a(config-subif)#atm esi-address 00000C750501.01
```

You can see the registered NSAP address in Example 10-52.

Example 10-52 *Showing the Registered NSAP Address from the CPE*

```
C7505-7a#sh atm ilmi-status

Interface : ATM0/0 Interface Type : Private UNI (User-side)
ILMI VCC : (0, 16) ILMI Keepalive : Enabled/Up (5 Sec 4 Retries)
ILMI State:      UpAndNormal
Peer IP Addr:     0.0.0.0         Peer IF Name:     atmVirtual.02.1.1.01
Peer MaxVPIbits:  0               Peer MaxVCIbits:  16
Active Prefix(s) :
47.0000.0000.0000.0100.0200.8850
End-System Registered Address(s) :
47.0000.0000.0000.0100.0200.8850.0000.0c75.0501.01(Confirmed)
C7505-7a#
```

The two CPEs are successfully connected to the network and are ready to be configured. ILMI is a great aid in parameter autonegotiation.

SVCs

You use the setup shown in Figure 10-11 for the SVC configuration. For the SVC setup, multipoint ATM subinterfaces .10 are used in the routers, configuring the IP subnetwork 172.18.1.0/24.

Figure 10-11 *Network Setup for SVCs*

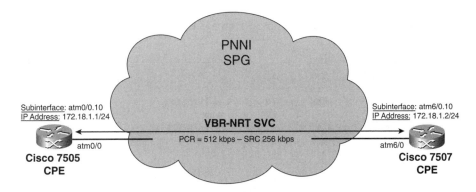

You configured the subinterface for SVCs in the preceding section. It is important to note that the ESI was configured in the subinterface, so you can have multiple registered addresses per physical interface. If you use multiple subinterfaces for SVCs, each subinterface should have its own AESA. If no ILMI is running in the link, you need to manually configure the AESA using the command **atm nsap** in the router and the command **addaddr** in the PNNI node.

To configure SVCs, you need to set up the destination IP address-to-destination AESA address mapping. You first create a multipoint subinterface and configure an ATM SVC and corresponding ATM AESA destination, as shown in Example 10-53.

Example 10-53 *Subinterface Configuration for SVCs*

```
C7507-1a(config)#interface ATM 6/0.10 multipoint
C7507-1a(config-subif)#svc SVC_test nsap
  47.0000.0000.0000.0100.0200.8850.0000.0c75.0501.01
```

You enter ATM-VC configuration mode (see Example 10-54). In this mode, you configure the destination Layer 3 address, as well as encapsulation, OAM characteristics, and traffic and QoS parameters.

Example 10-54 *ATM-VC Configuration Mode*

```
C7507-1a(config-if-atm-vc)#encapsulation aal5snap
C7507-1a(config-if-atm-vc)#vbr-nrt 512 256
C7507-1a(config-if-atm-vc)#oam-svc manage
C7507-1a(config-if-atm-vc)#protocol ip 172.18.1.2 broadcast
C7507-1a(config-if-atm-vc)#^Z
C7507-1a#
```

The broadcast keyword in the IP mapping allows broadcasts such as routing updates to traverse this SVC.

You do the same on the other end, as shown in Example 10-55.

Example 10-55 *Remote CPE Configuration for SVCs*

```
C7505-7a#conf t
Enter configuration commands, one per line.  End with CNTL/Z.
C7505-7a(config)#interface ATM 0/0.10 multipoint
C7505-7a(config-subif)# svc SVC_test nsap
  47.0000.0000.0000.0100.0100.8600.0000.0c75.0701.01
C7505-7a(config-if-atm-vc)#encapsulation aal5snap
C7505-7a(config-if-atm-vc)#vbr-nrt 512 256
```

Example 10-55 *Remote CPE Configuration for SVCs (Continued)*

```
C7505-7a(config-if-atm-vc)#oam-svc manage
C7505-7a(config-if-atm-vc)#protocol ip 172.18.1.1 broadcast
C7505-7a(config-if-atm-vc)#^Z
C7505-7a#
```

You should have IP connectivity across your ATM SVC. We use the **ping** command as shown in Example 10-56. The first ping is lost because of the SVC setup.

Example 10-56 *Using IP PING to Check Connectivity*

```
C7505-7a#ping 172.18.1.1

Type escape sequence to abort.
Sending 5, 100-byte ICMP Echos to 172.18.1.1, timeout is 2 seconds:
.!!!!
Success rate is 80 percent (4/5), round-trip min/avg/max = 8/9/12 ms
C7505-7a#sh atm vc
               VCD /                                  Peak  Avg/Min Burst
Interface   Name      VPI  VCI  Type  Encaps  SC  Kbps    Kbps  Cells  Sts
0/0         Sig        0    5   PVC   SAAL   VBR  149760  2000    16   UP
0/0         ILMI       0   16   PVC   ILMI   VBR  1498    1498    16   UP
0/0.10      SVC_test   0   37   SVC   SNAP   VBR   477     238   128   UP
C7505-7a#
```

You can see this SVC from the PNNI control plane (SES or PXM-45) using the command **dsppncons**, as shown in Example 10-57. With this command, you can see the calling and called addresses as well as cross-connect details. This command displays cross-connect port, VPI, and VCI information, not end-to-end values. In the BPX-SES, pnport 12.2 is used as the UNI link connected to the router, and pnport 4.3 is a PNNI link toward the MGX-8850.

Example 10-57 *Displaying SVC Connections from the PNNI Controller*

```
SES-3a.1.PXM.a > dsppncons
         Port    VPI  VCI  CallRef:Flag      X-Port   VPI  VCI  CallRef:Flag
  Type OAM-Type  Pri
         4.3      0   42     6: 1            12.2      0   37      1: 0
  PTP    No    8
    Calling-Addr: 47.0000000000000010002008850.00000c750501.01
    Called-Addr: 47.0000000000000010001008600.00000c750701.01
         12.2     0   37     1: 0            4.3       0   42      6: 1
  PTP    Yes   8
    Calling-Addr: 47.0000000000000010002008850.00000c750501.01
    Called-Addr: 47.0000000000000010001008600.00000c750701.01

SES-3a.1.PXM.a >
```

The command **dsppncon** displays more details regarding the cross-connect, as shown in Example 10-58.

Example 10-58 *Using* **dsppncon** *to See Call Details*

```
SES-3a.1.PXM.a > dsppncon 12.2 0 37


 CallRef:        1 CallRefFlag:   0  CallLeafRef :        0
 Calling-address: 47.0000000000000010002008850.00000c750501.01    <= Cisco 7505 to
   MGX-8850
 Calling-subaddress #1: N/A
 Calling-subaddress #2: N/A
 Called-address: 47.0000000000000010001008600.00000c750701.01     <= Cisco 7507 to
   BPX-SES
 Called-subaddress #1: N/A
 Called-subaddress #2: N/A
 OE Port :           4.3 OE VPI :    0  OE VCI :     42
 OE CallRef:        6 OE CallRefFlag:   1
 OAM-Type : OAM Endpoint
 Routing Priority : 8
 Connection-type : SVC   Cast-type : point-to-point   Bearer-class :BCOBX
 Service-category :VBR-NRT   Call-clipping-susceptibility:no
 Tx conformance :VBR.1  Rx conformance :VBR.1
 Tx pcr :     1208         Rx pcr :       1208
 Tx scr :     604 Rx scr :   604
 Tx Per Util :  100         Rx Per Util :   100
 Tx mbs :      0 Rx mbs :     0
 Tx cdvt : 250000
 Tx frame-discard-option :enable  Rx frame-discard-option :enable

Type <CR> to continue, Q<CR> to stop:
 Max ctd :    N/A
 Max Tx cdv :    N/A    Max Rx cdv :    N/A
 Max Tx clr :    N/A    Max Rx clr :    N/A
 NCCI value: no record found

SES-3a.1.PXM.a >
```

One difference between router and switch configuration is the units in which traffic parameters are specified. In particular, router traffic rates are specified in kilobits per second (Kbps), and switch traffic rates are given in cells per second (CPS). The relationship between both units is shown in Equation 10-1.

Equation 10-1 *Relationship Between Kbps and CPS*

$$CPS = \text{roundup}\left(\frac{kbps \cdot 125}{53}\right)$$

Using the formula shown in Equation 10-1, Table 10-3 shows the values used in the SVC. The Kbps are configured in the router, and the CPS are signaled in the PNNI information elements. On a router, rates are always configured in Kbps; however, the ATM traffic descriptor information element (Q.2931, Section 4.5.6) specifies rates in CPS.

Table 10-3 *Rate Parameters in Kbps and CPS*

SVC Parameter	Kbps	CPS
PCR	512	1208
SCR	256	604

SPVCs

For the SPVC provisioning, you use the network configuration shown in Figure 10-12. For SPVC configuration, multipoint ATM subinterfaces .20 are used in the routers. The IP subnetwork to be configured in the SPVC scenarios is 172.18.2.0/24.

Figure 10-12 *Network Setup for SPVCs*

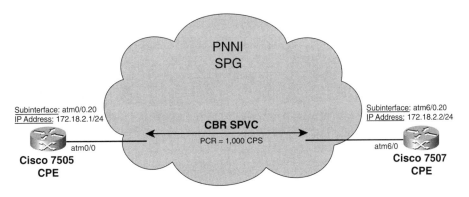

As we will detail, the CPE segment of SPVCs is permanent. From the router's perspective, you configure a PVC (see Example 10-59). You use 1000 CPS in the network. That corresponds to 424 Kbps in the router using the formula shown in Equation 10-1.

Example 10-59 *CPE Configuration for SPVCs*

```
C7507-1a(config)#interface ATM 6/0.20 multipoint
C7507-1a(config-subif)#description Subinterface for SPVCs towards BPX-SES switch.
C7507-1a(config-subif)#ip address 172.18.2.1 255.255.255.0
C7507-1a(config-subif)#pvc PVC_test 100/100
C7507-1a(config-if-atm-vc)#protocol ip 172.18.2.2 broadcast
C7507-1a(config-if-atm-vc)#oam-pvc manage
C7507-1a(config-if-atm-vc)#cbr 424
```

On the network, you configure a dual-endpoint SPVC. A dual-endpoint SPVC has a master endpoint and a slave persistent endpoint. The master endpoint is responsible for originating the SPVC call. Some of the advantages of a dual-endpoint SPVC model, such as slave endpoint alarm indication and slave port load model, are described in the later section "Notes on SPVCs."

You start by configuring a slave SPVC endpoint, as shown in Example 10-60.

Example 10-60 *Configuring a Slave SPVC Endpoint from the Controller*

```
SES-3a.1.PXM.a > addcon 12.2 100 100 1 2 -lpcr 1000 -rpcr 1000
LOCAL ADDR: 4700000000000001000100860000000000C020000.100.100

SES-3a.1.PXM.a >
```

The output of this command is the endpoint identifier formed with the port's AESA SPVC endpoint plus VPI and VCI values. This output is used in the master endpoint addcon.

You can display the AESA SPVC endpoint using the command **dspspvcaddr**. See Example 10-61.

Example 10-61 *Displaying the AESA SPVC Endpoint*

```
SES-3a.1.PXM.a > dspspvcaddr

Interface Id     Soft VC Address(es)
-----------      -------------------
       11.2      47.0000.0000.0000.0100.0100.8600.0000.000b.0200.00
       12.2      47.0000.0000.0000.0100.0100.8600.0000.000c.0200.00
        4.3      47.0000.0000.0000.0100.0100.8600.0000.0004.0300.00

SES-3a.1.PXM.a >
```

These SPVC AESAs are formed by prepending the SPVC prefix to 2 bytes equal to 0 plus the LIN for the VSI slave in hexadecimal format and a selector of 0 (see Figure 10-13).

Figure 10-13 *SPVC Endpoint AESA*

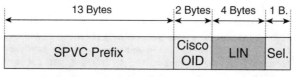

• Cisco OID = 0x0000
• Selector = 0x00

You can now configure the SPVC master endpoint in the MGX-8850 PNNI node. This configuration is performed in the AXSM card, as shown in Example 10-62.

Example 10-62 *Creating a Master SPVC Endpoint in the MGX-8850*

```
m8850-7a.2.AXSM.a > addcon 1 100 100 1 1 -slave
   470000000000000010001008600000000000C020000.100.100 -lpcr 1000 -rpcr 1000
master endpoint added successfully
master endpoint id : 47000000000000001000200885000000102180100.100.100

m8850-7a.2.AXSM.a >
And we can check the AXSM connection endpoints.
m8850-7a.2.AXSM.a > dspcons
record    Identifier    Type    SrvcType    M/S    Upld      Admn    Alarm
------    ----------    ----    --------    ---    ----      ----    -----
     0    01 0100 00100    VCC        cbr1    M    013d4b18      UP     none

m8850-7a.2.AXSM.a > dspvsicons
 LCN    Type    lLin    lVpi    lVci    rLin    rVpi    rVci    cksmVal    pCref
======================================================================
00003 s/svc    01021801    0000    000005    01073b22    0001    000041    030d546c    0000
00004 p/spvc    01021801    0100    000100    01011802    0000    000043    00481867    0000

m8850-7a.2.AXSM.a >
```

NOTE It is important to note that ILMI connection using VPI/VCI = 0/16 is not displayed in the output of the command **dspvsicons**. This is because it is not a connection set up by VSI. It is configured directly by slave platform software because of ILMI's distributed nature. ILMI messages are extended to the VSI master using VSI protocol passthrough.

You also can check the SPVC from the control plane, as shown in Example 10-63. That would be a PXM-45 card in the MGX-8850 PNNI node and SES in the BPX-SES PNNI node.

Example 10-63 *Displaying SPVC Information from the Controller*

```
m8850-7a.7.PXM.a > dspcons

Local Port    Vpi.Vci    Remote Port    Vpi.Vci    State    Owner    Pri    Persistency
--------------------+--------------------+---------+-------+---+-----------
2:1.1:1    100 100    Routed    100 100    OK    MASTER    8    Persistent
Local  Addr: 47.00000000000001000200885000.000001021801.00
Remote Addr: 47.00000000000001000100086000.0000000c0200.00
Preferred Route ID:-

m8850-7a.7.PXM.a >
```

Because the remote port resides in a remote node, the remote port is displayed as routed for a successful call. You can see the local cross-connect information using the command **dsppncons**, as shown in Example 10-64.

Example 10-64 *Displaying PNNI Node Cross-Connects*

```
SES-3a.1.PXM.a > dsppncons
          Port  VPI  VCI  CallRef:Flag       X-Port  VPI  VCI  CallRef:Flag
  Type OAM-Type  Pri
          4.3    0    43      7: 1           12.2   100  100      1: 0
  PTP    No     8
    Calling-Addr: 47.000000000000001000200880.000001021801.00
    Called-Addr: 47.000000000000001000100880.0000000c0200.00
          12.2  100  100      1: 0           4.3     0    43      7: 1
  PTP    No     8
    Calling-Addr: 47.000000000000001000200880.000001021801.00
    Called-Addr: 47.000000000000001000100880.0000000c0200.00

SES-3a.1.PXM.a >
```

Finally, you configure the other end's CPE, and you have IP connectivity. See Example 10-65.

Example 10-65 *Configuring the Rremote CPE for SPVCs*

```
C7505-7a(config)#interface ATM 0/0.20 multipoint
C7505-7a(config-subif)#description Subinterface for SPVCs towards MGX-8850 switch.
C7505-7a(config-subif)#ip address 172.18.2.2 255.255.255.0
C7505-7a(config-subif)#pvc PVC_test 100/100
C7505-7a(config-if-atm-vc)#protocol ip 172.18.2.1 broadcast
C7505-7a(config-if-atm-vc)#oam-pvc manage
C7505-7a(config-if-atm-vc)#cbr 424
C7505-7a(config-if-atm-vc)#^Z
C7505-7a#ping 172.18.2.1

Type escape sequence to abort.
Sending 5, 100-byte ICMP Echos to 172.18.2.1, timeout is 2 seconds:
!!!!!
Success rate is 100 percent (5/5), round-trip min/avg/max = 1/2/4 ms
C7505-7a#
```

In the SPVC case, the first ping does not fail because the SPVC was already signaled and set up in the PNNI network.

SPVCs Using RPM-PR as CPE

Figure 10-14 shows the setup using an RPM-PR card as CPE for SPVC endpoints in the MGX-8850-based PNNI node. The RPM-PR card currently does not support SVCs, only PVC and SPVC endpoints.

Figure 10-14 *Network Setup for SPVCs Using RPM-PR*

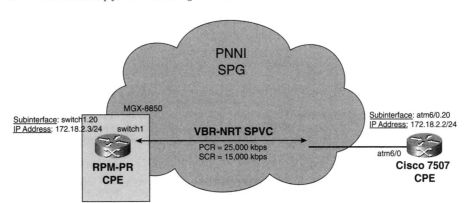

You configure a new PVC in the Cisco 7507 SPVC subinterface with the corresponding
Layer 3 mapping (see Example 10-66). A variable bit rate (VBR-NRT) PVC is configured
with a Peak Cell Rate (PCR) of 25,000 Kbps, a Sustained Cell Rate (SCR) of 15,000 Kbps,
and a Maximum Burst Size (MBS) of 1000 cells.

Example 10-66 *Configuring the CPE for a New SPVC*

```
C7507-1a#conf t
Enter configuration commands, one per line.  End with CNTL/Z.
C7507-1a(config)#interface ATM 6/0.20
C7507-1a(config-subif)#pvc PVC_RPM 100/200
C7507-1a(config-if-atm-vc)#protocol ip 172.18.2.3 broadcast
C7507-1a(config-if-atm-vc)#oam-pvc manage
C7507-1a(config-if-atm-vc)#vbr-nrt 25000 15000 1000^Z
C7507-1a#
```

You can perform the RPM-PR configuration. First of all, the RPM-PR internal ATM port
must be seen from the PNNI control plane. To achieve that, you need to configure a resource
partition in the RPM-PR's switch interface that is managed from the PXM-45 proxy VSI
slave. Refer to Example 10-67. You create a VCC resource partition using controller-id 2,
in which you configure the VPI and VCI ranges and guaranteed and maximum bandwidth.

Example 10-67 *Creating an RPM-PR VSI Resource Partition*

```
RPM_10_8850#conf t
RPM_10_8850(config)#interface switch 1
RPM_10_8850(config-if)#switch partition vcc 1 2
RPM_10_8850(config-if-swpart)#vpi 0 0
RPM_10_8850(config-if-swpart)#vci 32 3808
RPM_10_8850(config-if-swpart)#ingress-percentage-bandwidth 100 100
RPM_10_8850(config-if-swpart)#egress-percentage-bandwidth 100 100
RPM_10_8850(config-if-swpart)#exit
```

With this configuration, the pnport and corresponding SPVC AESA are visible from the PNNI control plane. See Example 10-68.

Example 10-68 *Displaying the RPM-PR SPVC Endpoint AESA*

```
m8850-7a.7.PXM.a > dspspvcaddr

Interface Id      Soft VC Address(es)
------------      -------------------
    1:1.2:2       47.0000.0000.0000.0100.0200.8850.0000.0101.1802.00
    2:1.1:1       47.0000.0000.0000.0100.0200.8850.0000.0102.1801.00
       7.36       47.0000.0000.0000.0100.0200.8850.0000.0107.3b24.00
       7.37       47.0000.0000.0000.0100.0200.8850.0000.0107.3b25.00
       7.38       47.0000.0000.0000.0100.0200.8850.0000.0107.3b26.00
       7.35       47.0000.0000.0000.0100.0200.8850.0000.0107.3b23.00
       10.1       47.0000.0000.0000.0100.0200.8850.0000.0107.5301.00
   12:1.1:1       47.0000.0000.0000.0100.0200.8850.0000.010c.1801.00

m8850-7a.7.PXM.a >
```

The RPM SPVC AESA is formed with the MGX-8850 SPVC prefix, the PnPort LIN in hexadecimal, and a selector of 0, as shown in Figure 10-15.

Figure 10-15 *RPM-PR SPVC AESA*

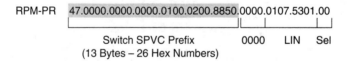

RPM-PR SPVC Endpoint AESA:

RPM-PR 47.0000.0000.0000.0100.0200.8850.0000.0107.5301.00

Switch SPVC Prefix   0000   LIN   Sel
(13 Bytes – 26 Hex Numbers)

Now you configure a multipoint subinterface for SPVCs with its IP address and ATM PVC (see Example 10-69). The ATM PVC is configured with the same traffic characteristics as the previous one.

Example 10-69 *Configuring the RPM-PR for SPVCs*

```
RPM_10_8850(config)#interface switch 1.20 multipoint
RPM_10_8850(config-subif)#description Subinterface for SPVCs.
RPM_10_8850(config-subif)#ip address 172.18.2.3 255.255.255.0
RPM_10_8850(config-subif)#pvc 0/100
RPM_10_8850(config-if-atm-vc)#oam-pvc manage
RPM_10_8850(config-if-atm-vc)#protocol ip 172.18.2.1
RPM_10_8850(config-if-atm-vc)#encapsulation aal5snap
RPM_10_8850(config-if-atm-vc)#vbr-nrt 25000 15000 1000
RPM_10_8850(config-if-atm-vc)#exit
```

At this point, the CPEs are configured for SPVCs, and you need to configure the SPVC in the network. To do that, you start by configuring the RPM-PR switch connection that is the MGX-8850 PNNI node's SPVC endpoint. The remote AESA is the BPX-SES SPVC AESA for port 12.2. See Example 10-70.

Example 10-70 *Configuring a Slave SPVC Connection Endpoint in an RPM-PR*

```
RPM_10_8850(config-subif)#switch connection vcc 0 100 master remote raddr ?
  XX.XXXX. ... .XXX.XX  remote NSAP address

RPM_10_8850(config-subif)#$0000.0100.0100.8600.0000.000c.0200.00 100 200
RPM_10_8850(config-if-swconn)#^Z
RPM_10_8850#
```

You configure the RPM-PR SPVC endpoint as a slave endpoint with the parameter **master remote** in the **switch connection** command. As shown in Example 10-71, the SPVC endpoint appears in the PNNI control plane.

Example 10-71 *The RPM-PR SPVC Endpoint from the Control Plane*

```
m8850-7a.7.PXM.a > dspcon 10.1 0 100
Port                     Vpi Vci            Owner     State    Persistency
---------------------------------------------------------------------------
Local  10:-1.1:-1      0.100              SLAVE     FAIL     Persistent
       Address: 47.0000000000000010002008850.000001075301.00
       Node name: m8850-7a
Remote Routed           0.0                MASTER    --       Persistent
       Address: 00.00000000000000000000000000.000000000000.00
       Node name:

------------------- Provisioning Parameters -------------------
Connection Type: VCC        Cast Type: Point-to-Point
Service Category: nrt-VBR    Conformance: nrt-VBR.3
Bearer Class: BCOB-X
Last Fail Cause: N/A                         Attempts: 0

Continuity Check: Disabled   Frame Discard: Disabled
L-Utils: 100    R-Utils: 100   Max Cost: 0     Routing Cost: 0
OAM Segment Ep: Enabled
Priority: -

---------- Traffic Parameters ----------
Values: Configured (Signalled)
Tx PCR:  58963      (-)      Rx PCR:  58963      (-)

Type <CR> to continue, Q<CR> to stop:
Tx SCR:  35378      (-)      Rx SCR:  35378      (-)
Tx MBS:  1000       (-)      Rx MBS:  1000       (-)
Tx CDVT: 250000     (-)
Tx CDV:  N/A                 Rx CDV:  N/A
```

continues

Example 10-71 *The RPM-PR SPVC Endpoint from the Control Plane (Continued)*

```
Tx CTD:  N/A              Rx CTD:  N/A

------------------- Preferred Route Parameters-----------------
Preferred Route ID: -
Currently on preferred route: N/A

m8850-7a.7.PXM.a >
```

From the **dspcon** output, you can see that the remote information is still empty because it has not yet been configured. No attempts have been made to set up the SPVC because this is the slave endpoint.

From the same output, you can also verify the relationship between units expressed in Kbps and CPS. As you know, you configured the SPVC endpoint with a PCR = 25,000 Kbps and an SCR of 15,000 Kbps. Using the formula shown in Equation 10-1, the values expressed in CPS are PCR = 58,963 CPS and SCR = 35,378 CPS. These values in CPS are shown in the **dspcon** command output.

To conclude, you configure the master SPVC endpoint with traffic parameters matching the aforementioned values and expressed in CPS, as shown in Example 10-72. For the SPVC call to connect, all QoS and traffic parameters need to match in both ends.

Example 10-72 *Adding the Master SPVC Endpoint*

```
SES-3a.1.PXM.a > addcon 12.2 100 200 7 1
  47000000000000001000200885000000107530100.0.100 -lpcr 58963 -lscr 35378 -lmbs
  1000 -rpcr 58963 -rscr 35378 -rmbs 1000

SES-3a.1.PXM.a >
```

You can see that the connection establishment is successful in Example 10-73.

Example 10-73 *Displaying SPVC Connections and Checking Their Status*

```
m8850-7a.7.PXM.a > dspcons

Local Port  Vpi.Vci   Remote Port  Vpi.Vci    State    Owner  Pri Persistency
-------------------------+-------------------------+---------+-------+---+-----------
10.1          0 100    Routed       100 200     OK      SLAVE  -   Persistent
Local  Addr: 47.000000000000001000200885000000107530100.0.100
Remote Addr: 47.000000000000001000100886000000000000c0200.00
Preferred Route ID:-
2:1.1:1     100 100    Routed       100 100     OK      MASTER 8   Persistent
Local  Addr: 47.000000000000001000200885000000001021801.00
Remote Addr: 47.000000000000001000100886000000000000c0200.00
Preferred Route ID:-

m8850-7a.7.PXM.a >
```

Furthermore, as shown in Example 10-74, you have IP connectivity.

Example 10-74 *Checking IP Connectivity Through the SPVC*

```
C7507-1a#ping 172.18.2.3

Type escape sequence to abort.
Sending 5, 100-byte ICMP Echos to 172.18.2.3, timeout is 2 seconds:
!!!!!
Success rate is 100 percent (5/5), round-trip min/avg/max = 1/2/4 ms
C7507-1a#
```

Using the command **dsppncons**, you can see the cross-connecting port, VPI, and VCI, as shown in Example 10-75.

Example 10-75 *Displaying PNNI Node Cross-Connects*

```
m8850-7a.7.PXM.a > dsppncons
          Port   VPI   VCI  CallRef:Flag        X-Port    VPI   VCI
   CallRef:Flag  Type OAM-Type  Pri
          10.1    0    100        10: 0          1:1.2:2    0    54
   12: 1     PTP    No      8
     Calling-Addr: 47.000000000000010001008600.0000000c0200.00
     Called-Addr: 47.000000000000010002008850.000001075301.00
...
```

You can also see that the calling address is the BPX-SES SPVC AESA. This is because the BPX-SES endpoint was configured as the master connection endpoint. To revisit this concept, the connection endpoint configured as master is responsible for initializing the connection setup.

Summary of PNNI Configuration Commands

This section details in a concise format all the configuration commands used. All the steps described in the "Generic Configuration Model" section are listed in the summary tables.

Generic PNNI Node Bringup

Table 10-4 recaps all the commands used to bring up a PNNI node. Also refer to Figure 10-3 for the corresponding flow chart.

Table 10-4 *Summary of PNNI Node Bringup Commands*

Task	BPX-8600 and SES	MGX-8850 and MGX-8950
1	(1) **uptrk** **cnfrsrc** **cnfvsiif**	None
2	(1) **addshelf** **addctrlr**	(2) **addcontroller**
3	(3) **cnfpnni-node**	

(1) In the BPX-8600 switch CLI

(2) In the PXM-45 CLI

(3) In the PNNI controller CLI (PXM-45, PXM-1E or SES)

Generic Link Configuration

Last but not least, the link configuration commands are summarized in Table 10-5. The tasks or steps can be matched with the ones shown in Figure 10-4.

Table 10-5 *Summary of PNNI Link and Configuration*

Task	BPX-8600 and SES	MGX-8850 and MGX-8950
1	(1) **upln** **addport** **upport** **cnfrsrc** **cnfvsiif**	(2) **cnfcdsct** **upln** **addport** **addpart**
2	(1) **cnfport**	(2) **upilmi**
3	(3) **cnfpnportsig**	
4	(4) **pvc sig 0/5 qsaal** **pvc ILMI 0/16 ilmi**	

Table 10-5 *Summary of PNNI Link and Configuration (Continued)*

Task	BPX-8600 and SES	MGX-8850 and MGX-8950
5	(4) **interface atm** *x/x.x* {**multipoint** | **point-to-point**} **ip address**	
6 (SPVC)	(5) **addcon**	(2)(6) **addcon**
7 (SPVC)	(4) **pvc** [*name*] **VPI/VCI**	
8 (SVC)	None	
9 (SVC)	(4) atm {esi-address | nsap-address} **svc** [*name*] **nsap** {*20-byte AESA*}	

(1) In the BPX-8600 switch CLI

(2) In the AXSM card CLI

(3) In the PNNI controller CLI

(4) In the router CLI

(5) In the SES CLI

(6) In the RPM-PR CLI

To configure a PNNI link facing a PNNI neighbor, Steps 1 through 3 in Table 10-5 are performed. Step 3 specifies PNNI1.0 signaling.

For a link facing a router, Step 3 specifies UNI signaling, followed by Steps 4 and 5. To run SPVCs, Steps 6 and 7 are performed, and to run SVCs, Step 9 is required.

Notes on SPVCs

The original Cisco multiservice switching SPVC implementation follows the dual-endpoint SPVC model. In this model, an SPVC connection consists of two endpoints: an SPVC master endpoint and an SPVC slave endpoint.

The following list details their functions:

- The master endpoint owns and is responsible for routing and signaling. It controls the SPVC setup.

- The slave endpoint is a passive endpoint that is responsible for accepting or rejecting the SPVC setup.

This behavior is shown in Figure 10-16.

Figure 10-16 *Dual-Endpoint SPVC Model*

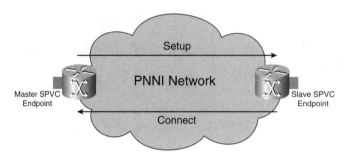

In a dual-endpoint SPVC model, both endpoints must be provisioned with the same service category and traffic parameters for the slave endpoint to accept the call. The slave endpoint is persistent.

This model is more robust than a single-endpoint SPVC model and has the following advantages:

- **AIS generation on slave endpoint**—A persistent slave endpoint sends an alarm indication to the attached CPE if the SPVC is derouted.

- **Statistics at the slave endpoint**—Statistics and counters can be kept because the slave endpoint is persistent.

- **SVCs and SPVCs on the same interface**—Because the VPI/VCI pair is reserved at a persistent slave endpoint, an incoming SVC setup cannot use them.

- **ATM-FR and ATM-CE service interworking**—If a slave endpoint is on a Frame Relay interface, Frame Relay traffic parameters are explicitly specified. Currently, no standard exists for signaling Frame Relay parameters across ATM, so a single-endpoint SPVC model must use proprietary signaling. The same holds true for ATM-to-circuit emulation interworking.

- **Protection against configuration errors**—A dual-endpoint SPVC model protects users from making an SPVC call to the wrong ATM switch or port because the slave endpoint needs to be configured for the call to be accepted.

Given these advantages, a dual-endpoint SPVC model is normally preferred. In the case of interworking between a multiservice switching platform and a platform supporting only the single-endpoint SPVC model, there are different options. The multiservice switching platform can be configured with a master SPVC endpoint, and no configuration is required in the other platform. The multiservice switching platform can also be configured with a slave SPVC endpoint. However, the other platform needs to be configured, and all service categories and traffic parameters need to match between both endpoints. You can use the formula shown in Equation 10-1 to translate traffic parameter values in different units.

In particular, Table 10-6 summarizes the service category mappings to use when you configure SPVCs between a multiservice switching platform and a Cisco IOS Software-based platform.

Table 10-6   *Service Category Mappings for SPVCs*

Cisco IOS Software Platform Service Class	MSS Platform Service Class
cbr	cbr2(11)
vbr-rt	vbr1rt(2)
vbr-nrt	vbr1nrt(5)
ubr	ubr1(8)
abr	abrstd(10)

Current PNNI software in multiservice switching platforms also supports single-endpoint SPVC model and nonpersistent SPVC slave endpoints. In this case, a PNNI node can be configured so that an SPVC or SPVP request can override existing SVCs and SVPs using the requested VPI/VCI. This is because an SVC can take a different VPI/VCI pair but an SPVC endpoint cannot. The command **cnfsvcoverride** is provided for that functionality; it defaults to no override. On the other end of the spectrum, pnports can be configured to block incoming SPVC requests for nonpersistent endpoints. This is achieved with the parameter **-nonpersblock** in the command **cnfpnportcc,** which configures the PnPort call control. Single-endpoint SPVCs can be released with the command **clrspvcnonpers**.

Another new feature allows persistent slave endpoints in a dual-endpoint SPVC model to have a nodal traffic conformance tolerance factor for allowing SPVC request acceptance with different traffic parameters. This tolerance factor is configured as a percentage from the configured master endpoint's traffic parameters. The command **cnftrftolerance** lets you specify this nodal percentage, which defaults to 5 percent.

Configuring Filters

PNNI software allows the configuration of call filters. These filters are similar to access lists (ACLs) in Cisco IOS Software both in functionality and provisioning sequence.

Configurable filters include address filters to reject or allow (deny or permit) specific calling or called parties in the setup message, or a combination of calling and called parties. Calling parties, called parties, or combinations can be rejected or allowed based on a beginning or ending set of digits in each field.

Each address filter can contain multiple entries sorted by an index, so a single address filter can contain numerous rules permitting or denying complete and partial addresses.

Finally, these filters are applied to pnports in either the ingress or egress direction.

You can configure an example of address filtering. In this example, you allow SVCs to be set up from one CPE but not from the other.

You start by creating the address filter in the MGX-8850, called Filter_test (see Example 10-76). The default absent action is permit. In this filter, you deny the Cisco 7507 AESA as a calling AESA in a setup message.

Example 10-76 *Creating a Filter in a PNNI Node*

```
m8850-7a.7.PXM.a > addfltset Filter_test

m8850-7a.7.PXM.a > cnffltset Filter_test -address
  47.00000000000010001008600.00000c750701.01 -length 160 -list calling
  -accessMode deny

m8850-7a.7.PXM.a > dspfltset -name Filter_test
FilterName: Filter_test
Index: 1
Address: 4700000000000001000100860000000c75070101
AddrLen: 160 bits
AddrPlan: Nsap
AccessMode: Deny
AddrList: Calling Party List
----------------------------------------

m8850-7a.7.PXM.a >
```

For partial address matches, you can use the parameter **-address**. The parameter **–address** can be followed by these:

- Digits before three periods—An address beginning with those digits (such as **-address 470091...**)

- Digits after three periods—An address ending with those digits (such as **-address ...75070101**).

You now apply the filter to pnport 2:1.1:1 connected to the Cisco 7505 router in the outbound direction, as shown in Example 10-77.

Example 10-77 *Applying the Access Filter to a PnPort*

```
m8850-7a.7.PXM.a > cnfpnportacc 2:1.1:1 -out Filter_test

m8850-7a.7.PXM.a >
```

You try to set up a connection from the Cisco 7507. That setup message has the Cisco 7507 AESA as the calling party and is rejected in the MGX-8850 at port 2:1.1:1. See Example 10-78.

Example 10-78 *Verifying the Deny Filter Functionality*

```
C7507-1a#ping 172.18.1.2

Type escape sequence to abort.
Sending 5, 100-byte ICMP Echos to 172.18.1.2, timeout is 2 seconds:
.....
Success rate is 0 percent (0/5)
C7507-1a#
```

This behavior is shown in Figure 10-17.

Figure 10-17 *Address Filtering Example*

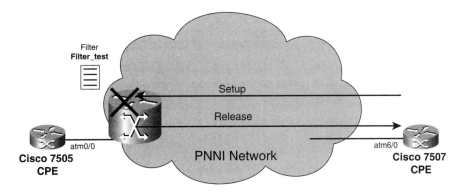

An ATM signaling debug in the Cisco 7507 router shows how the call is not completed. Note that the calling party address corresponds to the Cisco 7507 AESA router connected to the BPX-SES PNNI node. See Example 10-79.

Example 10-79 *ATM Signaling Debug Showing the Call Released*

```
17:43:13: ATMSIG(ATM6/0 0,0 - 0022/00): (vcnum:0) API - alloc_connection_id 16
17:43:13: ATMSIG: Called Party Addr:
  47.00000000000001000200850.00000C750501.01
17:43:13: ATMSIG: Calling Party Addr:
  47.00000000000001000100860.00000C750701.01
17:43:13: ATMSIG(ATM6/0 0,0 - 0022/00): (vcnum:29) Null(U0) -> Call Initiated(U1)
17:43:13: ATMSIG(ATM6/0 0,0 - 0022/00): (vcnum:29) Input event: Rcvd Call
  Proceeding in Call Initiated(U1)
17:43:13: ATMSIG(ATM6/0 0,56 - 0022/00): (vcnum:29) Connection Identifier IE:
  associated sig = 88    vpi = 0    vci = 56
17:43:13: ATMSIG(ATM6/0 0,56 - 0022/00): (vcnum:29) Call Initiated(U1)
  -> Outgoing Call Proceeding(U3)
```

continues

Example 10-79 *ATM Signaling Debug Showing the Call Released (Continued)*

```
17:43:13: ATMSIG(ATM6/0 0,56 - 0022/00): (vcnum:29) Input event: Rcvd Release in
   Outgoing Call Proceeding(U3)
17:43:13: ATMSIG(ATM6/0 0,56 - 0022/00): (vcnum:29)cause = temporary failure,
   location = Private Network
17:43:13: ATMSIG(ATM6/0 0,56 - 0022/00): (vcnum:29) Outgoing Call Proceeding(U3)
   -> Release Indication(U12)
ATMAPI: (c<-s): RELEASEv2 ci: 0x16, cause: 0x29
17:43:13:
ATMAPI: (c->s): RELEASE_COMPv2 ci: 0x16 cause: 0x29
17:43:13: ATMSIG(ATM6/0 0,56 - 0022/00): (vcnum:29) building cause code - cause =
   (0x1B)destination out of order, IE_cause = (0x1B)destination out of
   order, location = User
17:43:13: ATMSIG(ATM6/0 0,56 - 0022/00): (vcnum:29) Output Release Complete msg,
   Release Indication(U12) state
17:43:13: ATMSIG(ATM6/0 0,56 - 0022/00): (vcnum:29) Release Indication(U12) ->
   Dead
```

However, if the call is initiated from the Cisco 7505 router, it is successful. In this case, the calling party address matches the Cisco 7505 AESA connected to the MGX-8850 PNNI node. See Example 10-80.

Example 10-80 *Initiating the Call from the 7505 Router*

```
ATMSIG_API: Called Party Addr:
   47.0000000000000010001008600.00000C750701.01
ATMSIG_API: Calling Party Addr:
   47.0000000000000010002008850.00000C750501.01
ATMSIG_API:(ATM0/0 0,0 - 0004/00): (vcnum:9) Null(U0) -> Call Initiated(U1)
ATMSIG_API:(ATM0/0 0,0 - 0004/00): (vcnum:9) Input event : Rcvd Call Proceeding
   in Call Initiated(U1)
ATMSIG_API:(ATM0/0 0,54 - 0004/00): (vcnum:9) Connection Identifier IE:
   associated sig = 88   vpi = 0   vci = 54
ATMSIG_API:(ATM0/0 0,54 - 0004/00): (vcnum:9) Call Initiated(U1) -> Outgoing Call
   Proceeding(U3)
ATMSIG_API:(ATM0/0 0,54 - 0004/00): (vcnum:9) Input event : Rcvd Connect in
   Outgoing Call Proceeding(U3)
12:46:31: ProcessBLLI: IE length = 1.
ATMSIG_API:(ATM0/0 0,54 - 0004/00): (vcnum:9) Connection Identifier IE: associated
   sig = 88   vpi = 0   vci = 54
ATMSIG_API:(ATM0/0 0,54 - 0004/00): (vcnum:9) Input event : Req Connect Ack in
   Outgoing Call Proceeding(U3)
ATMSIG_API:(ATM0/0 0,54 - 0004/00): (vcnum:9) Output Connect Ack msg, Outgoing
   Call Proceeding(U3) state
ATMSIG_API:(ATM0/0 0,54 - 0004/00): (vcnum:9) Outgoing Call Proceeding(U3) ->
   Active(U10)
12:46:31:
ATMAPI: (c<-s): CONNECTv2 ci: 0x4 ei: 0xFFFFFFFF
```

You delete the filter assignment to port 1:1.2:2, and you are back to normal. See Example 10-81.

Example 10-81 *Deleting a PnPort Access Filter*

```
m8850-7a.7.PXM.a > delpnportacc 2:1.1:1 out

m8850-7a.7.PXM.a >
```

Summary

This chapter explored PNNI configuration in multiservice switching networks. The specific configuration commands in three different POPs were covered, and a generic configuration model and a summary of configuration commands for all platforms were presented. Multiple PNNI verification commands aid in the operational aspects.

One important point to remember is that these configurations are orthogonal with the MPLS configurations covered in Chapters 6 and 7 in the sense that both MPLS and PNNI configurations can be present concurrently, not interfering with each other. This explicitly demonstrates ships in the night (SIN) mode. In the more general case, a multiservice switching network can run MPLS and PNNI as presented concurrently, along with other control planes such as NCDP, or even control planes not yet developed. This concept provides investment protection as well as incremental service provisioning.

The scalability of the multiservice switching network is multiplied by the use of feeder shelves.

CHAPTER 11

Advanced PNNI Configuration

From a routing protocol perspective, the PNNI protocol includes highly developed routing features. Hierarchical routing and constrained-based routing are some of the sophisticated routing elements. This chapter covers advanced PNNI configuration in multiservice switching networks, including these two topics.

Building on the PNNI network from Chapter 10, "PNNI Implementation and Provision," this chapter describes hierarchical PNNI configuration. Multiple Peer Group PNNI provides great scalability for PNNI networks. This chapter also includes the concepts and configuration of Interim Interswitch Signaling Protocol (IISP) and traffic engineering.

Hierarchical PNNI Configuration

This section covers the migration of a single peer group (SPG) PNNI network into a hierarchical multiple peer group (MPG) network.

There are two fundamental MPG migration models—adding a lower PNNI level, and adding a higher PNNI level:

- **Adding a lower PNNI level**—This growth model consists of replacing a lowest-level physical node in an SPG with a logical PNNI node and moving the lowest level down the hierarchy. It can be used as the SPG grows or when you add new lower-level nodes to the PNNI network. The original SPG level is higher than the MPG lowest level.

- **Adding a higher PNNI level**—This growth model is more often used to join unpeered or separate PNNI networks forming two different peer groups that meet as logical group nodes (LGNs) at a higher hierarchical level. The original SPG level is the same as the lowest MPG level.

This chapter goes through a complete example of MPG migration, following the model of adding a lower PNNI level. Starting from the SPG PNNI network built in Chapter 10, you will configure a new lowest hierarchical plane at level 80. As shown in Figure 11-1, one of the main enablers of this migration is the addressing scheme you choose, stressing the importance of a planned and thorough addressing plan.

Figure 11-1 *PNNI Node's AESAs for Hierarchical PNNI*

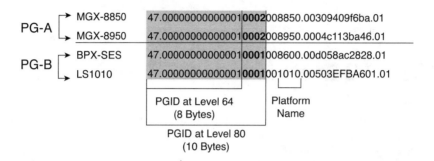

PNNI Node's ATM Addresses (Switch AESAs):

In Figure 11-1, you can see that level 64 has a common peer group ID (PGID) among all nodes. That is why the initial network is an SPG. However, the lower level 80 has two different PGIDs and thus two different regions. The MGX-8850 and MGX-8950 PNNI nodes are within a peer group called PG-A, and the BPX-SES and LS1010 PNNI nodes are in a different peer group called PG-B.

Initial SPG PNNI Network

The final state of the PNNI network from Chapter 10 is the initial stage of the MPG migration. The original PNNI network is shown in Figure 11-2.

Figure 11-2 *Initial SPG PNNI Network*

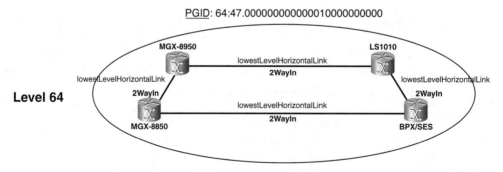

PGID: 64:47.0000000000000010000000000

In Figure 11-2, you can see that the original state has a level 64 SPG PNNI network with four PNNI nodes. These PNNI nodes are physical nodes that are interconnected by lowest-level horizontal PNNI links in **twoWayInside** state.

Adding a Lower Level to the First Node in Peer Group A

The first step in the MPG configuration is to move down the hierarchy in the MGX-8850 PNNI node to the lower level 80 and replace the level 64 PNNI node with a logical PNNI node. This logical node is an LGN, as shown in Figure 11-3.

Figure 11-3 *Moving the First Node Down the Hierarchy*

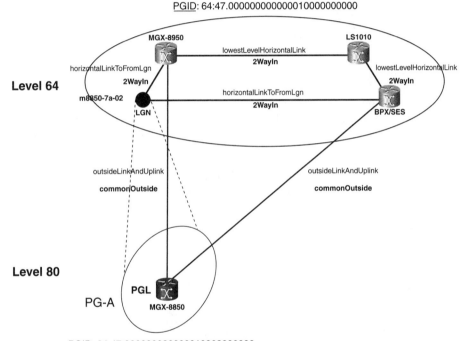

PGID: 64:47.00000000000010000000000

Level 64

Level 80

PG-A

PGID: 64:47.00000000000010002000000

From a configuration perspective, before you change the PNNI configuration, you need to administratively down the PNNI node. This step does not affect existing calls, but it prevents new switched connections from being successfully established. After the PNNI node is administratively down, the lowest-level PNNI node (node index 1) is configured with the new level of 80, the new node ID, and the new peer group ID. In addition, a non-null election priority is configured. Example 11-1 shows these configuration steps.

Example 11-1 *Lowering the MGX-8850 PNNI Node Level*

```
m8850-7a.7.PXM.a > cnfpnni-node 1 -enable false

m8850-7a.7.PXM.a > cnfpnni-node 1 -level 80

m8850-7a.7.PXM.a > cnfpnni-node 1 -nodeId
  80:160:47.00000000000010002008850.00309409f6ba.01

m8850-7a.7.PXM.a > cnfpnni-node 1 -pgId 80:47.00000000000010002000000

m8850-7a.7.PXM.a > cnfpnni-election 1 -priority 50
```

Last, a logical PNNI node at level 64 is created, and the two PNNI nodes are administratively upped. See Example 11-2.

Example 11-2 *Adding a LGN*

```
m8850-7a.7.PXM.a > addpnni-node 64

Get MAC Address from ID PROM NV...OK: 00 30 94 09 F6 BA

m8850-7a.7.PXM.a > cnfpnni-node 1 -enable true

m8850-7a.7.PXM.a > cnfpnni-node 2 -enable true
```

NOTE   The higher-level PNNI node's name is formed by taking the lowest-level PNNI node's name and appending a dash and the node index. For example, for a lowest-level PNNI node named m8850-7a, the higher-level node with index 2 is named m8850-7a-02.

The first thing to check is that the PNNI node at level 80 is the peer group leader (PGL) of the LGN at level 64. You do this by checking the PNNI election, as Example 11-3 shows.

Example 11-3 *Checking the Peer Group Leadership*

```
m8850-7a.7.PXM.a > dsppnni-election 1

node index: 1
    PGL state......        OperPgl       Init time(sec).......      15
    Priority.......           100        Override delay(sec)..      30
                                         Re-election time(sec)      15
    Pref PGL..............80:160:47.0000000000000010002008850.00309409f6ba.01
    PGL..................80:160:47.0000000000000010002008850.00309409f6ba.01
    Active parent node id..64:80:47.0000000000000010002000000.00309409f6ba.00
```

You can see that it is operational and acting as PGL. It is important to note that 50 priority units were added to the configured 50 to make 100. This is done in PNNI to provide stability for the elected PGL.

The second item you need to look up are the PNNI links. Figure 11-3 shows that each of the two PNNI nodes has two PNNI links. You can check this using the command **dsppnni-link**, as shown in Example 11-4.

Example 11-4 *Using* **dsppnni-link**

```
m8850-7a.7.PXM.a > dsppnni-link

node index    : 1
Local port id:    16848898         Remote port id:      262912
Local Phy Port Id: 1:1.2:2
    Type.        outsideLinkAndUplink    Hello state.......commonOutside
```

Example 11-4 *Using* **dsppnni-link** *(Continued)*

```
      Derive agg..........        0    Intf index..........   16848898
      SVC RCC index.......        0    Hello pkt RX........        571
                                       Hello pkt TX........        584
      Remote node name.......SES-3a
      Remote node id........64:160:47.00000000000010001008600.00d058ac2828.01
      Upnode id.............64:160:47.00000000000010001008600.00d058ac2828.01
      Upnode ATM addr.......47.00000000000010001008600.00d058ac2828.01
      Common peer group id...64:47.00.0000.0000.0001.0000.0000.00

 node index  : 1
 Local port id:   17569793       Remote port id:   17111041
 Local Phy Port Id: 12:1.1:1
      Type.      outsideLinkAndUplink    Hello state......commonOutside
      Derive agg..........        0    Intf index..........   17569793
      SVC RCC index.......        0    Hello pkt RX........        578

 Type <CR> to continue, Q<CR> to stop:
                                       Hello pkt TX........        580
      Remote node name.......m8950-7b
      Remote node id........64:160:47.00000000000010002008950.0004c113ba46.01
      Upnode id.............64:160:47.00000000000010002008950.0004c113ba46.01
      Upnode ATM addr........47.00000000000010002008950.0004c113ba46.01
      Common peer group id...64:47.00.0000.0000.0001.0000.0000.00

 node index  : 2
 Local port id:           1       Remote port id:           1
 Local Phy Port Id: n/a
      Type.   horizontalLinkToFromLgn    Hello state....... twoWayInside
      Derive agg..........        0    Intf index..........          0
      SVC RCC index.......        1    Hello pkt RX........          7
                                       Hello pkt TX........          8
      Remote node name.......SES-3a
      Remote node id........64:160:47.00000000000010001008600.00d058ac2828.01
      Upnode id.............0:0:00.00000000000000000000000.000000000000.00
      Upnode ATM addr........00.00000000000000000000000.000000000000.00
      Common peer group id...00:00.00.0000.0000.0000.0000.0000.00

 node index  : 2
 Local port id:           2       Remote port id:           1
 Local Phy Port Id: n/a
      Type.   horizontalLinkToFromLgn    Hello state....... twoWayInside
      Derive agg..........        0    Intf index..........          0
      SVC RCC index.......        2    Hello pkt RX........          5
                                       Hello pkt TX........          8
      Remote node name.......m8950-7b
      Remote node id........64:160:47.00000000000010002008950.0004c113ba46.01
      Upnode id.............0:0:00.00000000000000000000000.000000000000.00
      Upnode ATM addr........00.00000000000000000000000.000000000000.00
      Common peer group id...00:00.00.0000.0000.0000.0000.0000.00

 m8850-7a.7.PXM.a >
```

To begin with, we will analyze the links of the PNNI node at the lowest level (level 80), which has node index 1. This PNNI node has two links (which are the physical links) of type **outsideLinkAndUplink**. One of them has the SES/BPX as a remote node, and the other has the MGX-8950 as a remote node. These links are *outside links* because the remote nodes are outside the local node's peer group, but they are also *uplinks* because the remote nodes are at the higher level 64. Hence, the links are of type **outsideLinkAndUplink**. The hello state of these links is **commonOutside**, indicating that the node finds a common level of the routing hierarchy and achieves full bidirectional communication with a neighbor node. A switch can advertise links that reach the **commonOutside** state in PTSEs as uplinks to the upnode.

Second, let's look at the logical links from the highest-level PNNI node, which has a node index of 2. These two links are horizontal links because the peer PNNI node is at the same hierarchical level of 64, but they are not lowest-level. They receive the type **horizontalLinkToFromLgn**, indicating that they are horizontal links from a logical group node. These links achieve a **twoWayInside** state, indicating bidirectional communication. Database summary, PNNI Topology State Element (PTSE) request, PNNI Topology State Packets (PTSPs), and PTSE acknowledgment packets can only be transmitted over links in **twoWayInside** state. If the link is in **twoWayInside** state, the local node receives the remote node ID and port ID.

Finally, it's also worth noting that the links from the LGN have a non-null SVC-RCC index. They use SVC-based routing control channels (RCCs). You can see this using the command **dsppnni-svcc-rcc**, as shown in Example 11-5.

Example 11-5 *Using* **dsppnni-svcc-rcc**

```
m8850-7a.7.PXM.a > dsppnni-svcc-rcc

node index: 2                        svc index: 1
   Hello pkt RX........     15      SVCC VPI............      0
   Hello pkt TX........     14      SVCC VCI............      66
   Hello state...........twoWayInside
   Remote node id........64:160:47.00000000000001000100860000.00d058ac2828.01
   Remote node ATM addr...47.00000000000001000100860000.00d058ac2828.01

node index: 2                        svc index: 2
   Hello pkt RX........     14      SVCC VPI............      10
   Hello pkt TX........     14      SVCC VCI............      37
   Hello state...........twoWayInside
   Remote node id........64:160:47.00000000000001000200895000.0004c113ba46.01
   Remote node ATM addr...47.00000000000001000200895000.0004c113ba46.01

m8850-7a.7.PXM.a >
```

Adding a Lower Level to the Second Node in Peer Group A

The second step in the MPG migration is to lower the hierarchical level of the MGX-8950 PNNI node to level 80 and add its logical PNNI node at level 64, as shown in Figure 11-4. At level 80, the MGX-8950 PNNI node belongs to the same peer group as the MGX-8850 at level 80. That peer group is called PG-A.

Figure 11-4 *Moving the Second Node Down the Hierarchy*

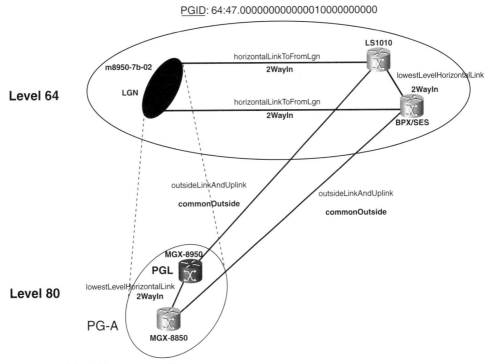

Using the same procedure, as Example 11-6 shows, you lower the lowest-level node to level 80 in the MGX-8950 and add a logical node at level 64. You also configure the peer group leader election priority so that the MGX-8950 lowest-level node is elected PGL.

Example 11-6 *Adding a Lower Level to the MGX-8950 PNNI Node*

```
m8950-7b.7.PXM.a > cnfpnni-node 1 -enable false

m8950-7b.7.PXM.a > cnfpnni-node 1 -level 80

m8950-7b.7.PXM.a > cnfpnni-node 1 -nodeId
  80:160:47.00000000000010002008950.0004c113ba46.01

m8950-7b.7.PXM.a > cnfpnni-node 1 -pgId 80:47.00000000000010002000000
```

continues

Example 11-6 *Adding a Lower Level to the MGX-8950 PNNI Node (Continued)*

```
m8950-7b.7.PXM.a > cnfpnni-election 1 -priority 150

m8950-7b.7.PXM.a > addpnni-node 64

Get MAC Address from ID PROM NV...OK: 00 04 C1 13 BA 46

m8950-7b.7.PXM.a > cnfpnni-node 1 -enable true

m8950-7b.7.PXM.a > cnfpnni-node 2 -enable true
```

At this point, the MGX-8950 lowest level is elected PGL (see Example 11-7). This is indicated in Figure 11-4 with a darker node color.

Example 11-7 *Displaying PNNI Leadership Election*

```
m8950-7b.7.PXM.a > dsppnni-election

node index: 1
    PGL state......        OperPgl      Init time(sec).......         15
    Priority.......        200          Override delay(sec)..         30
                                        Re-election time(sec)         15
    Pref PGL..............80:160:47.000000000000010002008950.0004c113ba46.01
    PGL...................80:160:47.000000000000010002008950.0004c113ba46.01
    Active parent node id..64:80:47.000000000000010002000000.0004c113ba46.00


node index: 2
    PGL state......        OperNotPgl   Init time(sec).......         15
    Priority.......        0            Override delay(sec)..         30
                                        Re-election time(sec)         15
    Pref PGL..............0:0:00.000000000000000000000000.000000000000.00
    PGL...................0:0:00.000000000000000000000000.000000000000.00
    Active parent node id..0:0:00.000000000000000000000000.000000000000.00
```

As with the MGX-8850 PNNI node, the MGX-8950's election priority is increased by 50 the moment it is elected. See Example 11-8. On the MGX-8850, the election priority returns to 50 because it is no longer the PGL.

Example 11-8 *Checking the Election Priority*

```
m8850-7a.7.PXM.a > dsppnni-election

node index: 1
    PGL state......        OperNotPgl   Init time(sec).......         15
    Priority.......        50           Override delay(sec)..         30
                                        Re-election time(sec)         15
    Pref PGL..............80:160:47.000000000000010002008950.0004c113ba46.01
    PGL...................80:160:47.000000000000010002008950.0004c113ba46.01
    Active parent node id..0:0:00.000000000000000000000000.000000000000.00
```

Example 11-8 *Checking the Election Priority (Continued)*

```
node index: 2
    PGL state......        Starting     Init time(sec).......       15
    Priority.......               0     Override delay(sec)..       30
                                        Re-election time(sec)       15
    Pref PGL..............0:0:00.000000000000000000000000.000000000000.00
    PGL...................0:0:00.000000000000000000000000.000000000000.00
    Active parent node id..0:0:00.000000000000000000000000.000000000000.00


m8850-7a.7.PXM.a >
```

It is extremely important to note that the level-64 PNNI LGN runs in the MGX-8950
because it was elected PGL. The PNNI election at level 64 in the MGX-8850 stays in
starting state because the MGX-8850 is not PGL and does not run PNNI at level 64.
Similarly, PNNI neighbors and links at level 64 are visible only from the PGL—that is,
from the MGX-8950. PNNI neighbors and links at level 64 are not visible from the MGX-
8850, because it is not PGL.

From a design perspective, it is advisable to configure at least primary and secondary PGLs
in a peer group, because a single PGL is a single point of failure. The selection of the PGL
node is an important part of designing MPG PNNI networks. PGL functionality uses extra
CPU power, so you should avoid PNNI nodes with extra CPU load. These include border
nodes and nodes with many signaling links.

Example 11-9 explores the PNNI neighbors seen from the MGX-8950 PNNI node.

Example 11-9 *PNNI Neighbors to the MGX-8950 Node*

```
m8950-7b.7.PXM.a > dsppnni-neighbor

node index    : 1

node name     : m8850-7a
Remote node id: 80:160:47.00000000000001000208850.00309409f6ba.01
Neighbor state: FULL
    Port count.........          1    SVC RCC index.......        0
    RX DS pkts..........         3    TX DS pkts..........        2
    RX PTSP pkts........        37    TX PTSP pkts........       99
    RX PTSE req pkts....         1    TX PTSE req pkts....        1
    RX PTSE ack pkts....        49    TX PTSE ack pkts....       11


node index    : 2

node name     : LS-1010
Remote node id: 64:160:47.000000000000010001001010.00503efba601.00
Neighbor state: FULL
    Port count.........          1    SVC RCC index.......        2
```

continues

Example 11-9 *PNNI Neighbors to the MGX-8950 Node (Continued)*

```
       RX DS pkts.........      2     TX DS pkts.........      3
       RX PTSP pkts........     3     TX PTSP pkts........     1
       RX PTSE req pkts....     0     TX PTSE req pkts....     0
       RX PTSE ack pkts....     1     TX PTSE ack pkts....     3

   node index    : 2

   node name     : SES-3a
   Remote node id: 64:160:47.00000000000010001008600.00d058ac2828.01
   Neighbor state: FULL
       Port count.........      1     SVC RCC index.......     1
       RX DS pkts.........      2     TX DS pkts.........      3
       RX PTSP pkts........    10     TX PTSP pkts........    10
       RX PTSE req pkts....     1     TX PTSE req pkts....     2
       RX PTSE ack pkts....     4     TX PTSE ack pkts....     3

   m8950-7b.7.PXM.a >
```

The lowest-level PNNI node index 1 has the MGX-8850 as a PNNI neighbor in **FULL** state. The remote node ID has a level indicator of 80. The PNNI node index 2 at level 64 has two neighbors in **FULL** state: the nodes LS-1010 and SES-3a, both with a level indicator of 64. Note that these two neighbors at level 64 use SVC-based RCCs for the hello protocol.

Referring to Figure 11-4, you can verify the PNNI links at both levels.

First, you see the PNNI links in node index 1 at level 80. You use the command **dsppnni-link**, specifying a node index, as shown in Example 11-10.

Example 11-10 *Specifying a Node Index with* **dsppnni-link**

```
   m8950-7b.7.PXM.a > dsppnni-link 1

   node index   : 1
   Local port id:    16848897        Remote port id: 2148532224
   Local Phy Port Id: 1:1.1:1
       Type.       outsideLinkAndUplink    Hello state.......commonOutside
       Derive agg..........     0     Intf index..........  16848897
       SVC RCC index........     0     Hello pkt RX.........     4356
                                       Hello pkt TX.........     3646
       Remote node id.........64:160:47.00000000000010001001010.00503efba601.00
       Upnode id.............64:160:47.00000000000010001001010.00503efba601.00
       Upnode ATM addr........47.00000000000010001001010.00503efba601.01
       Common peer group id...64:47.00.0000.0000.0001.0000.0000.00

   node index   : 1
   Local port id:    17111041        Remote port id:    17569793
   Local Phy Port Id: 5:1.1:1
       Type. lowestLevelHorizontalLink    Hello state....... twoWayInside
       Derive agg..........     0     Intf index..........  17111041
```

Example 11-10 *Specifying a Node Index with* **dsppnni-link** *(Continued)*

```
     SVC RCC index........      0     Hello pkt RX.........      677
                                      Hello pkt TX.........      749
     Remote node name.......m8850-7a
     Remote node id........80:160:47.00000000000010002008850.00309409f6ba.01
     Upnode id.............0:0:00.00000000000000000000000.000000000000.00
     Upnode ATM addr........00.00000000000000000000000.000000000000.00
     Common peer group id...00:00.00.0000.0000.0000.0000.0000.00

m8950-7b.7.PXM.a >
```

You see two links: a **lowestLevelHorizontalLink** toward the level 80 MGX-8850 neighbor, and an **outsideLinkAndUplink** toward the level 64 LS-1010.

Second, there are also two PNNI logical links from the level 64 logical group node. These are shown in Example 11-11.

Example 11-11 *PNNI Links from the Level 64 Logical Group Node*

```
m8950-7b.7.PXM.a > dsppnni-link 2

node index    : 2
Local port id:          1        Remote port id:          1
Local Phy Port Id: n/a
    Type.   horizontalLinkToFromLgn      Hello state....... twoWayInside
    Derive agg..........      0        Intf index..........      0
    SVC RCC index........      1        Hello pkt RX.........      11
                                        Hello pkt TX.........      13
    Remote node name.......SES-3a
    Remote node id........64:160:47.00000000000010001008600.00d058ac2828.01
    Upnode id.............0:0:00.00000000000000000000000.000000000000.00
    Upnode ATM addr........00.00000000000000000000000.000000000000.00
    Common peer group id...00:00.00.0000.0000.0000.0000.0000.00

node index    : 2
Local port id:          2        Remote port id:    33476608
Local Phy Port Id: n/a
    Type.   horizontalLinkToFromLgn      Hello state....... twoWayInside
    Derive agg..........      0        Intf index..........      0
    SVC RCC index........      2        Hello pkt RX.........      12
                                        Hello pkt TX.........      14
    Remote node name.......LS-1010
    Remote node id........64:160:47.00000000000010001001010.00503efba601.00
    Upnode id.............0:0:00.00000000000000000000000.000000000000.00
    Upnode ATM addr........00.00000000000000000000000.000000000000.00
    Common peer group id...00:00.00.0000.0000.0000.0000.0000.00

m8950-7b.7.PXM.a >
```

Node index 2 has two **horizontalLinkToFromLgn** logical links in **twoWayInside** state toward the level 64 PNNI nodes SES-3a and LS-1010. At this point, the peer group PG-A is configured and operational.

Adding a Lower Level to the Third Node in Peer Group B

The third step in the PNNI hierarchical configuration is to add a lower level to the SES-BPX PNNI node. At level 80, this PNNI node is in a different peer group than the previous two PNNI nodes. This peer group is called PG-B, as shown in Figure 11-5.

Figure 11-5 *Moving the Third Node Down the Hierarchy*

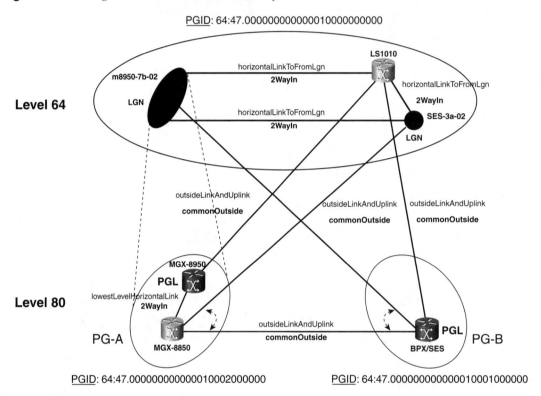

As with the previous two nodes, the lowest level is lowered to level 80. The peer group ID at level 80 is different from the previous two PNNI nodes, so this level 80 PNNI node is in a different peer group. A non-null election priority is also configured for the lowest-level node. You can see the configuration in Example 11-12.

Example 11-12 *Adding a Lower Level to the BPX-SES PNNI Node*

```
SES-3a.1.PXM.a > cnfpnni-node 1 -enable false

SES-3a.1.PXM.a > cnfpnni-node 1 -level 80

SES-3a.1.PXM.a > cnfpnni-node 1 -pgId 80:47.000000000000010001000000

SES-3a.1.PXM.a > cnfpnni-node 1 -nodeId
```

Example 11-12 *Adding a Lower Level to the BPX-SES PNNI Node (Continued)*

```
  80:160:47.000000000000010001008600.00d058ac2828.01

SES-3a.1.PXM.a > cnfpnni-election 1 -priority 1

SES-3a.1.PXM.a > addpnni-node 64

SES-3a.1.PXM.a > cnfpnni-node 1 -enable true

SES-3a.1.PXM.a > cnfpnni-node 2 -enable true

SES-3a.1.PXM.a >
```

The SES PNNI node is elected peer group leader with a priority of 51 (the configured value of 1 plus 50 to provide hysteresis).

It is important to note that at level 80 (the lowest level), the PNNI node does not have any PNNI neighbors. The other two nodes at level 80 are in a different peer group. You can see this with the command **dsppnni-neighbor**, as shown in Example 11-13.

Example 11-13 *Displaying PNNI Neighbors Using* **dsppnni-neighbor**

```
SES-3a.1.PXM.a > dsppnni-neighbor

node index    : 2

node name    : m8950-7b-02
Remote node id: 64:80:47.000000000000010002000000.0004c113ba46.00
Neighbor state: FULL
      Port count.........        1    SVC RCC index.......        1
      RX DS pkts.........        2    TX DS pkts.........        3
      RX PTSP pkts.......        8    TX PTSP pkts.......        6
      RX PTSE req pkts....       1    TX PTSE req pkts....       1
      RX PTSE ack pkts....       3    TX PTSE ack pkts....       1

node index    : 2

node name    : LS-1010
Remote node id: 64:160:47.000000000000010001001010.00503efba601.00
Neighbor state: FULL
      Port count.........        1    SVC RCC index.......        2
      RX DS pkts.........        2    TX DS pkts.........        3
      RX PTSP pkts.......        1    TX PTSP pkts.......        1
      RX PTSE req pkts....       0    TX PTSE req pkts....       0
      RX PTSE ack pkts....       1    TX PTSE ack pkts....       1

SES-3a.1.PXM.a >
```

Note that the first neighbor (m8950-7b-02) is an LGN and the second neighbor is a physical node.

The link type and status of the SES PNNI nodes correspond to the ones the MGX-8850 had in the first configuration step. The lowest-level PNNI node has two **outsideLinkAndUplink** links in **commonOutside** state, and the higher-level node has two **horizontalLinkToFromLgn** links in **twoWayInside** state.

In Figure 11-5, a dotted arc between two links from a level-80 node means that it is the same link.

Adding a Lower Level to the Fourth Node

The fourth and final step in the hierarchical configuration involves adding a lower level to the Cisco IOS-based PNNI node. Figure 11-6 shows the end result of this step and the complete exercise.

Figure 11-6 *Final MPG PNNI Network*

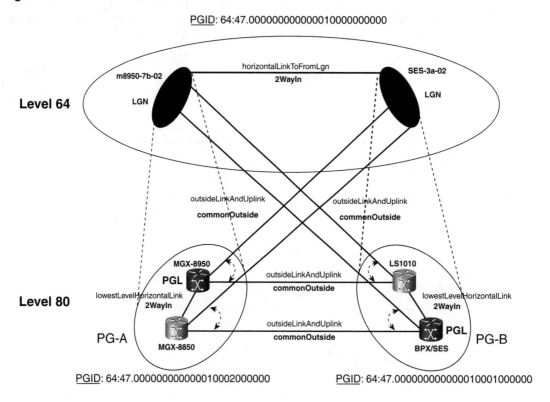

As Example 11-14 shows, when using a different set of commands, you need to execute the same tasks to create a lower level in the Cisco IOS Software-based PNNI node. You need to administratively down the PNNI node, lower the level to 80, add a parent PNNI node at level 64, and enable both PNNI nodes.

Example 11-14 *Adding a Lower PNNI Level to the IOS PNNI Node*

```
LS-1010#conf t
Enter configuration commands, one per line.  End with CNTL/Z.
LS-1010(config)#atm router pnni
LS-1010(config-atm-router)#node 1 disable
LS-1010(config-pnni-node)#node 1 level 80
LS-1010(config-pnni-node)#node 2 level 64
LS-1010(config-pnni-node)#node 1
LS-1010(config-pnni-node)#parent 2
LS-1010(config-pnni-node)#node 1 enable
LS-1010(config-pnni-node)#node 2 enable
```

Because you did not configure an election priority in the lowest-level node, the SES-based node is still PGL. The higher-level PNNI node is not running in the Cisco IOS Software-based PNNI node. See Example 11-15.

Example 11-15 *Displaying PNNI Node Details*

```
LS-1010#show atm pnni local-node

PNNI node 1 is enabled and running
  Node name: LS-1010
  System address         47.00000000000001000100101.00503EFBA601.01
  Node ID        80:160:47.00000000000001000100101.00503EFBA601.00
  Peer group ID       80:47.0000.0000.0000.0100.0100.0000
  Level 80, Priority 0 0, No. of interfaces 2, No. of neighbors 1
  Parent Node Index: 2
  Node Allows Transit Calls
  Node Representation: simple
<snip>
PNNI node 2 is enabled and not running
  Node name: LS-1010.2.64
  System address         47.00000000000001000100101.00503EFBA601.02
  Node ID        64:80:47.00000000000001000100000.00503EFBA601.00
  Peer group ID       64:47.0000.0000.0000.0100.0000.0000
  Level 64, Priority 0 0, No. of interfaces 0, No. of neighbors 0
  Parent Node Index: NONE
  Node Allows Transit Calls
  Node Representation: simple
<snip>
```

You can see in Example 11-16 that the SES-3a PNNI node is still PGL.

Example 11-16 *Checking the PNNI Election Priority*

```
LS-1010#show atm pnni election

PGL Status.............: Not PGL
Preferred PGL..........: (9) SES-3a
Preferred PGL Priority.: 51
Active PGL.............: (9) SES-3a
Active PGL Priority....: 51
Active PGL For.........: 00:02:27
Current FSM State......: PGLE Operating: Not PGL
Last FSM State.........: PGLE Calculating
Last FSM Event.........: Preferred PGL Is Not Self

Configured Priority....: 0
Advertised Priority....: 0
Conf. Parent Node Index: NONE
PGL Init Interval......: 15 secs
Search Peer Interval...: 75 secs
Re-election Interval...: 15 secs
Override Delay.........: 30 secs
LS-1010#
```

This is the final stage in the hierarchical level. You have achieved hierarchical abstraction. Figure 11-7 shows the topology picture that the BPX-SES PNNI node has—that is, the PNNI network as seen from peer group PG-B.

Figure 11-7 *Hierarchical Abstraction*

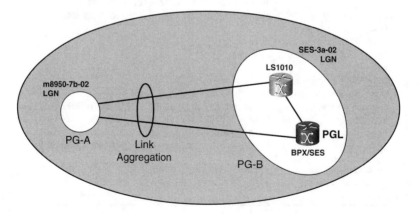

The BPX-SES PNNI node has one PNNI neighbor at level 80 (LS-1010) and one neighbor at level 64 (MGX-8950-02), as shown in Figures 11-6 and 11-7 as well as Example 11-17.

Example 11-17 *Displaying SES-BPX Node's PNNI Neighbors*

```
SES-3a.1.PXM.a > dsppnni-neighbor

node index    : 1

node name     : LS-1010
Remote node id: 80:160:47.000000000000010001001010.00503efba601.00
Neighbor state: FULL
    Port count.........      1    SVC RCC index.......      0
    RX DS pkts.........      2    TX DS pkts.........      2
    RX PTSP pkts.......      2    TX PTSP pkts.......      9
    RX PTSE req pkts....     1    TX PTSE req pkts....     1
    RX PTSE ack pkts....     2    TX PTSE ack pkts....     1


node index    : 2

node name     : m8950-7b-02
Remote node id: 64:80:47.000000000000010002000000.0004c113ba46.00
Neighbor state: FULL
    Port count.........      1    SVC RCC index.......      1
    RX DS pkts.........      2    TX DS pkts.........      3
    RX PTSP pkts.......     13    TX PTSP pkts.......     11
    RX PTSE req pkts....     1    TX PTSE req pkts....     1
    RX PTSE ack pkts....     5    TX PTSE ack pkts....     3
SES-3a.1.PXM.a >
```

Each node at the lowest level sees four PNNI nodes: two lowest-level nodes, one upnode at the higher level, and one upnode's neighbor.

All the PNNI nodes at level 80 have two links: one link of type **lowestLevelHorizontal-Link** to the PNNI neighbor within the peer group, and a second link of type **outsideLink-AndUpLink** to the other peer group's LGN.

At level 64, both LGNs have only one PNNI link. This one PNNI link is a **horizontalLink-ToFromLgn** in **twoWayInside** state. The following command in Example 11-18 shows the PNNI links at the highest-level node of index 2.

Example 11-18 *Displaying PNNI Links at the Highest Level*

```
SES-3a.1.PXM.a > dsppnni-link 2

node index   : 2
Local port id:          1         Remote port id:          1
Local Phy Port Id: n/a
    Type.    horizontalLinkToFromLgn    Hello state....... twoWayInside
    Derive agg..........      0    Intf index..........      0
    SVC RCC index.......      1    Hello pkt RX.........     24
                                   Hello pkt TX.........     29
    Remote node name.......m8950-7b-02
```

continues

Example 11-18 *Displaying PNNI Links at the Highest Level (Continued)*

```
       Remote node id.........64:80:47.00000000000010002000000.0004c113ba46.00
       Upnode id.............0:0:00.00000000000000000000000000.000000000000.00
       Upnode ATM addr.......00.00000000000000000000000000.000000000000.00
       Common peer group id...00:00.00.0000.0000.0000.0000.0000.00

SES-3a.1.PXM.a >
```

Because you know the topology at all levels, you know that two physical links at level 80 connect the LGNs at level 64. Those two links are seen as a single logical link at level 64. The reason for this is that the two links at the lower level have the same aggregation token, so they are aggregated into a single logical link. The aggregation token defaults to 0. Thus, the default behavior is to aggregate links into a single logical link at a higher hierarchical level (a summary link between peer groups).

You can see the aggregation token in Example 11-19, a 32-bit number used for link aggregation, using the command **dsppnni-intf**. You can modify it with the command **cnfpnni-intf**. The aggregation token identifies uplinks to be aggregated into a single link such that uplinks with the same token are the single link.

Example 11-19 *Displaying the PNNI Link Aggregation Token*

```
m8950-7b.7.PXM.a > dsppnni-intf 1:1.1:1

Physical port id: 1:1.1:1          Logical port id:   16848897
    Aggr token.........       0     AW-NRTVBR..........       5040
    AW-CBR.............    5040     AW-ABR.............       5040
    AW-RTVBR...........    5040     AW-UBR.............       5040

m8950-7b.7.PXM.a >
```

With link aggregation, a single link represents multiple links between child peer groups (PGs) the same way that with node aggregation a simple LGN represents the complete child PG. Both processes are summarized under the concept of PNNI topology aggregation, which gives the PNNI protocol great scalability.

Link aggregation is shown in Figure 11-8.

With link aggregation, metrics (additive parameters) from individual links are added to determine the metric of the aggregated link. The worst attribute (nonadditive parameters) from individual links becomes the attribute of the aggregated link.

Figure 11-8 *Link Aggregation Example*

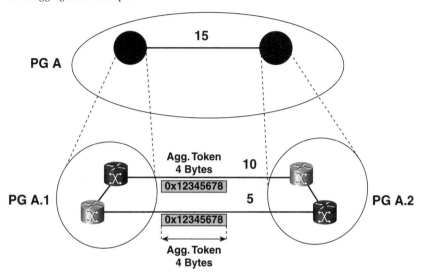

Abstraction Achieved

As pointed out earlier, you have achieved hierarchical abstraction. Many display commands have proven the multiple-level concept. This section shows some more commands to wrap up this idea.

You use the same **aesa_ping** command used in Chapter 10 in the "PNNI Network Verification" section with the same source PNNI node and AESA destination. See Example 11-20. The difference is that, in Chapter 10, the PNNI network was an SPG network, and now it is an MPG network.

Example 11-20 *Using the Command* **aesa_ping**

```
m8950-7b.7.PXM.a > aesa_ping 47.000000000000010001008600.00d058ac2828.01

Ping Got CLI message, index=0

PING: from PNNI - SOURCE ROUTE
DTL    1 : Number of (Node/port)elements    1

DTL 1:NODE 1: :80:160:71:0:0::2:0:137:80:

 Port 1:16848897
DTL    2 : Number of (Node/port)elements    2

DTL 2:NODE 1: :64:80:71:0:0::2:0:0:0:
```

continues

Example 11-20 *Using the Command* **aesa_ping** *(Continued)*

```
 Port 1:0

DTL 2:NODE 2: :64:80:71:0:0::1:0:0:0:

 Port 2:0

Port List : no of ports =    1

Port ID    1:16848897

m8950-7b.7.PXM.a >
```

The setup message has a designated transit list (DTL) stack. This is because, given the hierarchical topology aggregation, topology information is summarized outside the peer group.

Table 11-1 summarizes the nodes listed in the two DTLs. You can compare it to Table 10-2 in Chapter 10.

Table 11-1   *PNNI Node IDs in the* **aesa_ping** *Path for MPG*

DTL	Node	PNNI Node ID	Node
1	Node 1	80:160:47.0000...02008950...	MGX-8950
2	Node 1	64:80:47.0000...02000000...	MGX-8950-02
2	Node 2	64:80:47.0000...01000000...	BPX-8600-02

DTL 1 shows the nodes listed at level 80. On the other hand, DTL 2 lists the nodes at level 64, which are the two LGNs at the highest level.

Another useful command is **dsppnni-path**, which displays the precalculated routes to a specified node as Example 11-21 shows. You use this command to see the routes to all nodes for the administrative weight parameter for the CBR service type.

Example 11-21 *Using the Command* **dsppnni-path** *for AW with CBR*

```
m8950-7b.7.PXM.a > dsppnni-path aw cbr

node #/PortId   node id                                                  node name
--------------  --------------------------------------------------------  ----------
D  2/        0 80:160:47.00000000000001000200 8850.00309409f6ba.01  m8850-7a
S  1/ 17111041 80:160:47.00000000000001000200 8950.0004c113ba46.01  m8950-7b

node #/PortId   node id                                                  node name
--------------  --------------------------------------------------------  ----------
D  5/        0 64:80:47.000000000000010001000000.00d058ac2828.00  SES-3a-02
   7/        0 64:80:47.000000000000010002000000.0004c113ba46.00  m8950-7b-02
S  1/ 16848897 80:160:47.00000000000001000200 8950.0004c113ba46.01  m8950-7b

m8950-7b.7.PXM.a >
```

You can see in Example 11-22 that from the MGX-8950 to the MGX-8850 is a path at the lowest level 80 across physical nodes. However, the routes to the node SES-3a-02 need to go across the level 64 MGX-8950-02 node. If you change the election priority of the nodes in PG-A using the command **cnfpnni-election** so that the MGX-8850 is elected PGL, the output of the command **dsppnni-path** changes to show the MGX-8850-02 node at level 64.

Example 11-22 *Using the Command* **dsppnni-path**

```
m8950-7b.7.PXM.a > dsppnni-path aw cbr

node #/PortId   node id                                            node name
--------------  -------------------------------------------------- ----------
D  2/        0 80:160:47.00000000000001000200880.00309409f6ba.01  m8850-7a
S  1/ 17111041 80:160:47.0000000000000100020088950.0004c113ba46.01 m8950-7b


node #/PortId   node id                                            node name
--------------  -------------------------------------------------- ----------
D  5/        0 64:80:47.000000000000001000100000.00d058ac2828.00  SES-3a-02
   6/        0 64:80:47.000000000000001000200000.00309409f6ba.00  m8850-7a-02
S  1/ 16848897 80:160:47.00000000000001000200880.0004c113ba46.01  m8950-7b

m8950-7b.7.PXM.a >
```

IISP

This section covers IISP configuration. A new PNNI node that is part of a PNNI private network will be joined to your network using an IISP interface. The topology is shown in Figure 11-9.

Figure 11-9 *IISP Configuration*

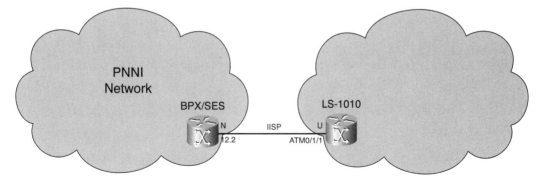

The LS-1010 has the following ATM address, as Example 11-23 shows.

Example 11-23 *Changing the ATM Address*

```
IOS_pnni_node(config)#atm address 49.FEE4DECAFC0FFEE4BED2FEED.DAD000001010.00
```

The first configuration step is to bring up a line, port, and resource partition in interface 12.1 in BPX-SES PNNI node. An ATMF service class template (SCT) also needs to be assigned to the VSI resource partition.

As soon as the physical and ATM layers are up, you can start the IISP-specific configuration. Essentially, two tasks are required for IISP configuration:

- Configure the PnPort signaling to IISP, selecting a side (IISP is asymmetric. It has network and user sides).

- Configure a static route pointing to the remote prefix out of the IISP interface.

In IISP, the network side allocates VPI/VCI to avoid collisions. IISP's routing face supports crankback generation and load balancing over multiple IISP links and multiple addresses. IISP's signaling side supports simple call-control mechanisms to provide QoS support.

To carry out the first step and configure IISP signaling, you use the command **cnfpnportsig** in the SES controller with the PnPort administratively down to select the IISP version 3.1 side network. See Example 11-24.

Example 11-24 *Configuring IISP Signaling*

```
SES-3a.1.PXM.a > dnpnport 12.1

SES-3a.1.PXM.a > cnfpnportsig 12.1 -nniver iisp31 -side network

SES-3a.1.PXM.a > uppnport 12.1

SES-3a.1.PXM.a > dsppnportsig 12.1

provisioned IF-type: nni      version:      iisp31
sigType:  private             side:         network
addrPlan:  aesa
VpiVciAllocator:  n/a         HopCounterGen:  n/a
PassAlongCapab:  enable
sigVpi:          0            sigVci:              5
rccVpi:          n/a          rccVci:              n/a
svc routing priority: 8

SES-3a.1.PXM.a >
```

On the Cisco IOS Software PNNI node, you need to disable autoconfiguration and use the interface-level command **atm iisp**, as Example 11-25 shows.

Example 11-25 *Configuring IISP Signaling*

```
IOS_pnni_node#conf t
Enter configuration commands, one per line.  End with CNTL/Z.
IOS_pnni_node(config)#interface ATM0/0/0
```

Example 11-25 *Configuring IISP Signaling (Continued)*

```
IOS_pnni_node(config-if)#no atm auto-configuration
IOS_pnni_node(config-if)#atm iisp side user version 3.1
IOS_pnni_node(config-if)#exit
```

The first task is complete.

In the second task, you need to configure a static route. In the SES controller, this task is accomplished with the command **addaddr**. For IISP configuration, the configured protocol needs to be **static**, and the type needs to be **external**. Also, the static route needs to be redistributed into the PNNI routing process so that remote PNNI nodes learn this route. You do this by setting the parameter **redst** to **yes**. See Example 11-26. In this case, 8 bits is enough to make this prefix unique, because the remote network uses a different address plan.

Example 11-26 *Configuring Static Routes for IISP*

```
SES-3a.1.PXM.a > addaddr 12.1 49 8 -proto static -type ext -redst yes

SES-3a.1.PXM.a > dspaddr 12.1
49
length: 8        type: exterior      proto: static
scope: 0         plan: nsap_e164     redistribute: true
transit network id:


SES-3a.1.PXM.a >
```

The second task is performed in the Cisco IOS Software platform using the command **atm route** (see Example 11-27). The static ATM route also needs to be redistributed under the PNNI process. You can see the redistribution status of static routes within the PNNI routing process using the command **show atm pnni local-node**.

Example 11-27 *Configuring Static Routes for IISP*

```
IOS_pnni_node(config)#atm route 47 ATM 0/0/0
IOS_pnni_node(config)#atm router pnni
IOS_pnni_node(config-atm-route)#node 1
IOS_pnni_node(config-pnni-node)#redistribute atm-static
IOS_pnni_node(config-pnni-node)#^Z
IOS_pnni_node#
IOS_pnni_node#show atm pnni local-node 1 | incl static
  Redistributing static routes: Yes
IOS_pnni_node#
```

The IISP link is now ready. You can go to the MGX-8950 and check the PNNI routing table as well as reachability to the IISP linked PNNI network. See Example 11-28.

Example 11-28 *Checking the PNNI Routing Table*

```
m8950-7b.7.PXM.a > dsppnni-reachable-addr network

scope..............        0     Advertising node number     5
Exterior...........     false
ATM addr prefix.....47.0000.0000.0000.0100.01/80
Transit network id..
Advertising nodeid..64:80:47.00000000000010001000000.00d058ac2828.00
Node name..........SES-3a-02

scope..............        0     Advertising node number     2
Exterior...........     false
ATM addr prefix.....47.0000.0000.0000.0100.0200.8850/104
Transit network id..
Advertising nodeid..80:160:47.00000000000010002008850.00309409f6ba.01
Node name..........m8850-7a

scope..............        0     Advertising node number     5
Exterior...........     true
ATM addr prefix.....49/8
Transit network id..
Advertising nodeid..64:80:47.00000000000010001000000.00d058ac2828.00
Node name..........SES-3a-02

m8950-7b.7.PXM.a >
```

The output of the command **dsppnni-reachable-addr** is also a good way to check address summarization.

You can also use the command **aesa_ping** to verify connectivity with the other network.

The closing portion of the IISP configuration includes creating an SPVC with each endpoint in a different PNNI network traversing the IISP link. SPVC signaling also goes across the IISP link. As Example 11-29 shows, you need only a master SPVC endpoint, because the remote PNNI node implements a single-endpoint SPVC model.

Example 11-29 *Adding an SPVC Traversing the IISP Link*

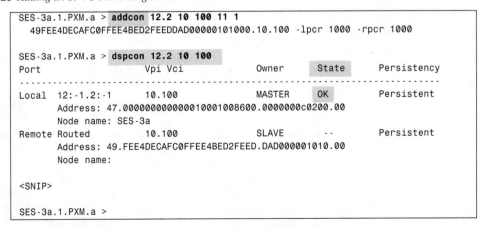

```
SES-3a.1.PXM.a > addcon 12.2 10 100 11 1
    49FEE4DECAFC0FFEE4BED2FEEDDAD00000101000.10.100 -lpcr 1000 -rpcr 1000

SES-3a.1.PXM.a > dspcon 12.2 10 100
Port                    Vpi Vci        Owner       State     Persistency
- - - - - - - - - - - - - - - - - - - - - - - - - - - - - - - - - - - - - - - - - -
Local 12:-1.2:-1     10.100          MASTER      OK        Persistent
        Address: 47.00000000000010001008600.0000000c0200.00
        Node name: SES-3a
Remote Routed        10.100          SLAVE       --        Persistent
        Address: 49.FEE4DECAFC0FFEE4BED2FEED.DAD000001010.00
        Node name:

<SNIP>

SES-3a.1.PXM.a >
```

The remote SPVC endpoint can be seen from the remote node, as shown in Example 11-30.

Example 11-30 *Looking at the SPVC Endpoint*

```
IOS_pnni_node#show atm vc interface ATM 0/0/0 10 100

Interface: ATM0/0/0, Type: oc3suni
VPI = 10   VCI = 100
Status: UP
Time-since-last-status-change: 00:01:21
Connection-type: SoftVC
Cast-type: point-to-point
Hold-priority: none
 Soft vc location: Destination
 Remote ATM address: 47.0000.0000.0000.0100.0100.8600.0000.000c.0200.00
 Remote VPI: 10
 Remote VCI: 100
 Soft vc call state: Active
Packet-discard-option: disabled
Usage-Parameter-Control (UPC): pass
Number of OAM-configured connections: 0
OAM-configuration: disabled
OAM-states:  Not-applicable
Cross-connect-interface: ATM0/0/0, Type: oc3suni
Cross-connect-VPI = 0
Cross-connect-VCI = 90
Cross-connect-UPC: pass
Cross-connect OAM-configuration: disabled
Cross-connect OAM-state:  Not-applicable
Rx cells: 0, Tx cells: 0
Rx connection-traffic-table-index: 2147483641
Rx service-category: CBR (Constant Bit Rate)
Rx pcr-clp01: 424
Rx scr-clp01: none
Rx mcr-clp01: none
Rx      cdvt: 1024 (from default for interface)
Rx       mbs: none
Tx connection-traffic-table-index: 2147483641
Tx service-category: CBR (Constant Bit Rate)
Tx pcr-clp01: 424
Tx scr-clp01: none
Tx mcr-clp01: none
Tx      cdvt: none
Tx       mbs: none

IOS_pnni_node#
```

Because you configured 1000 cells per second (CPS) for the local and remote PCR values, the Cisco IOS Software PNNI node shows 424 kilobits per second (Kbps).

Traffic Engineering

As mentioned at the beginning of this chapter, PNNI is a constrained-based routing protocol. It advertises and keeps in its database information about the status of the links and nodes using metrics and attributes. PNNI supports different link metrics (such as AW, CTD, and CDV) and different attributes (such as AvCR, MaxCR, CLR, and so on) for different classes of service. The support for different link metrics and attributes is one of the enablers of the advanced traffic engineering and QoS routing capabilities.

Different classes of service support different traffic parameters. The traffic parameter administrative weight (AW) is required for all classes of service. Others, such as maximum cell rate (MaxCR), are required for some classes of service but are optional for others. Table 11-2 summarizes the required and optional traffic parameters.

Table 11-2   *PNNI Classes of Service and Traffic Parameters*

	CBR	VBR-RT	VBR-NRT	ABR	UBR
AW	Required	Required	Required	Required	Required
MaxCR	Optional	Optional	Optional	Required	Required
AvCR	Required	Required	Required	Required	N/A
MaxCTD	Required	Required	Required	N/A	N/A
CDV	Required	Required	N/A	N/A	N/A
CLR_0	Required	Required	Required	N/A	N/A
CLR_{0+1}	Required	Required	Required	N/A	N/A
CRM	N/A	Optional	Optional	N/A	N/A
VF	N/A	Optional	Optional	N/A	N/A

Class-Based Routing and Constrained-Based Routing

Evidence of the class-based nature of the PNNI routing protocol can be seen with different commands. The command **dsppnni-path** displays precalculated routes to specific nodes or to all nodes, for different classes of service, based on AW, cell transfer delay, or cell delay variation.

First, as Example 11-31 shows, you can display the precomputed path from the MGX-8850 toward the two adjacent nodes for the CBR service category.

Example 11-31 *Displaying the PNNI Precomputed Paths*

```
m8850-7a.7.PXM.a > dsppnni-path aw cbr 2

node #/PortId   node id                                              node name
--------------  ----------------------------------------------------  ----------
D  2/         0 64:160:47.0000000000000010001008600.00d058ac2828.01 SES-3a
S  1/  16848898 64:160:47.0000000000000010002008850.00309409f6ba.01 m8850-7a

m8850-7a.7.PXM.a > dsppnni-path aw cbr 3

node #/PortId   node id                                              node name
--------------  ----------------------------------------------------  ----------
D  3/         0 64:160:47.0000000000000010002008950.0004c113ba46.01 m8950-7b
S  1/  17569793 64:160:47.0000000000000010002008850.00309409f6ba.01 m8850-7a

m8850-7a.7.PXM.a >
```

You can see that CBR connections terminating in either of these two nodes take a unique path.

The precalculated path from the MGX-8850 PNNI node toward the LS-1010 PNNI node based on administrative weight for the CBR class is as follows (see Example 11-32).

Example 11-32 *Displaying Load Sharing Paths*

```
m8850-7a.7.PXM.a > dsppnni-path aw cbr 4

node #/PortId   node id                                              node name
--------------  ----------------------------------------------------  ----------
D  4/         0 64:160:47.0000000000000010001001010.00503efba601.00 LS-1010
   2/    721408 64:160:47.0000000000000010001008600.00d058ac2828.01 SES-3a
S  1/  16848898 64:160:47.0000000000000010002008850.00309409f6ba.01 m8850-7a

node #/PortId   node id                                              node name
--------------  ----------------------------------------------------  ----------
D  4/         0 64:160:47.0000000000000010001001010.00503efba601.00 LS-1010
   3/  16848897 64:160:47.0000000000000010002008950.0004c113ba46.01 m8950-7b
S  1/  17569793 64:160:47.0000000000000010002008850.00309409f6ba.01 m8850-7a

m8850-7a.7.PXM.a >
```

You can see in Example 11-32 that the source node is the MGX-8850 and the destination node is the LS-1010. There are two via-nodes in two equal-cost paths. One path traverses the SES-3a PNNI node, and the other path goes across the MGX-8950 PNNI node.

With the command **cnfpnni-intf** you can configure the class of service-based AW on a PNNI interface (see Example 11-33). You can increase the AW of the CBR service category in PnPort 1:1.2:2, connecting the MGX-8850 with the BPX-SES, and then check the path again.

Example 11-33 *Configuring the AW for CBR Traffic on a Link*

```
m8850-7a.7.PXM.a > cnfpnni-intf 1:1.2:2 -awcbr 10000

m8850-7a.7.PXM.a > dsppnni-path aw cbr 4

node #/PortId   node id                                              node name
--------------  ---------------------------------------------------- ----------
D  4/           0 64:160:47.0000000000000010001001010.00503efba601.00 LS-1010
   3/    16848897 64:160:47.0000000000000010002008950.0004c113ba46.01 m8950-7b
S  1/    17569793 64:160:47.0000000000000010002008850.00309409f6ba.01 m8850-7a

m8850-7a.7.PXM.a >
```

You can see in Example 11-33 that CBR connections now take a unique path through the via-node MGX-8950. The two paths toward the LS-1010 are now of unequal cost.

Finally, as Example 11-34 shows, you can decrease the AW of the CBR service category on the same interface and check the precomputed paths.

Example 11-34 *Changing the AW for CBR Traffic*

```
m8850-7a.7.PXM.a > cnfpnni-intf 1:1.2:2 -awcbr 10

m8850-7a.7.PXM.a > dsppnni-path aw cbr 4

node #/PortId   node id                                              node name
--------------  ---------------------------------------------------- ----------
D  4/           0 64:160:47.0000000000000010001001010.00503efba601.00 LS-1010
   2/     721408 64:160:47.0000000000000010001008600.00d058ac2828.01 SES-3a
S  1/    16848898 64:160:47.0000000000000010002008850.00309409f6ba.01 m8850-7a

m8850-7a.7.PXM.a >
```

The unique and different path that connections now take is through the SES-3a node.

All the routing information, including nodes, link states, and reachability information, is kept in the PNNI internal database. You can see the contents of the PNNI internal database using the command **dsppnni-idb**, as Example 11-35 shows. To narrow the command's output, you can see information for a specific node and a specific link. You gather the port ID in decimal notation corresponding to a specific PnPort using the command **dsppnportidmaps**.

Example 11-35 *Viewing the PNNI Internal Database Using* **dsppnni-idb**

```
29

node index: 1
    Local port id....... 16848898    Remote port id....... 262912
    Local link index....        1    Remote link index....      1
    Local node number...        1    Remote node number...      2
```

Example 11-35 *Viewing the PNNI Internal Database Using* **dsppnni-idb** *(Continued)*

```
PGL node index......        0      LGN node index.......        0
Transit restricted..      off      Complex node........      off
Branching restricted       on      PGL.................    false
Ancestor...........     false      Border node.........    false
VP capable.........      true      Link type...........horizontal
Non-transit for PGL election..      off
node id..............64:160:47.0000000000000010002008850.00309409f6ba.01
node name............m8850-7a

                     forward direction
            CBR    RTVBR   NRTVBR    ABR     UBR
           ------  ------  ------  ------  ------
     AW      5040    5040    5040    5040    5040
     MaxCR  96000   96000   96000   96000   96000
     AvCR   93293   93293   93293   93293   93293
     CTD       56      56      56     n/a     n/a
     CDV       10      10     n/a     n/a     n/a
     CLR0      10       8       6     n/a     n/a
     CLR0+1     8       8       8     n/a     n/a
     CRM      n/a     n/a     n/a     n/a     n/a
     VF       n/a     n/a     n/a     n/a     n/a

m8850-7a.7.PXM.a >
```

The output of the command **dsppnni-idb** specified on a PNNI link shows the routing protocol's class-based nature, as well as its metric and QoS-based characteristics.

You can specify different node characteristics and link colors. To name a couple, a PNNI node has a transit-restricted flag advertised in the nodal info PTSE, and a PNNI link has a VP-capable flag.

Connection Tracing

The PNNI nodes provide connection-tracing capabilities for existing active connections. The connection trace information includes all nodes, cross-connecting ports, VPI/VCI — in other words, the complete path. Path trace is a sibling tool that allows the user to see the path of a connection in the process of being established. The path-tracing tool also shows crankback information.

The connection-tracing tool is used with the combination of the commands **conntrace** and **dspconntracebuffers**, as you can see in Example 11-36.

Example 11-36 *Using the Commands* **conntrace** *and* **dspconntracebuffers**

```
m8850-7a.7.PXM.a > conntrace 2:1.1:1 -vpi 100 -vci 100

m8850-7a.7.PXM.a > dspconntracebuffers

- - - - - - - - - - - - - - - - - - - - - - - - - - - - -
```

continues

Example 11-36 *Using the Commands* **conntrace** *and* **dspconntracebuffers** *(Continued)*

```
dspconntracebuffers: next record
--------------------------------

Last update time: Mar 7 2002 19:46:31
Result: SUCCESS      Reason: N/A

Incoming Port: 16914433    Physical PortId: 2:1.1:1
VPI   : 100   VCI: 100   CallRef: 8093
Node Name: m8850-7a NodeId: 80:160:47.00000000000010002008850.00309409f6ba.01
Outgoing Port: 16848898    Physical PortId: 1:1.2:2

VPI   : 0   VCI: 36   CallRef: 1
Node Name:  NodeId: 80:160:47.00000000000010001008600.00d058ac2828.01
Outgoing Port: 786944   VPI   : 100   VCI: 100   CallRef: 10   Physical PortId: 12.2


m8850-7a.7.PXM.a >
```

You can display symmetric information by going to the other end of the connection and performing a connection trace. See Example 11-37.

Example 11-37 *Connection Tracing*

```
SES-3a.1.PXM.a > conntrace 12.2 -vpi 100 -vci 100

SES-3a.1.PXM.a > dspconntracebuffer 12.2 100 100

Last update time: Mar 7 2002 7:17:20
Result: SUCCESS      Reason: N/A

Incoming Port: 786944   Physical PortId: 12.2
VPI   : 100   VCI: 100   CallRef: 10
Node Name: SES-3a NodeId: 80:160:47.00000000000010001008600.00d058ac2828.01
Outgoing Port: 262912   Physical PortId: 4.3

VPI   : 0   VCI: 36   CallRef: 1
Node Name:  NodeId: 80:160:47.00000000000010002008850.00309409f6ba.01
Outgoing Port: 16914433   VPI   : 100   VCI: 100   CallRef: 8093   Physical PortId:
2:1.1:1


SES-3a.1.PXM.a >
```

NOTE   In this section, we identified an SPVC connection for tracing by providing a port, VPI and VCI. For SVC connection tracing, however, we need to identify the SVC call using the **call reference**.

Route Optimization

PNNI provides per-link and per-VPI/VCI range schedulable route optimization features. The command **cnfrteopt** (see Example 11-38) configures the route optimization task. You can see this configuration with the command **dsprteoptcnf**.

Example 11-38 *Route Optimization Using* **cnfrteopt**

```
m8850-7a.7.PXM.a > cnfrteopt 2:1.1:1 enable -tod 01:00..03:00

m8850-7a.7.PXM.a > dsprteoptcnf
Configuration of Route Optimization:
Percentage Reduction Threshold: 30
Port           Enable  VPI/VCI Range    Interval   Time Range
7.35           no
7.36           no
7.37           no
7.38           no
10.1           no
2:1.1:1        yes     all              60         01:00..03:00

m8850-7a.7.PXM.a >
```

Alternatively, you can manually trigger the route optimization on a per-port, per-VPI/VCI range, or per-connection basis using the command **optrte**.

Preferred Routes

One of the most advanced traffic engineering features is the ability to specify a preferred route for an SPVC or SPVP connection. In essence, the user manually specifies the DTL to be used in the setup message. This allows the user to specify the nodes and links a connection should take.

As soon as a preferred route is specified, routing attempts for the connection try to route the connection via the preferred route before attempting any other routes. Generic CAC (GCAC) is checked for the preferred route to verify that the bandwidth requirements can be satisfied at that point in time. Depending on the directed route flag, further routing attempts might take place. When a route is marked as a directed route, the connection tries to route only on the specified preferred route. For nondirected routes, further routing attempts take place if the preferred route fails. This functionality is equivalent to AutoRoute preferred and directed routes.

The routes are stated as the full node/port combination that makes up the complete and fully specified path and are stored in the database. Because you specify the DTL, the preferred route can be specified within only a single peer group.

From a command-line interface perspective, the **pref** abbreviation creates the preferred route family of commands. The commands **addpref**, **modpref**, **delpref**, **dsppref**, and

dspprefs manipulate the preferred routes. These commands are self-explanatory. The command **cnfconpref** associates and dissociates the created preferred routes with SPVC and SPVP connections such that many SPVCs can be associated with a single preferred route.

Connection Admission Control and Oversubscription

Connection Admission Control (CAC) happens at the VSI slave in the resource manager module. There are several CAC algorithms, including CAC based on bandwidth, logical connection numbers (LCNs), and enhanced CAC algorithms. Basic CAC is the default CAC algorithm for all service types except the ATM Forum UBR service category and MPLS service types. For these, the LCN CAC algorithm is used by default. Table 11-3 describes the effective cell rate (ECR) calculation used in basic CAC.

Table 11-3   *ECR Values for Basic CAC*

Service Category	ECR Value
CBR	PCR
VBR-RT, VBR-NRT	SCR
ABR	MCR
UBR	N/A

Enhanced CAC takes into the ECR calculation various QoS requirements.

It is important not to mistake CAC performed at the VSI slave with generic CAC or GCAC performed by PNNI at the controller.

Multiservice switching PNNI implementations have a set of CAC parameters that can be modified on a per-class basis. You can use the command **dsppnportcac** executed at the PNNI controller CLI to display those parameters and their default values. See Example 11-39.

Example 11-39   *Displaying the Default CAC Parameters*

```
m8850-7a.7.PXM.a > dsppnportcac 1:1.2:2


                  cbr:      rt-vbr:     nrt-vbr:         ubr:          abr:          sig:
bookFactor:       100%         100%         100%         100%         100%         100%
maxBw:       100.0000%    100.0000%    100.0000%    100.0000%    100.0000%    100.0000%
minBw:         0.0000%      0.0000%      0.0000%      0.0000%      0.0000%      0.9928%
maxVc:           100%         100%         100%         100%         100%         100%
minVc:             0%           0%           0%           0%           0%           1%
maxVcBw:           0            0            0            0            0            0


m8850-7a.7.PXM.a >
```

The minimum values are guaranteed quantities, and the maximum values describe the upper limit based on shared resources, for both bandwidth and LCNs.

On the VSI slave side, you can display the load model showing available bandwidth and LCN values on a port's partition, as well as the per-service category, using the command **dspload**, as Example 11-40 shows.

Example 11-40 *Using* **dspload**

```
m8850-7a.1.AXSM.a > dspload 2 1
       +---------------------------------------------+
       ¦  I N T E R F A C E   L O A D   I N F O  ¦
       +---------------------------------------------+
       ¦ Maximum Channels         : 0004000        ¦
       ¦ Guaranteed Channels      : 0001000        ¦
       ¦ Igr Maximum Bandwidth    : 0096000        ¦
       ¦ Igr Guaranteed Bandwidth : 0096000        ¦
       ¦ Egr Maximum Bandwidth    : 0096000        ¦
       ¦ Egr Guaranteed Bandwidth : 0096000        ¦
       ¦ Available Igr Channels   : 0003995        ¦
       ¦ Available Egr Channels   : 0003995        ¦
       ¦ Available Igr Bandwidth  : 0058519        ¦
       ¦ Available Egr Bandwidth  : 0058519        ¦
       +---------------------------------------------+
       ¦         E X C E P T -- V A L U E S        ¦
       +---------------------------------------------+
       ¦ SERV-CATEG ¦ VAR-TYPE ¦ INGRESS ¦ EGRESS  ¦
       ¦ VSI-SIG    ¦ Avl Chnl ¦ 0003995 ¦ 0003995 ¦
       ¦ CBR        ¦ Avl Chnl ¦ 0003995 ¦ 0003995 ¦
       ¦ VBR-RT     ¦ Avl Chnl ¦ 0003995 ¦ 0003995 ¦
       ¦ VBR-nRT    ¦ Avl Chnl ¦ 0003995 ¦ 0003995 ¦
       ¦ UBR        ¦ Avl Chnl ¦ 0003995 ¦ 0003995 ¦
       ¦ ABR        ¦ Avl Chnl ¦ 0003995 ¦ 0003995 ¦
       +---------------------------------------------+
       ¦ VSI-SIG    ¦ Avl Bw   ¦ 0058519 ¦ 0058519 ¦
       ¦ CBR        ¦ Avl Bw   ¦ 0058519 ¦ 0058519 ¦
       ¦ VBR-RT     ¦ Avl Bw   ¦ 0058519 ¦ 0058519 ¦
       ¦ VBR-nRT    ¦ Avl Bw   ¦ 0058519 ¦ 0058519 ¦
       ¦ UBR        ¦ Avl Bw   ¦ 0058519 ¦ 0058519 ¦
       ¦ ABR        ¦ Avl Bw   ¦ 0058519 ¦ 0058519 ¦
       +---------------------------------------------+

m8850-7a.1.AXSM.a >
```

The interface load information shows per-partition quantities, and the except values show per-service category values within the VSI partition. The **dspload** command output shows that, by default, the values for all the service categories are the same and equal the per-partition values for both bandwidth and LCNs. All available channels are equal to 3995 LCNs, and the available bandwidth is 58,519 CPS. This is because the command **dsppnportcac** shows all minimum values equal to 0, except for the signaling service category, the resources of which are used.

The following examples demonstrate the relationship between the PNNI VSI master pnportcac values and the VSI slave load values.

CAC Example 1

In the first example, the maximum bandwidth for CAC for VBR-NRT service types is made null. The command **cnfpnportcac** in the PNNI controller is used while the PnPort is administratively down. See Example 11-41.

Example 11-41 *Changing the Maximum CAC Bandwidth for nrt-VBR*

```
m8850-7a.7.PXM.a > dnpnport 1:1.2:2

m8850-7a.7.PXM.a > cnfpnportcac 1:1.2:2 nrtvbr -maxbw 0
WARNING: New CAC parameters apply to existing connections also

m8850-7a.7.PXM.a > uppnport 1:1.2:2
```

All the VBR-NRT connections that were traversing the PnPort 1:1.2:2 in the MGX-8850 were derouted (as you can see using the command **dsppncons**), and the load model looks as Example 11-42 shows.

Example 11-42 *Checking the Load Model*

```
m8850-7a.1.AXSM.a > dspload 2 1
    +--------------------------------------------+
    ¦    I N T E R F A C E   L O A D   I N F O    ¦
    +--------------------------------------------+
    ¦ Maximum Channels          : 0004000         ¦
    ¦ Guaranteed Channels       : 0001000         ¦
    ¦ Igr Maximum Bandwidth     : 0096000         ¦
    ¦ Igr Guaranteed Bandwidth  : 0096000         ¦
    ¦ Egr Maximum Bandwidth     : 0096000         ¦
    ¦ Egr Guaranteed Bandwidth  : 0096000         ¦
    ¦ Available Igr Channels    : 0003997         ¦
    ¦ Available Egr Channels    : 0003997         ¦
    ¦ Available Igr Bandwidth   : 0094897         ¦
    ¦ Available Egr Bandwidth   : 0094897         ¦
    +--------------------------------------------+
    ¦          E X C E P T -- V A L U E S         ¦
    +--------------------------------------------+
    ¦ SERV-CATEG ¦ VAR-TYPE ¦ INGRESS ¦ EGRESS   ¦
    ¦ VSI-SIG    ¦ Avl Chnl ¦ 0003997 ¦ 0003997  ¦
    ¦ CBR        ¦ Avl Chnl ¦ 0003997 ¦ 0003997  ¦
    ¦ VBR-RT     ¦ Avl Chnl ¦ 0003997 ¦ 0003997  ¦
    ¦ VBR-nRT    ¦ Avl Chnl ¦ 0003997 ¦ 0003997  ¦
    ¦ UBR        ¦ Avl Chnl ¦ 0003997 ¦ 0003997  ¦
    ¦ ABR        ¦ Avl Chnl ¦ 0003997 ¦ 0003997  ¦
    +--------------------------------------------+
    ¦ VSI-SIG    ¦ Avl Bw   ¦ 0094897 ¦ 0094897   ¦
    ¦ CBR        ¦ Avl Bw   ¦ 0094897 ¦ 0094897   ¦
    ¦ VBR-RT     ¦ Avl Bw   ¦ 0094897 ¦ 0094897   ¦
    ¦ VBR-nRT    ¦ Avl Bw   ¦ 0000000 ¦ 0000000   ¦
```

Example 11-42 *Checking the Load Model (Continued)*

```
          ¦ UBR       ¦ Avl Bw  ¦ 0094897 ¦ 0094897 ¦
          ¦ ABR       ¦ Avl Bw  ¦ 0094897 ¦ 0094897 ¦
          +-------------------------------------------+

   m8850-7a.1.AXSM.a >
```

From the **dspload** command output, you can see that you now have more available bandwidth and LCNs from the connections that took an alternative path. In addition, the except values for VBR-NRT available bandwidth are null. The subpartitioned bandwidth is shown in Figure 11-10.

Figure 11-10 *Bandwidth Subpartitioning in CAC Example 1*

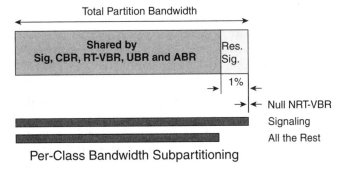

Per-Class Bandwidth Subpartitioning

NOTE	In a PNNI application, it is critical to match the per-class PNNI advertised link values used by GCAC (modified with the command **cnfpnni-intf**) with the VSI slave CAC values (modified with the command **cnfpnportcac**).

Performing only a modification in the CAC parameters can lead to unnecessary crankbacks or underutilized bandwidth. In particular, considering CAC Example 1, PNNI is unaware of the change in CAC. You configured no bandwidth for the VBR-NRT service types in the interface, but PNNI tries to route VBR-NRT connections over it, and GCAC passes. This scenario leads to crankbacks.

CAC Example 2

The second example involves changing the minimum bandwidth for CBR service types. The CAC values were set back to the defaults before beginning.

Using the command **cnfpnportcac**, the minimum bandwidth for CBR is changed to 40 percent (see Example 11-43).

Example 11-43 *Using **cnfpnportcac** to Change the Minimum Bandwidth for CBR*

```
   m8850-7a.7.PXM.a > cnfpnportcac 1:1.2:2 cbr -minbw 40
   WARNING: New CAC parameters apply to existing connections also
```

continues

Example 11-43 *Using* **cnfpnportcac** *to Change the Minimum Bandwidth for CBR (Continued)*

```
m8850-7a.7.PXM.a >

m8850-7a.7.PXM.a > dsppnportcac 1:1.2:2

                    cbr:      rt-vbr:    nrt-vbr:      ubr:       abr:       sig:
bookFactor:        100%        100%        100%        100%       100%       100%
maxBw:         100.0000%   100.0000%   100.0000%   100.0000%  100.0000%  100.0000%
minBw:          40.0000%     0.0000%     0.0000%     0.0000%    0.0000%    0.9928%
maxVc:             100%        100%        100%        100%       100%       100%
minVc:               0%          0%          0%          0%         0%         1%
maxVcBw:             0           0           0           0          0          0

m8850-7a.7.PXM.a >
```

NOTE

The command **dsppnportcac** shows a minimum guaranteed bandwidth for the signaling service category of approximately 1 percent. This is done to reserve some bandwidth for the signaling VCs, because the signaling service types are critical. A minimum LCN of 1 percent is also configured by default.

In real life, the signaling VCs are more active when the interface doesn't have much traffic. For example, after an interface flap, the signaling VCs set up all the SVCs and SPVCs that were previously torn down, but there is no user traffic. In contrast, the more VCs that are created (and, therefore, the more user traffic), the less traffic there is on the signaling VCs.

You can display the per-class subpartition and except CAC values from the VSI slave using the command **dspload**, as shown in Example 11-44.

Example 11-44 *Displaying the Subpartition with* **dspload**

```
m8850-7a.1.AXSM.a > dspload 2 1
        +-------------------------------------------------+
        ¦    I N T E R F A C E    L O A D    I N F O      ¦
        +-------------------------------------------------+
        ¦ Maximum Channels          : 0004000             ¦
        ¦ Guaranteed Channels       : 0001000             ¦
        ¦ Igr Maximum Bandwidth     : 0096000             ¦
        ¦ Igr Guaranteed Bandwidth  : 0096000             ¦
        ¦ Egr Maximum Bandwidth     : 0096000             ¦
        ¦ Egr Guaranteed Bandwidth  : 0096000             ¦
        ¦ Available Igr Channels    : 0003995             ¦
        ¦ Available Egr Channels    : 0003995             ¦
        ¦ Available Igr Bandwidth   : 0093293             ¦
        ¦ Available Egr Bandwidth   : 0093293             ¦
        +-------------------------------------------------+
        ¦           E X C E P T -- V A L U E S            ¦
        +-------------------------------------------------+
        ¦ SERV-CATEG ¦ VAR-TYPE ¦ INGRESS  ¦ EGRESS  ¦
        ¦ VSI-SIG    ¦ Avl Chnl ¦ 0003995  ¦ 0003995 ¦
        ¦ CBR        ¦ Avl Chnl ¦ 0003995  ¦ 0003995 ¦
```

Example 11-44 *Displaying the Subpartition with* **dspload** *(Continued)*

```
                ¦ VBR-RT    ¦ Avl Chnl ¦ 0003995 ¦ 0003995 ¦
                ¦ VBR-nRT   ¦ Avl Chnl ¦ 0003995 ¦ 0003995 ¦
                ¦ UBR       ¦ Avl Chnl ¦ 0003995 ¦ 0003995 ¦
                ¦ ABR       ¦ Avl Chnl ¦ 0003995 ¦ 0003995 ¦
                +------------------------------------------------+
                ¦ VSI-SIG   ¦ Avl Bw   ¦ 0055893 ¦ 0055893 ¦
                ¦ CBR       ¦ Avl Bw   ¦ 0093293 ¦ 0093293 ¦
                ¦ VBR-RT    ¦ Avl Bw   ¦ 0055893 ¦ 0055893 ¦
                ¦ VBR-nRT   ¦ Avl Bw   ¦ 0055893 ¦ 0055893 ¦
                ¦ UBR       ¦ Avl Bw   ¦ 0055893 ¦ 0055893 ¦
                ¦ ABR       ¦ Avl Bw   ¦ 0055893 ¦ 0055893 ¦
                +------------------------------------------------+

  m8850-7a.1.AXSM.a >
```

This example is summarized in Figure 11-11.

Figure 11-11 *Bandwidth Subpartitioning in CAC Example 2*

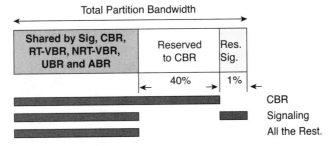

Total Partition Bandwidth

Per-Class Bandwidth Subpartitioning

CAC Example 3

In the third example, four CBR connections use 2100 CPS in the PnPort. The initial PnPort CAC configuration is as Example 11-45 shows.

Example 11-45 *Checking the PnPort CAC Initial Values*

```
m8850-7a.7.PXM.a > dsppnportcac 2:1.1:1

                 cbr:      rt-vbr:    nrt-vbr:      ubr:       abr:        sig:
bookFactor:      100%       100%        100%       100%       100%        100%
maxBw:       100.0000%  100.0000%   100.0000%  100.0000%  100.0000%  100.0000%
minBw:         0.0000%    0.0000%     0.0000%    0.0000%    0.0000%     0.6681%
maxVc:           100%       100%        100%       100%       100%        100%
minVc:             0%         0%          0%         0%         0%          1%
maxVcBw:           0          0           0          0          0           0

m8850-7a.7.PXM.a >
```

There is 0.6681 percent of guaranteed bandwidth and 1 percent of guaranteed LCN for the signaling service category. As Example 11-46 shows, using the command **dsppnportrsrc**, allows you to see the breakdown of actual values expressed in CPS and the number of channels.

Example 11-46 *Using the Command* **dsppnportrsrc** *to Check Per-Class Resource Allocation*

```
m8850-7a.7.PXM.a > dsppnportrsrc 2:1.1:1

                      cbr:    rt-vbr:   nrt-vbr:      ubr:      abr:      sig:
  Max TxCR CPS:      50000     50000      50000     50000     50000     50000
  Max RxCR CPS:      50000     50000      50000     50000     50000     50000
  MinGuar TxCR CPS:      0         0          0         0         0       334
  MinGuar RxCR CPS:      0         0          0         0         0       334
  Min Tx CLR:           10         8          6         6         6         8
  Min Rx CLR:           10         8          6         6         6         8
  Avl TxCR CPS:      47566     47566      47566     47566     47566     47566
  Avl RxCR CPS:      47566     47566      47566     47566     47566     47566
  OvSub AvTx CPS:    47566     47566      47566     47566     47566     47566
  OvSub AvRx CPS:    47566     47566      47566     47566     47566     47566
  # Avl Tx Chans:     1986      1986       1986      1986      1986      1995
  # Avl Rx Chans:     1986      1986       1986      1986      1986      1995

m8850-7a.7.PXM.a >
```

Regarding bandwidth subpartitioning per class of service in the partition, the guaranteed bandwidth for the signaling of 0.6681 percent out of 50,000 CPS equals 334 CPS. That is why the signaling service category has a guaranteed bandwidth of 334 CPS. You can see that the signaling service category uses 334 CPS because its available bandwidth equals the available bandwidth in other service categories.

Subtracting 334 CPS plus the used bandwidth (2100 CPS) from 50,000 CPS of total bandwidth gives you 47,566 CPS shared by all service categories.

Now the book factor is configured in this PnPort for the CBR service category. The book factor can be seen as the percentage utilization for the service category in the partition. A book factor of 50 percent means that the port is oversubscribed on a 2:1 ratio for the CBR service category. A book factor of 200 percent indicates an undersubscription factor of the same ratio. See Example 11-47.

Example 11-47 *Configuring the* **bookfactor** *to Oversubscribe*

```
m8850-7a.7.PXM.a > cnfpnportcac 2:1.1:1 cbr -bookfactor 50
WARNING: New CAC parameters apply to existing connections also

m8850-7a.7.PXM.a >
```

In this configuration, the CBR service types are oversubscribed by 2:1. You can see this in the output of the command **dsppnportrsrc**, as Example 11-48 shows.

Example 11-48 *Checking Oversubscription Using the Command* **dsppnportrsrc**

```
m8850-7a.7.PXM.a > dsppnportrsrc 2:1.1:1

                    cbr:    rt-vbr:   nrt-vbr:    ubr:     abr:     sig:
Max TxCR CPS:       50000    50000     50000     50000    50000    50000
Max RxCR CPS:       50000    50000     50000     50000    50000    50000
MinGuar TxCR CPS:       0        0         0         0        0      334
MinGuar RxCR CPS:       0        0         0         0        0      334
Min Tx CLR:            10        8         6         6        6        8
Min Rx CLR:            10        8         6         6        6        8
Avl TxCR CPS:       48616    48616     48616     48616    48616    48616
Avl RxCR CPS:       48616    48616     48616     48616    48616    48616
OvSub AvTx CPS:     97232    48616     48616     48616    48616    48616
OvSub AvRx CPS:     97232    48616     48616     48616    48616    48616
# Avl Tx Chans:      1986     1986      1986      1986     1986     1995
# Avl Rx Chans:      1986     1986      1986      1986     1986     1995

m8850-7a.7.PXM.a >
```

NOTE   This example shows per-partition and per-class overbooking. Overbooking can also be performed on a per-connection basis.

You can see that the available bandwidth for all service types increased by 1050 CPS. This is because the four CBR connections with a total of 2100 CPS now have an effective rate of 2100 CPS divided by 2 (1050 CPS). The CBR service category is oversubscribed by a factor of 2. See Figure 11-12.

Figure 11-12 *Oversubscription in CAC Example 3*

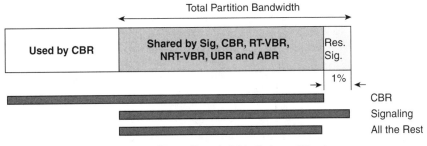

Per-Class Bandwidth Subpartitioning

In Figure 11-12, the switch does not create bandwidth for CBR type connections, but it applies the book factor to CBR call reservations. So if a CBR connection is configured with PCR = 1000 CPS, only 500 CPS is reserved in the load model, making the effective bandwidth twice as much.

NOTE

For SVC and SPVC calls, the actual connection bandwidth is used for the GCAC algorithm. However, PNNI advertises a link bandwidth equal to the actual link bandwidth divided by the book factor. This is done because a connection can span multiple PNNI links with a different book factor.

For example, on a 100 Mbps link with a configured book factor of 10 percent, PNNI advertises a link of 1000 Mbps. For a call of 20 Mbps, GCAC uses 20 Mbps, but the actual CAC performed by the VSI slave uses 2 Mbps.

In summary, the connection bandwidth is adjusted based on the book factor at the CAC level in the VSI slave, but not at the GCAC level in the PNNI process. At the PNNI level, the book factor is considered when links are advertised.

You now configure the minimum guaranteed bandwidth for the CBR service category equal to 10 percent of the total partition bandwidth. See Example 11-49.

Example 11-49 *Changing the CAC Minimum Bandwidth for CBR*

```
m8850-7a.7.PXM.a > cnfpnportcac 2:1.1:1 cbr -minbw 10
WARNING: New CAC parameters apply to existing connections also

m8850-7a.7.PXM.a >
```

You have the available bandwidth and channels for all service classes as Example 11-50 shows.

Example 11-50 *Checking Available Resources Using the Command* **dsppnportrsrc**

```
m8850-7a.7.PXM.a > dsppnportrsrc 2:1.1:1

                    cbr:    rt-vbr:   nrt-vbr:    ubr:      abr:      sig:
Max TxCR CPS:      50000     50000     50000     50000     50000     50000
Max RxCR CPS:      50000     50000     50000     50000     50000     50000
MinGuar TxCR CPS:   5000         0         0         0         0       334
MinGuar RxCR CPS:   5000         0         0         0         0       334
Min Tx CLR:           10         8         6         6         6         8
Min Rx CLR:           10         8         6         6         6         8
Avl TxCR CPS:      48616     44666     44666     44666     44666     44666
Avl RxCR CPS:      48616     44666     44666     44666     44666     44666
OvSub AvTx CPS:    97232     44666     44666     44666     44666     44666
OvSub AvRx CPS:    97232     44666     44666     44666     44666     44666
# Avl Tx Chans:     1986      1986      1986      1986      1986      1995
# Avl Rx Chans:     1986      1986      1986      1986      1986      1995

m8850-7a.7.PXM.a >
```

You can see that the CBR service category has a guaranteed bandwidth of 5000 CPS. This 5000 CPS was subtracted from the available bandwidth of all the service types. All service types (including CBR) share 44,666 CPS (which equals 50,000 CPS total partition bandwidth minus 334 CPS reserved for signaling service category minus 5000 CPS) set aside for CBR service category.

The CBR service category has a total of 50,000 CPS minus 334 CPS, or 49,666 CPS (from which 5000 CPS is guaranteed and the rest is shared). The four CBR connections effectively use 1050 CPS, so the available bandwidth equals 48,616 CPS. The bandwidth set aside for the four CBR connections is taken from the CBR reserved guaranteed bandwidth.

Summary

This chapter covered two practical implications and applications of PNNI's advanced features. First, node, link, and address summarization and hierarchical organization provide massive scalability to PNNI networks. Second, the class-based awareness present in both routing updates and decisions, as well as CAC subpartitioning, brings enhanced traffic engineering capabilities. This chapter also covered static ATM routing configuration suitable to join two ATM domains.

CHAPTER 12

Virtual Switch Review

This book began with a discussion of the multiservice switching network architecture. The chapters that followed delved into the specific realizations of the architecture. Now it is time to look at the Cisco virtual switch architecture (CVSA) as a whole.

CVSA

The CVSA, built on the Multiservice Switching Forum (MSF) framework, enables true multiservice switching capabilities in an open and standard way and lets the telecom industry grow in a modular and horizontal fashion. The MSF plays a key role in enabling these changes, gathering different vendors and service providers to standardize the multiservice switching framework.

In this architecture, the adaptation plane transforms services into a format that the forwarding plane can cross-connect. The control plane manages this forwarding plane, and the two planes work in tandem to create a virtual switch. Different control-plane implementations include MPLS, PNNI, and voice (through the Media Gateway Control Protocol [MGCP].) The control plane can in turn be directed by an application plane such as SS7 intelligent networks or RFC 2547 MPLS VPNs. (Future control planes, either vendor-developed or service provider-developed, will be integrated into an existing infrastructure.)

There also can be different forwarding planes, such as cell-based, frame-based, and lambda-based optical cross-connects (OXC). The key in allowing true multiservice functionality resides in the partitioning function.

The MSF architecture not only defines the partitioning and sharing of resources in a switch controlled in a standard way, but it also defines a set of open interfaces among the planes to enable interworking and interoperability. In general terms, CVSA enables the use of any and every control plane over any and every forwarding plane.

Another important consequence of the architecture decomposition is that the core transport can evolve independently of the edge services.

Cisco's architecture also enhances the network's high-availability features by separating the control plane and the forwarding plane.

Virtual Switch Interface

The Virtual Switch Interface (VSI) discussed in Chapter 2, "SCI: Virtual Switch Interface," allows all of the following:

- **Multiple simultaneous controllers**—The controllers do not need to reside on a common platform, and they can have different locations. Each controller can be upgraded separately.

- **Control plane independence**—Each control plane manages resources independently, and each service receives the QoS required.

- **Dynamic partitioning**—The resources allocated to each controller can be changed dynamically. This allows planning for future events, such as adding a service to the network or modifying its resources according to the number of users. It also enables incremental provisioning.

- **Quality of Service (QoS)**—Each partition receives a guaranteed number of resources, but it also can use a shared pool to achieve maximum use. Separate queues are assigned to different services. Each service receives native QoS support.

To provide pure separation between the control and forwarding planes, the interface between them needs to be purely messaging-based. To ensure no adverse interaction between the control and forwarding planes, no databases should be shared between them. This is one of VSI's strengths. In cases where the control and forwarding planes reside on the same card, VSI provides a clean, logical separation between the two planes.

VSI is a published and open interface that allows third parties to develop control planes to manage the switches. It also allows switches to be controlled by a Cisco-developed control plane.

What's Next?

Using CVSA, different service-optimized controlled planes executing different service profiles can manage a common forwarding infrastructure. This concept provides the best possible framework for multiple voice and data services. In all cases, a migration path is also provided. CVSA maximizes the service differentiation concept, allowing the flexible support of a mix of existing and yet-to-be-developed services with only incremental investment in the most cost-efficient way. Taking the existing infrastructure, new protocols providing next-generation services can be added in a way that does not affect existing services.

Service Traffic Groups, Types, and Categories

This appendix contains the definitions of service groups, service types, and service categories used by the Virtual Switch Interface (VSI) protocol. Refer to Table A-1 to select the appropriate traffic management type.

Service Groups

Service groups are the most general taxonomy of traffic management types. Different service types and categories belong to each one of the service groups.

Table A-1 *VSI Service Groups*

Service Group	Description
0	MPLS service categories
1	ATM Forum service categories
2	IP service categories (obsolete)
3 through 10	Vendor-specific service categories (used by local agreement between switch and controller)

NOTE The IP service group (2) and its corresponding service types and service categories are obsolete, because DiffServ has redefined the meaning of the Type of Service (ToS) bits of the IP packet header.

Service Types

Service types are defined to allow a controller to specify different traffic management types on VSI messages, as shown in Tables A-2 through A-6. The specific service type mapping into hardware programming is left to the VSI slave and is transparent to the VSI master. The VSI master specifies a type of traffic using the service types defined.

Table A-2   *VSI Special Service Types*

Type	Name	Description
0x0000	Null	No type is needed. Use for reservation without bandwidth only.
0x0001	Default	Use the default switch type.
0x0002	Signaling	Signaling for network topology, and so on.
0x0003 through 0x00FF	Reserved	Currently unused.

Table A-3   *ATM Forum Service Types*

Type	Name	Description
0x0100	CBR.1	Constant Bit Rate
0x0101	VBR-RT.1	Variable Bit Rate-Real Time
0x0102	VBR-RT.2	Variable Bit Rate-Real Time
0x0103	VBR-RT.3	Variable Bit Rate-Real Time
0x0104	VBR-NRT.1	Variable Bit Rate-Nonreal Time
0x0105	VBR-NRT.2	Variable Bit Rate-Nonreal Time
0x0106	VBR-NRT.3	Variable Bit Rate-Nonreal Time
0x0107	UBR.1	Unassigned Bit Rate
0x0108	UBR.2	Unassigned Bit Rate
0x0109	ABR	Available Bit Rate
0x010A	CBR.2	Constant Bit Rate
0x010B	CBR.3	Constant Bit Rate
0x010C through 0x01FF	Reserved	Currently unused

Table A-4   *DiffServ/Multiprotocol Label Switching (MPLS) Service Types*

Type	Name	Description
0x0200	DiffServ CS0	DiffServ Class Selector 0 (lowest priority)
0x0201	DiffServ CS1	DiffServ Class Selector 1
0x0202	DiffServ CS2	DiffServ Class Selector 2
0x0203	DiffServ CS3	DiffServ Class Selector 3
0x0204	DiffServ CS4	DiffServ Class Selector 4
0x0205	DiffServ CS5	DiffServ Class Selector 5
0x0206	DiffServ CS6	DiffServ Class Selector 6
0x0207	DiffServ CS7	DiffServ Class Selector 7
0x0208 through 0x020F	Reserved	Currently unused
0x0210	MPLS-ABR	MPLS LVC with ABR flow control
0x0211	Reserved	Currently unused
0x0212	DiffServ DF	DiffServ Default
0x0213	DiffServ EF	DiffServ Expedited Forwarding
0x0214	DiffServ AFC1	DiffServ Assured Forwarding 1
0x0215	DiffServ AFC2	DiffServ Assured Forwarding 2
0x0216	DiffServ AFC3	DiffServ Assured Forwarding 3
0x0217	DiffServ AFC4	DiffServ Assured Forwarding 4
0x0218 through 0x02FF	Reserved	Currently unused

NOTE   Previously, the DiffServ Class Selectors service types were called MPLS COS. For example, the old traffic type *MPLS COS 0* is now called *DiffServ CS 0*.

Table A-5   *Vendor-Specific Service Types*

Type	Name	Description
0x0300 through 0x07FF	Reserved	Reserved for vendor-specific types

Table A-6   *Cisco Proprietary Service Types (AutoRoute and Portable AutoRoute)*

Type	Name	Description
0x0400	AutoRoute HP	AutoRoute High-Priority data
0x0401	AutoRoute NTS	AutoRoute Non-Time-Stamped data
0x0402	AutoRoute ABR	AutoRoute Available Bit Rate (standard)
0x0403	AutoRoute VBR-NRT	AutoRoute Variable Bit Rate-Nonreal Time
0x0404	AutoRoute CBR	AutoRoute Constant Bit Rate
0x0405	AutoRoute BDA	AutoRoute Bursty Data A traffic
0x0406	AutoRoute BDB	AutoRoute Bursty Data B traffic (ForeSight)
0x0407	AutoRoute TS	AutoRoute Time-Stamped data
0x0408	AutoRoute Voice	AutoRoute Voice traffic
0x0409	AutoRoute ABR-F	AutoRoute Available Bit Rate (ForeSight)
0x040A	AutoRoute VBR-RT	AutoRoute Variable Bit Rate-Real Time
0x040B through 0x04FF	Reserved	Currently unused

Service Categories

To simplify certain functions such as resource provisioning and load advertisement, service types are grouped into service categories. For example, as shown in Table A-7, the ATM Forum service types CBR.1, CBR.2, and CBR.3 are grouped into the service category CBR.

Table A-7 *ATM Forum Service Categories*

Category	Name	Description
0x0100	CBR	Consists of CBR.1, CBR.2, and CBR.3
0x0101	VBR-RT	Consists of VBR-RT.1, VBR-RT.2, and VBR-RT.3
0x0102	VBR-NRT	Consists of VBR-NRT.1, VBR-NRT.2, and VBR-NRT.3
0x0103	UBR	Consists of UBR.1 and UBR.2
0x0104	ABR	Consists of ABR
0x0105 through 0x01FF	Reserved	Currently unused

Service categories are meant to accommodate ATM Forum service types.

For the rest of the service categories—VSI, MPLS, vendor-specific, and Cisco-proprietary—the categories are the same as the corresponding service types.

INDEX

Numerics

A

B

M

Train with authorized Cisco Learning Partners.

Discover all that's possible on the Internet.

One of the biggest challenges facing networking professionals is how to stay current with today's ever-changing technologies in the global Internet economy. Nobody understands this better than Cisco Learning Partners, the only companies that deliver training developed by Cisco Systems.

Just go to **www.cisco.com/go/training_ad**. You'll find more than 120 Cisco Learning Partners in over 90 countries worldwide.* Only Cisco Learning Partners have instructors that are certified by Cisco to provide recommended training on Cisco networks and to prepare you for certifications.

To get ahead in this world, you first have to be able to keep up. Insist on training that is developed and authorized by Cisco, as indicated by the Cisco Learning Partner or Cisco Learning Solutions Partner logo.

Visit **www.cisco.com/go/training_ad** today.

CISCO SYSTEMS

EMPOWERING THE
INTERNET GENERATION™

Cisco Press

Learning is serious business.

Invest wisely.

Cisco AVVID IP Telephony Solutions

Cisco IP Telephony
David Lovell
1-58705-050-1 • **Availabe Now**

Cisco IP Telephony is based on the successful CIPT training class taught by the author and other Cisco-certified training partners. This book provides networking professionals with the fundamentals to implement a Cisco AVVID IP Telephony solution that can be run over a data network, therefore reducing costs associated with running separate data and telephone networks. *Cisco IP Telephony* focuses on using Cisco CallManager and other IP telephony components connected in LANs and WANs. This book provides you with a foundation for working with Cisco IP Telephony products, specifically Cisco CallManager. If your task is to install, configure, support, and maintain a CIPT network, this is the book for you.

Cisco CallManager Fundamentals: A Cisco AVVID Solution
Anne Smith, John Alexander, Chris Pearce, and Delon Whetton
1-58705-008-0 • **Availabe Now**

Cisco CallManager Fundamentals provides examples and reference information about CallManager, the call processing component of the Cisco AVVID (Architecture for Voice, Video, and Integrated Data) IP Telephony solution. *Cisco CallManager Fundamentals* uses examples and architectural descriptions to explain how CallManager processes calls. This book details the inner workings of CallManager so that those responsible for designing and maintaining a Voice over IP (VoIP) solution from Cisco Systems can understand the role each component plays and how they interrelate. You will learn detailed information about hardware and software components, call routing, media processing, system management and monitoring, and call detail records. The authors, all members of the CallManager group at Cisco Systems, also provide a list of features and Cisco solutions that integrate with CallManager. This book is the perfect resource to supplement your understanding of CallManager.

Cisco Interactive Mentor

CIM Voice Internetworking:
VoIP Quality of Service

Cisco Systems, Inc.

1-58720-050-3 • Availabe Now

With *CIM Voice Internetworking: VoIP Quality of Service*, you acquire the skills needed to implement and fine tune QoS for Voice over IP routing on Cisco routers. From an overview of QoS concepts to detailed examination of Cisco IOS(r) Software QoS and VoIP routing commands, you will learn how to solve call quality problems caused by delay and inefficient packet compression in Voice over IP networks. Mastering techniques and methods developed by Cisco Technical Assistance Center engineers, you will enable and troubleshoot link efficiency mechanisms and QoS queuing components on multilink PPP and Frame Relay networks.

CIM Voice Internetworking: Basic Voice over IP

Cisco Systems, Inc.

1-58720-023-6 • Availabe Now

With *CIM Voice Internetworking: Basic Voice over IP*, you can master the telephony and voice internetworking knowledge you need to enhance the versatility and value of your communications infrastructure. Offering self-paced instruction and practice, this robust learning tool gives you a quick and cost-effective way to acquire Cisco knowledge and expertise. From an overview of traditional telephony and voice transmission concepts to the basics of routing voice and fax packets over a data network, you'll learn how to configure typical software features of a VoIP network and perform operational application tasks with interactive voice response (IVR). Using techniques developed by Cisco Technical Assistance Center engineers, you'll practice configuring and troubleshooting both analog and digital voice calls over IP networks. An excellent preparation tool for the Cisco Certified Network Professionals (CCNP) Voice Access Specialization and Cisco Certified Internetwork Expert (CCIE) exam.

Voice Technology

Voice over IP Fundamentals
Jonathan Davidson
1-57870-168-6 • **Availabe Now**

Voice over IP (VoIP), which integrates voice and data transmission, is quickly becoming an important factor in network
communications. It promises lower operational costs, greater
flexibility, and a variety of enhanced applications. *Voice
over IP Fundamentals* provides a thorough introduction to
this new technology to help experts in both the data and
telephone industries plan for the new networks.

Cisco Voice over Frame Relay, ATM, and IP
Steve McQuerry and Kelly McGrew
1-57870-227-5 • **Availabe Now**

Cisco Voice over Frame Relay, ATM, and IP is a direct
complement to the Cisco authorized training course of the
same name. Based on the content of the CVoice course, this
book provides an intermediate-level treatment of Cisco
voice technologies. The overall objective of the book is to
teach engineers how to design, integrate, and configure voice
over Frame Relay, ATM, and IP of enterprise or managed
network services using various Cisco 2600, 3600, 3810,
and 5300 multiservice access devices.

Deploying Cisco Voice over IP Solutions

Jonathan Davidson

1-58705-030-7 • **Availabe Now**

Deploying Cisco Voice over IP Solutions provides networking professionals the knowledge, advice, and insight necessary to design and deploy Voice over IP (VoIP) networks that meet customers' needs for scalability, services, and security. Beginning with an introduction to the important preliminary design elements that need to be considered before implementing VoIP, Deploying Cisco Voice over IP Solutions also demonstrates the basic tasks involved in designing an effective service provider-based VoIP network. It concludes with design and implementation guidelines for some of the more popular and widely requested VoIP services, such as prepaid services, fax services, and Virtual Private Networks (VPNs).

Integrating Voice and Data Networks

Scott Keagy

1-57870-196-1 • **Availabe Now**

The integration of voice and data networks is creating a fundamental change in the telecommunications and data industries. This change means better phone service, lower prices, new features, less maintenance, and more choices, leading to a complete convergence of the data networking and telecommunications industries into a single community. With all of this growth and change, network engineers and managers need specific information on how to integrate and configure packetized voice networks. *Integrating Voice and Data Networks* is designed as your one-stop shop for learning how to integrate traditional voice technology into existing Cisco data networks.

CCIE Professional Development

Cisco OSPF Command and Configuration Handbook
William R. Parkhurst Ph.D., CCIE

1-58705-071-4 • **Availabe Now**

Cisco OSPF Command and Configuration Handbook is a
clear, concise, and complete source of documentation for all
Cisco IOS Software OSPF commands. The way you use this
book will depend on your objectives. If you are preparing
for the CCIE written and lab exams, then this book can be
used as a laboratory guide to learn the purpose and use of
every OSPF command. If you are a network designer, then this
book can be used as a ready reference for any OSPF command.

Cisco BGP-4 Command and Configuration Handbook
William R. Parkhurst, Ph.D, CCIE

1-58705-017-X • **Availabe Now**

Cisco BGP-4 Command and Configuration Handbook is an
exhaustive practical reference to the commands contained
within BGP-4. For each command/subcommand, author
Bill Parkhurst explains the intended use or function and
how to properly configure it. Then he presents scenarios to
demonstrate every facet of the command and its use, along
with appropriate show and debug commands. Through the
discussion of functionality and the scenario-based configuration
examples, *Cisco BGP-4 Command and Configuration
Handbook* will help you gain a thorough understanding of
the practical side of BGP-4.

CCIE Practical Studies, Volume I
Karl Solie, CCIE

1-58720-002-3 • **Availabe Now**

CCIE Practical Studies, Volume I, provides the knowledge
to assemble and configure all the necessary hardware and
software components required to model complex, Cisco-
driven internetworks based on the OSI reference mode
from Layer 1 on up. Each chapter focuses on one or more
specific technologies or protocols and follows up with a
battery of CCIE exam-like labs for you to configure that
challenges your understanding of the chapter topics and
measures your aptitude as a CCIE candidate. The final
chapter of the book provides five CCIE "Simulation Labs."
These labs not only test your knowledge but your speed as
well. You will find *CCIE Practical Studies*, Volume I, to be
an indispensable preparation tool.